The Role of MicroRNAs in Plants

The Role of MicroRNAs in Plants

Special Issue Editor
Anthony A. Millar

MDPI • Basel • Beijing • Wuhan • Barcelona • Belgrade

Special Issue Editor
Anthony A. Millar
Research School of Biology,
Australian National University
Australia

Editorial Office
MDPI
St. Alban-Anlage 66
4052 Basel, Switzerland

This is a reprint of articles from the Special Issue published online in the open access journal *Plants* (ISSN 2223-7747) from 2018 to 2020 (available at: https://www.mdpi.com/journal/plants/special_issues/mciro_RNA_plant).

For citation purposes, cite each article independently as indicated on the article page online and as indicated below:

LastName, A.A.; LastName, B.B.; LastName, C.C. Article Title. *Journal Name* **Year**, *Article Number*, Page Range.

ISBN 978-3-03928-730-7 (Pbk)
ISBN 978-3-03928-731-4 (PDF)

Cover image courtesy of Anthony A. Millar, Maria Alonso-Peral and Junyan Li.

© 2020 by the authors. Articles in this book are Open Access and distributed under the Creative Commons Attribution (CC BY) license, which allows users to download, copy and build upon published articles, as long as the author and publisher are properly credited, which ensures maximum dissemination and a wider impact of our publications.

The book as a whole is distributed by MDPI under the terms and conditions of the Creative Commons license CC BY-NC-ND.

Contents

About the Special Issue Editor vii

Anthony A Millar
The Function of miRNAs in Plants
Reprinted from: *Plants* 2020, 9, 198, doi:10.3390/plants9020198 1

Érika Frydrych Capelari, Guilherme Cordenonsi da Fonseca, Frank Guzman and Rogerio Margis
Circular and Micro RNAs from *Arabidopsis thaliana* Flowers Are Simultaneously Isolated from AGO-IP Libraries
Reprinted from: *Plants* 2019, 8, 302, doi:10.3390/plants8090302 5

María José López-Galiano, Inmaculada García-Robles, Ana I. González-Hernández, Gemma Camañes, Begonya Vicedo, M. Dolores Real and Carolina Rausell
Expression of miR159 Is Altered in Tomato Plants Undergoing Drought Stress
Reprinted from: *Plants* 2019, 8, 201, doi:10.3390/plants8070201 18

Muhammad Shahbaz and Marinus Pilon
Conserved Cu-MicroRNAs in *Arabidopsis thaliana* Function in Copper Economy under Deficiency
Reprinted from: *Plants* 2019, 8, 141, doi:10.3390/plants8060141 29

Joseph L. Pegler, Jackson M. J. Oultram, Christopher P. L. Grof and Andrew L. Eamens
DRB1, DRB2 and DRB4 Are Required for Appropriate Regulation of the microRNA399/PHOSPHATE2 Expression Module in *Arabidopsis thaliana*
Reprinted from: *Plants* 2019, 8, 124, doi:10.3390/plants8050124 43

Joseph L Pegler, Jackson MJ Oultram, Christopher PL Grof and Andrew L Eamens
Profiling the Abiotic Stress Responsive microRNA Landscape of *Arabidopsis thaliana*
Reprinted from: *Plants* 2019, 8, 58, doi:10.3390/plants8030058 69

Michael Kravchik, Ran Stav, Eduard Belausov and Tzahi Arazi
Functional Characterization of microRNA171 Family in Tomato
Reprinted from: *Plants* 2019, 8, 10, doi:10.3390/plants8010010 87

Isaac Njaci, Brett Williams, Claudia Castillo-González, Martin B. Dickman, Xiuren Zhang and Sagadevan Mundree
Genome-Wide Investigation of the Role of MicroRNAs in Desiccation Tolerance in the Resurrection Grass *Tripogon loliiformis*
Reprinted from: *Plants* 2018, 7, 68, doi:10.3390/plants7030068 103

Anthony A. Millar, Allan Lohe and Gigi Wong
Biology and Function of miR159 in Plants
Reprinted from: *Plants* 2019, 8, 255, doi:10.3390/plants8050255 116

Zhanhui Zhang, Sachin Teotia, Jihua Tang and Guiliang Tang
Perspectives on microRNAs and Phased Small Interfering RNAs in Maize (*Zea mays* L.): Functions and Big Impact on Agronomic Traits Enhancement
Reprinted from: *Plants* 2019, 8, 170, doi:10.3390/plants8060170 133

Felipe Fenselau de Felippes
Gene Regulation Mediated by microRNA-Triggered Secondary Small RNAs in Plants
Reprinted from: *Plants* **2019**, *8*, 112, doi:10.3390/plants8050112 . . . . **150**

About the Special Issue Editor

Anthony Millar, Associate Professor, completed his PhD at the CSIRO Division of Plant Industry, Canberra, on the anaerobic response of cotton. He then carried out his postdoctoral studies at the University of British Columbia, Vancouver Canada, on the on the molecular biology of seed oil and cuticular wax biosynthesis in Arabidopsis. He then returned to CSIRO Canberra as a Research Scientist studying the hormone response of Arabidopsis and cereal germination as it related to seed dormancy and pre-harvest sprouting. He has been at the Australian National University since 2006, where his laboratory at the Division of Plant Sciences, Research School of Biology, studies gene silencing and plant RNA biology.

Editorial

The Function of miRNAs in Plants

Anthony A Millar

Division of Plant Science, Research School of Biology, The Australian National University, Canberra ACT 2601, Australia; tony.millar@anu.edu.au

Received: 9 January 2020; Accepted: 20 January 2020; Published: 5 February 2020

Abstract: MicroRNAs (miRNAs) are a class of small RNAs (sRNAs) that repress gene expression via high complementary binding sites in target mRNAs (messenger RNAs). Many miRNAs are ancient, and their intricate integration into gene expression programs have been fundamental for plant life, controlling developmental programs and executing responses to biotic/abiotic cues. Additionally, there are many less conserved miRNAs in each plant species, raising the possibility that the functional impact of miRNAs extends into virtually every aspect of plant biology. This Special Issue of *Plants* presents papers that investigate the function and mechanism of miRNAs in controlling development and abiotic stress response. This includes how miRNAs adapt plants to nutrient availability, and the silencing machinery that is responsible for this. Several papers profile changes in miRNA abundances during stress, and another study raises the possibility of circular RNAs acting as endogenous decoys to sequester and inhibit plant miRNA function. These papers act as foundational studies for the more difficult task ahead of determining the functional significance of these changes to miRNA abundances, or the presence of these circular RNAs. Finally, how miRNAs trigger the production of secondary sRNAs is reviewed, along with the potential agricultural impact of miRNAs and these secondary sRNA in the exemplar crop maize.

Keywords: miRNAs; development; abiotic stress; nutrient availability; circular RNAs; tasiRNA; phasiRNA

---

1. Introduction

MicroRNAs (miRNAs) have now been linked to most aspects of plant biology. They were first identified in plants less than 20 years ago, but they have been shown to be critical regulators of developmental process such as leaf morphogenesis, vegetative phase change, flowering time and response to environmental cues. This Special Issue presents a collection of papers that continues the molecular and functional characterization of plant miRNAs, as well as reviews that reflect on past achievements and outline the challenges and opportunities that lie ahead.

2. Development

Plant miRNAs are probably best known for their role in development. Many of the ancient miRNAs regulate highly conserved transcription factors or other regulatory genes that are fundamental in the development of terrestrial plants. MiR171 is one of these ancient miRNAs, being present in all lineages of land plants, where they negatively regulate genes encoding GRAS-domain SCARECROW-like transcription factors, but the functional outcome of this regulation is yet to be determined in many plant species. Part of the reason is that most plant miRNAs correspond to families of multiple redundant genes, and this is the case for miR171 in tomato, which has 11 family members. Kravchik et al. [1] investigate miR171 function in tomato using short tandem target mimic (STTM) technology, expressing a decoy that binds and sequesters all miR171 isoforms to inhibit the entire family [2]. They show miR171 in tomato is involved not only in shoot branching and leaf morphogenesis, but also in male development, as *STTM171* tomato plants had altered tapetal development and, consequently, altered pollen ontogenesis. Consistent with other species, miR171 appears to have diverse roles in tomato development.

3. Environmental Response

3.1. Abiotic Stress

To gain insights into whether an miRNA is involved in a stress response, the most obvious experiment is to determine whether its abundance changes under a certain stress. First, Pegler et al. [3] performed RNA-seq to determine the abundance of miRNAs in *Arabidopsis* under heat, drought, and salt stress conditions. This global survey identified many miRNAs with high-level fold changes under these conditions, thus identifying miRNAs that are candidates for playing important functional roles during these stresses. That study will act as a fundamental resource for future studies. Drought stress and water use efficiency will be a future key crop trait. Njaci et al. [4] identified miRNAs that alter in abundance under extreme water deficit in *Tripogon loliiformis*, a plant that can resurrect from a desiccated state. They found many conserved miRNAs differentially accumulated in roots and shoots during dehydration, likely reflecting the broad changes to metabolism and physiology during this extreme stress. Again, this study will act as a foundation for the investigation of miRNAs in desiccation tolerance, with the ultimate goal to utilize these for introducing tolerance into important crop species. In a more targeted study, López-Galiano et al. [5] focused on examining the change in miR159 abundance in tomato during drought, where they found miR159 to be downregulated during stress, while the mRNA levels of its corresponding target gene were derepressed. All these studies represent the start of the investigation of how miRNAs respond to stress and how they could be possibly utilized for developing stress tolerance. The much more challenging process lies ahead of determining what is the functional impact of these miRNA abundance changes and whether this information can be used to engineer stress tolerance into crop species. Indeed, despite miR159 being one of the earliest identified and most extensively studied miRNA, no clear conserved functional role for this miRNA has been identified. Millar et al. [6] summarize the literature concerning the biology of miR159, discussing the various potential functions that have been identified, and the questions that need to be addressed concerning this miRNA.

3.2. Nutrient Availability

MiRNA abundances also respond to nutrient availability. One such nutrient is copper (Cu), and four different miRNAs are known to respond to Cu levels, namely, miR397, miR398, miR408, and miR857, all highly conserved plant miRNAs. In this Special Issue, Shahbaz and Pilon [7] present an STTM which inhibited miR397, miR398, and miR408 simultaneously, resulting in higher levels of their target mRNAs under Cu-limiting conditions. The targets are all Cu-containing proteins, and failure to repress these targets under Cu-limiting conditions indirectly leads to reduced levels of an unrelated Cu-containing protein, plastocyanin, in the STTM transgenic plant lines. As plastocyanin is key for photosynthesis, this likely explains a decrease in photosynthetic electron transport activity of the STTM lines under Cu-limiting conditions, leading to a decrease in plant biomass. Therefore, the authors have superbly shown how these miRNAs regulate the Cu economy, channeling Cu to the most important Cu-containing proteins during Cu limitation.

One of the most limiting nutrients worldwide is phosphorous (P). Peglar et al. [8], investigate the machinery that is needed for the miR399-mediated regulation of *PHOSPHATE2* (*PHO2*). They show that in *Arabidopsis*, of the four members of DOUBLE-STRANDED RNA BINDING (DRB) protein family, DRB1 is the main player involved in the miR399 regulation of *PHO2*, but that DRB2 and DRB4 also play minor roles, and this regulation involves both an mRNA cleavage and translational repression mechanism. All these mechanisms are required to maintain *Arabidopsis* P homeostasis and highlight the complexity of this process.

4. Complexity of miRNA Regulation

MiRNA function can itself be regulated by RNAs where, in plants, noncoding RNA transcripts containing miRNA binding sites have been shown to act as decoys or miRNA target *MIMICs*, to sequester and inhibit miRNA function [9]. In animals, such RNAs are called competitive endogenous RNAs (ceRNAs), and some of the first identified were circular in form and contained multiple miRNA binding sites. It was thought that being circular increased stability and the effectiveness of being a decoy, but whether such RNAs exist in plants is unknown. In this Special Issue, Capelari et al. [10] bioinformatically mined publicly available RNA-seq data from ARGONAUTE-immunoprecipitation libraries (AGO-IP) and identified 1000s of potential circular RNAs, many of which contain potential miRNA binding sites. As many of the corresponding target mRNAs were found to be enriched in these AGO-IP libraries, this suggested that circular ceRNAs could be operating in plants. Obviously, more work is needed in confirming this, but in this intriguing paper, the authors have identified strong candidates to pursue.

In addition, miRNAs have been found not only to silence target transcripts through slicing, but in some instances slicing triggers the production of secondary siRNAs, known as trans-acting siRNAs (tasiRNA) or phased siRNAs (phasiRNAs). De Felippes [11] reviews the complex mechanisms and hypotheses by which tasiRNAs and phasiRNAs are generated, the factors involved, the regulatory advantages of transitivity, and the potential use of the natural amplification process to result in strong artificial gene silencing.

5. Application of miRNAs to Agriculture

Given miRNAs' central role in plant development via target key regulatory genes, and their potential role in stress response, their manipulation has the potential to alter key agronomic traits. Zhang et al. [12] summarizes the different gene silencing pathways and core machinery in maize, and their function in maize biology, detailing the traits that miRNAs, phasiRNAs, and tasiRNAs regulate and their potential use in agronomic improvement of maize, be it developmental timing, plant architecture, sex determination, fertility, or abiotic stress resistance. This gives an overview to the many potential applications in just one plant species. Given our extensive knowledge on the fundamental biology of plant miRNAs in model species, the future trajectory of this field will be their application in important crop species, where understanding their role and applying this knowledge have real potential for important agronomic outcomes.

Funding: This research received no external funding.

Acknowledgments: I wish to thank all colleagues for contributing articles to this Special Issue.

Conflicts of Interest: The author declares no conflict of interest.

References

1. Kravchik, M.; Stav, R.; Belausov, E.; Arazi, T. Functional Characterization of microRNA171 Family in Tomato. *Plants* **2019**, *8*, 10. [CrossRef] [PubMed]
2. Yan, J.; Gu, Y.; Jia, X.; Kang, W.; Pan, S.; Tang, X.; Chen, X.; Tang, G. Effective small RNA destruction by the expression of a Short Tandem Target Mimic in Arabidopsis. *Plant Cell* **2012**, *24*, 415–427. [CrossRef] [PubMed]
3. Pegler, J.L.; Oultram, J.M.J.; Grof, C.P.L.; Eamens, A.L. Profiling the Abiotic Stress Responsive microRNA Landscape of *Arabidopsis thaliana*. *Plants* **2019**, *8*, 58. [CrossRef] [PubMed]
4. Njaci, I.; Williams, B.; Castillo-González, C.; Dickman, M.B.; Zhang, X.; Mundree, S. Genome-Wide Investigation of the Role of MicroRNAs in Desiccation Tolerance in the Resurrection Grass *Tripogon loliiformis*. *Plants* **2018**, *7*, 68. [CrossRef] [PubMed]
5. López-Galiano, M.J.; García-Robles, I.; González-Hernández, A.I.; Camañes, G.; Vicedo, B.; Real, M.D.; Rausell, C. Expression of miR159 Is Altered in Tomato Plants Undergoing Drought Stress. *Plants* **2019**, *8*, 201.

6. Millar, A.A.; Lohe, A.; Wong, G. Biology and Function of miR159 in Plants. *Plants* **2019**, *8*, 255. [CrossRef] [PubMed]
7. Shahbaz, M.; Pilon, M. Conserved Cu-MicroRNAs in *Arabidopsis thaliana* Function in Copper Economy under Deficiency. *Plants* **2019**, *8*, 141. [CrossRef] [PubMed]
8. Pegler, J.L.; Oultram, J.M.J.; Grof, C.P.L.; Eamens, A.L. DRB1, DRB2 and DRB4 Are Required for Appropriate Regulation of the microRNA399/*PHOSPHATE2* Expression Module in *Arabidopsis thaliana*. *Plants* **2019**, *8*, 124. [CrossRef] [PubMed]
9. Franco-Zorrilla, J.M.; Valli, A.; Todesco, M.; Mateos, I.; Puga, M.I.; Rubio-Somoza, I.; Leyva, A.; Weigel, D.; García, J.A.; Paz-Ares, J. Target mimicry provides a new mechanism for regulation of microRNA activity. *Nat. Genet.* **2007**, *39*, 1033–1037. [CrossRef]
10. Frydrych Capelari, É.; da Fonseca, G.C.; Guzman, F.; Margis, R. Circular and Micro RNAs from *Arabidopsis thaliana* Flowers Are Simultaneously Isolated from AGO-IP Libraries. *Plants* **2019**, *8*, 302. [CrossRef] [PubMed]
11. De Felippes, F.F. Gene Regulation Mediated by microRNA-Triggered Secondary Small RNAs in Plants. *Plants* **2019**, *8*, 112. [CrossRef]
12. Zhang, Z.; Teotia, S.; Tang, J.; Tang, G. Perspectives on microRNAs and Phased Small Interfering RNAs in Maize (*Zea mays* L.): Functions and Big Impact on Agronomic Traits Enhancement. *Plants* **2019**, *8*, 170. [CrossRef]

© 2020 by the author. Licensee MDPI, Basel, Switzerland. This article is an open access article distributed under the terms and conditions of the Creative Commons Attribution (CC BY) license (http://creativecommons.org/licenses/by/4.0/).

Article

Circular and Micro RNAs from *Arabidopsis thaliana* Flowers Are Simultaneously Isolated from AGO-IP Libraries

Érika Frydrych Capelari [1,2,†], Guilherme Cordenonsi da Fonseca [1,†], Frank Guzman [1] and Rogerio Margis [1,2,3,*]

1. Programa de Pós-graduação em Biologia Celular e Molecular (PPGBCM), Centro de Biotecnologia, Universidade Federal do Rio Grande do Sul, Porto Alegre 91501-970, Brazil
2. Programa de Pós-graduação em Genética e Biologia Molecular (PPGBM), Universidade Federal do Rio Grande do Sul, Porto Alegre 91501-970, Brazil
3. Centro de Biotecnologia, Laboratório de Genomas e Populações de Plantas (LGPP), Universidade Federal do Rio Grande do Sul, Av. Bento Gonçalves, 9500—Laboratório 206 Prédio 43422, Porto Alegre 91501-970, Brazil
* Correspondence: rogerio.margis@ufrgs.br
† These authors contributed equally to this work.

Received: 26 April 2019; Accepted: 20 August 2019; Published: 26 August 2019

Abstract: Competing endogenous RNAs (ceRNAs) are natural transcripts that can act as endogenous sponges of microRNAs (miRNAs), modulating miRNA action upon target mRNAs. Circular RNAs (circRNAs) are one among the various classes of ceRNAs. They are produced from a process called back-splicing and have been identified in many eukaryotes. In plants, their effective action as a miRNA sponge was not yet demonstrated. To address this question, public mRNAseq data from Argonaute-immunoprecipitation libraries (AGO-IP) of *Arabidopsis thaliana* flowers were used in association with a bioinformatics comparative multi-method to identify putative circular RNAs. A total of 27,812 circRNAs, with at least two reads at the back-splicing junction, were identified. Further analyses were used to select those circRNAs with potential miRNAs binding sites. As AGO forms a ternary complex with miRNA and target mRNA, targets count in AGO-IP and input libraries were compared, demonstrating that mRNA targets of these miRNAs are enriched in AGO-IP libraries. Through this work, five circRNAs that may function as miRNA sponges were identified and one of them were validated by PCR and sequencing. Our findings indicate that this post-transcriptional regulation can also occur in plants.

Keywords: circRNA; microRNA; non-coding RNA; argonaute; immunoprecipitation; plant

1. Introduction

The advancements in high throughput sequencing technologies and the development of new bioinformatics tools expanded the knowledge about non-coding RNAs (ncRNAs) and their functions as regulators of gene expression. The ncRNAs can be subdivided into two major classes: (i) small non-coding RNAs (sncRNAs) and (ii) long non-coding RNAs (lncRNAs) [1]. The lncRNAs are usually more than 300 nucleotides in length and can be regulated by microRNAs (miRNAs) [2,3]. miRNAs represent small RNAs, with approximately 19–24 nucleotides, and their main function is to act as a post-transcriptional gene regulator, through the RNA Induced Silencing Complex (RISC). Argonaute (AGO) is the main protein involved in this regulatory complex. It harbors small RNAs in its active site and promotes the interaction between the miRNA sequence and the target messenger RNA (mRNA), forming a ternary miRNA:AGO:mRNA complex. It leads to a repression in gene expression either by the mRNA cleavage or by translational repression [4]. The regulation mediated by miRNA occurs through the base pairing of complementary sequences, known as miRNA response elements (MREs), between mRNA and miRNA [5].

Competing endogenous RNAs (ceRNAs) are transcripts of coding or non-coding genes that have MREs and can compete with mRNA targets for miRNAs binding. They can promote a reduction in miRNA action by decreasing their availability in the cytoplasm [6–9]. A typical example of this mechanism is represented by ncRNA IPS1, which interacts with miR-399 by mimicking the MRE of its target mRNA PHO2, in a mechanism called target mimicry [10]. It has been suggested that an interaction network exists among ceRNAs, which communicate and co-regulate themselves through competition for a limited set of miRNA [7]. Therefore, all transcripts that share similar MREs can potentially compete for a specific miRNA.

A distinct class of newly discovered endogenous non-coding RNA was denominated as circular RNA (circRNA) [11,12]. In the early 1990s, due to their low levels of expression, circRNAs were considered as being splicing artifacts, corresponding to transcripts with scrambled exon order and splicing errors [13–15]. circRNAs were also associated with pathologic agents like hepatitis delta virus (HDV) [16] and plant viroids [17]. With the advent of next-generation sequencing technology and bioinformatic tools, the identification, biogenesis, and functions of circRNAs have been described, allowing a better understanding of these molecules [18,19]. Thus, many circRNAs were shown to be expressed as abundant and stable molecules [18] in different organisms, like humans [20,21], animals [12,22], yeast [23], bacteria [24], and plants [19,23,25–29]. In addition, circRNAs exhibited development-specific, tissue-specific and cell type-specific expression in animals, suggesting a regulatory role [18,30,31].

CircRNAs are characterized by the lack of 5' caps and 3' poly-A tails. Instead, they form a covalently closed loop structure originated by back-splicing circularization in a mechanism mediated by the spliceosomes. In this process, the 3' region of a downstream exon of a given gene is linked to the 5' region of an upstream exon of that same gene. The circularization enhances the RNA stability, making circRNAs resistant to RNase R, an exonuclease that degrades linear RNAs [32]. Due to this stability, some exonic circRNAs have been shown to be at higher concentrations than their linear counterparts [18,33,34]. circRNAs can be originated from exons [15,34], introns [12,35] or both [36]. However, most of the circRNAs are originated from exons of protein-coding genes [37]. Thus, circRNAs may comprise a single or multiples exons.

Another feature of circRNAs that has aroused great interest is its multi-functionality. circRNAs have been implicated in: (i) regulation of RNA processing [22,38], (ii) transcription regulation [39], (iii) interaction with RNA binding proteins and ribonucleoproteins complexes [40,41], and (iv) acting as microRNA sponges, preventing miRNAs to bind their target mRNAs [12]. Furthermore, miRNA binding sites in circRNAs are less likely to have polymorphisms than flanking sequences or random sites, suggesting an important role of circRNAs in the regulation of miRNA activities [42]. Up to now, the study of circRNAs in plants has received much less attention, compared to the wide comprehensive knowledge of circRNAs in mammals, in which a large number of circRNAs have been identified and characterized [43].

Recent studies have shown that circRNA are present in many species of plants [19,23,25–27,29]. However, it was not yet demonstrated whether they could effectively act as miRNA sponges. To address this question, we used a publicly available sequencing data from an Argonaute-immunoprecipitation experiment (AGO-IP) from *Arabidopsis thaliana* flowers followed by sequencing of the associated RNAs [44] to screen for circRNAs with miRNA binding sites. In the present work, five putative circRNAs that may function as miRNA sponges were found, with one of them being validated by PCR and sequencing. Our findings suggest the existence of AGO-miRNA-circRNAs complexes, and contribute another step in the understanding of post-transcriptional regulation mechanisms in plants.

2. Results

2.1. Identification of circRNAs in AGO-IP Libraries

Circular RNAs with potential to act as sponges for miRNAs were identified in RNAseq data from libraries prepared from total RNA extracted from flowers A. thaliana. In a previous study, Carbonel and coworkers produced three independent libraries corresponding to Argonaute immunoprecipitation (AGO-IP) libraries [44]. These libraries were used in our analyses. Two lines of A. thaliana overexpressing the Argonaute wild type (DDH) and another overexpressing a mutant line with no ability to slice (DAH). Specific AGO-IP was carried using monoclonal antibodies directed against the human influenza hemagglutinin (HA) sequence tag present in the recombinant AGO (Figure 1).

Figure 1. Flowchart for identification of circRNAs, miRNAs and target mRNAs in AGO-IP and control libraries. The total RNA from *A. thaliana* flowers was divided in two fractions. One of them went through Argonaute immunoprecipitation (IP fraction) and the other was used as control (Input fraction). Different methodologies are represented by rhombus, while the outputs are represented by ellipses. Filled ellipses correspond to results also presented in tables.

The use of CirComPara allowed the identification of up to 29.358 circRNAs in AGO-IP RNAseq libraries (Table 1). Using the CircExplorer2 with the Segemehl anchor 86 putative circRNAs were identified, while using the Star anchor 15 and with TopHat, 23. The number of predicted circRNAs identified by FindCirc algorithm was 1422 and by the TestRealign was 27.812. The number of circRNAs hits is reduced to only three when certain methods that are more stringent are used (Table 1).

Table 1. Number of circRNAs identified in AGO-IP libraries by 5 different methods.

Identification Method		CircExplorer2			FindCirc	TestRealign
		Segemehl	Star	Tophat	-	-
CircExplorer2	Segemehl	86	9	12	10	26
	Star	-	15	7	3	3
	Tophat	-	-	23	7	7
FindCirc		-	-	-	1422	198
TestRealign		-	-	-	-	27,812

So far, we decided to focus on those circRNAs identified by at least three different methods. The description of these 12 circRNAs, including the library from which they were identified, the locus and function of parental gene, their origin and length are described in Table 2. The coordinates of the 12 circRNAs in the A. thaliana genome is listed in Supplementary Table S3. The majority of circRNAs was originated from perfect exon back-splicing, while two were produced from introns and another resulted from an imperfect exon back-splicing. The number of exons that form the chosen circRNAs varied from one to four. Their sizes ranged from 49 nt (At5g16880) to 1063 nt (At2g42170).

Table 2. Description of 12 putative circRNAs predicted by at least three methods.

Library	Gene_ID	Circ_ID	Parental Gene Function	Origin	Exons	Length (nt) ***	Methods
DDH-IP	At1g02560	circ_At1g02560	Nuclear encoded CLP protease 5	exonic	2	123	5
DDH-IP	At1g12080 **	circ_At1g12080	Vacuolar calcium-binding protein-related	exonic *	1	95	4
DDH-IP	At1g31810	circ_At1g31810	Formin Homology 14	exonic	1	50	3
DDH-IP	At1g52360	circ_At1g52360	Coatomer beta subunit	intronic	1	224	3
DDH-IP	At2g02410	circ_At2g02410	K06962—uncharacterized protein (K06962)	exonic	1	71	4
DDH-IP	At2g35940 **	circ_At2g35940	BEL1-like homeodomain 1	exonic	1	930	4
DDH-IP	At2g42170 **	circ_At2g42170	Actin family protein	exonic	4	1063	5
DDH-IP	At5g16880	circ_At5g16880	Target of Myb protein 1	exonic *	1	49	4
DDH-IP	At5g56950	circ_At5g56950	NAP-1 Nucleosome assembly protein	intronic *	1	118	4
DAH-IP	At3g01800	circ_At3g01800	Ribosome recycling factor	exonic	1	68	4
DAH-IP	At3g13990 **	circ_At3g13990	Kinase-related protein (DUF1296)	exonic	3	349	3
DAH-IP	At5g27720 **	circ_At5g27720	Small nuclear ribonucleoprotein family protein	exonic	4	321	5

* circRNA originated from an imperfect back-splicing; ** circRNAs with miRNA binding site; *** only exons considered.

2.2. circRNAs with miRNA Binding Sites

In order to identify those plants circRNAs that can function as miRNA sponges, only the five circRNAs that have binding sites to miRNAs were selected, among the 12 previous circRNAs (Table 3). The read count of each of the 12 circRNA, matching the back-splicing junction, was analyzed in both AGO-IP and control libraries, in order to identify the enrichment in the AGO-IP (Table 3). In total, AGO-IP libraries had 284,490,887 reads, while the control library had 594,458,195. From the 428 mature A. thaliana miRNAs, 14 miRNAs were predicted as having at least one of the five circRNAs as targets. 10 from these miRNAs were more abundant in AGO-IP libraries (highlighted with an *) in comparison to input library (Table 3). All the miRNAs that have predicted sites of translational inhibition were enriched. Those with cleavage sites were poorly represented or not detected at any library.

Table 3. Read counts of circRNA and microRNAs that are potentially associated.

circRNA	circRNA Read Counts ***		miRNA	miRNA Read Counts			Inhibiton By
	AGO-IP	Total RNA		AGO1-DDH	AGO1-DAH	Empty Vector	
circ_At1g12080 **	175	13	miR4221-5p *	265	171	7	Cleavage
			miR838-3p *	226	59	41	Translation
-	-	-	miR397a-5p *	1014	520	29	Translation
circ_At2g35940	8	0	miR5654-3p	2	0	0	Cleavage
			miR8182-5p *	2	9	0	Translation

Table 3. Cont.

circRNA	circRNA Read Counts ***		miRNA	miRNA Read Counts			Inhibiton By
	AGO-IP	Total RNA		AGO1-DDH	AGO1-DAH	Empty Vector	
-	-	-	miR830-3p *	29	13	0	Cleavage
-	-	-	miR833a-5p *	50	35	17	Translation
-	-	-	miR8174-3p	-	-	-	Cleavage
circ_At2g42170	8	0	miR831-3p *	81	11	3	Translation
-	-	-	miR838-3p *	226	59	41	Cleavage
-	-	-	miR4239-5p *	17	7	0	Translation
circ_At3g13990 **	17	0	miR5637-5p	-	-	-	Cleavage
-	-	-	miR780.2-3p	-	-	-	Cleavage
circ_At5g27720	17	0	miR838-3p *	226	59	41	Cleavage

* miRNA considered enriched in AGO-IP libraries; ** circRNAs validated; *** Read count normalized by the library size and with difference between assembled AGO-IPs (AGO-DAH/DDH) and Input libraries ($p < 0.05$); Expectation value ≤5.

2.3. The circRNAs Harbor Reverse Complementary Sequences of miRNAs which Targeted mRNAs Present in AGO-IP Libraries

The enriched miRNAs were selected to evaluate if their predicted target mRNAs were also present and enriched in AGO-IP libraries. In total, 260 mRNA targets were identified with an expectation range from 0.5 to 3 (Supplementary Table S2). Six out of the 10 enriched miRNAs presented mRNA targets with reads that were significantly more frequent in AGO-IP libraries than in the control input, reducing the number of predicted targets to 64 (Table 4).

Table 4. mRNAs targeted by miRNAs with circRNAs and enriched in AgoIP libraries.

Target_Access	miRNA	Expectation	Inhibition By	Lenght	Target Counts *		Function
					AgoIP	Input	
At2g38080.1		1	Cleavage	2021	58	21	Laccase/Diphenol oxidase
At5g60020.1	miR397a-5p	1	Cleavage	2049	33	18	Laccase 17
At3g06040.1		3	Cleavage	864	29	14	Ribosomal protein L12
At3g06470.1		3	Cleavage	1092	75	4	GNS1/SUR4 membrane protein
At3g54170.1		2.5	Cleavage	1262	22	10	FKBP12 interacting protein 37
At4g13070.1		2.5	Cleavage	1775	8	2	RNA-binding CRS1
At5g60040.1		2.5	Cleavage	4582	62	22	Nuclear RNA polymerase C1
At1g13350.1	miR4221-5p	3	Cleavage	2454	142	24	Protein kinase
At1g77660.1		3	Cleavage	1765	22	12	H3K4-specific methyltransferase
At2g33240.1		3	Cleavage	5313	36	12	Myosin XI D
At3g02170.1		3	Cleavage	3300	319	155	Longifolia2
At4g14510.1		3	Cleavage	2940	57	22	CRM family member 3B
At1g31650.1		3	Translation	2255	164	28	RHO guanyl-exchange factor 14
At2g38610.1		3	Translation	1452	56	26	RNA-binding KH protein
At2g35160.1	miR8182-5p	3	Cleavage	2798	20	9	SU(VAR)3-9 homolog 5
At4g22580.1		3	Cleavage	1628	39	10	Exostosin family protein
At1g23400.1		3	Cleavage	1822	81	24	RNA-binding CRS1
At1g49880.1	miR831-3p	2.5	Translation	803	50	2	FAD-linked sulfhydryl oxidase
At3g46060.1		3	Translation	1132	75	41	RAS-related protein RABE1C

Table 4. Cont.

Target_Access	miRNA	Expectation	Inhibition By	Lenght	Target Counts *		Function
					AgoIP	Input	
At2g36890.1	miR833a-5p	2.5	Cleavage	971	6	1	Myb-like DNA-binding domain
At3g12380.1		2.5	Cleavage	2323	33	16	Actin-related protein 5
At1g21740.1		3	Cleavage	2862	63	22	Protein of unknown function
At1g64180.1		3	Cleavage	2072	13	3	Intracellular transport protein
At1g70470.1		3	Cleavage	765	17	4	No annotated domains
At4g01080.1	miR838-3p	3	Cleavage	1583	98	33	Trichome-birefringence like 26
At5g09460.1		3	Cleavage	2546	124	41	Transcription Factor SAC51
At5g09461.1		3	Cleavage	2546	124	41	Conserved peptide upstream ORF
At5g20110.1		3	Cleavage	778	28	2	Dynein light chain type 1
At5g46030.1		2	Translation	732	26	12	Elongation factor Elf1 like
At2g44430.1		2.5	Translation	2196	98	19	DNA-binding protein
At5g22640.1		2.5	Translation	2814	247	115	MORN repeat-containing protein
At5g40340.1		2.5	Translation	3096	624	75	Tudor/PWWP/MBT protein
At5g56210.1		2.5	Translation	2004	22	5	WPP domain interacting protein 2
At5g62390.1		2.5	Translation	1859	349	152	BCL-2-associated athanogene 7
At5g17910.1		3	Translation	4532	178	77	No annotated domains
At5g41960.1		3	Translation	874	9	4	No annotated domains
At5g57790.1		3	Translation	1407	29	12	No annotated domains

* Read count normalized by the library size and with difference between assembled AGO-IPs (AGO-DAH/DDH) and Input libraries ($p < 0.05$).

2.4. circRNAs Validation by RT-PCR and Sequencing

PCR reactions with divergent primers were used in order to validate the back-splicing site of the five circRNAs presenting miRNA binding sites, all with more than two reads in AGO-IP libraries. Only one of the five circRNAs predicted by bioinformatics was amplified by RT-PCR using total RNAs extracted from A. thaliana flowers followed by RNase treatment and divergent primers (Figure 2). The circ_At3g13990 showed the expected electrophoretic band profile of 312 bp (Table 2). PCR negative and positive controls were done using genomic DNA (gDNA) and cDNAs from the parental gene with divergent and convergent primers, respectively. These amplification products were not detected in RNA samples from leaf, silique and steam (data not shown).

The total RT-PCR product from circ_At3g1399080 was purified and submitted to Sanger sequencing. The sequence resulted from back-splicing of At3g13990 exon 4 (E4) and exon 2 (E2) was obtained using the Primer circular Forward (PcF) (Figure 3). This result was also corroborated by 34 reads, present in AGO-IP libraries, that overhang with 3 or 4 nucleotides over the back-splicing site.

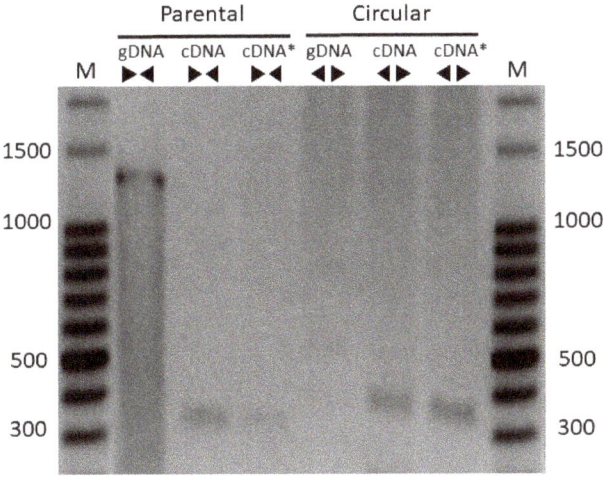

circRNA_At3g13990

Figure 2. Validation of circRNA by RT-PCR. PCR reactions were performed using divergent primers (◀▶) to amplify the circRNA_At3g13990. Convergent primers (▶◀) were used to amplify parental mRNA. Genomic DNA (gDNA) was used as control. Samples were analyzed on 1,5% agarose gel. (M) DNA size marker of 100 bp; cDNA: complementary DNA; cDNA*: complementary DNA produced from total RNA treated with RNase R previously to reverse transcription. bp: base pairs.

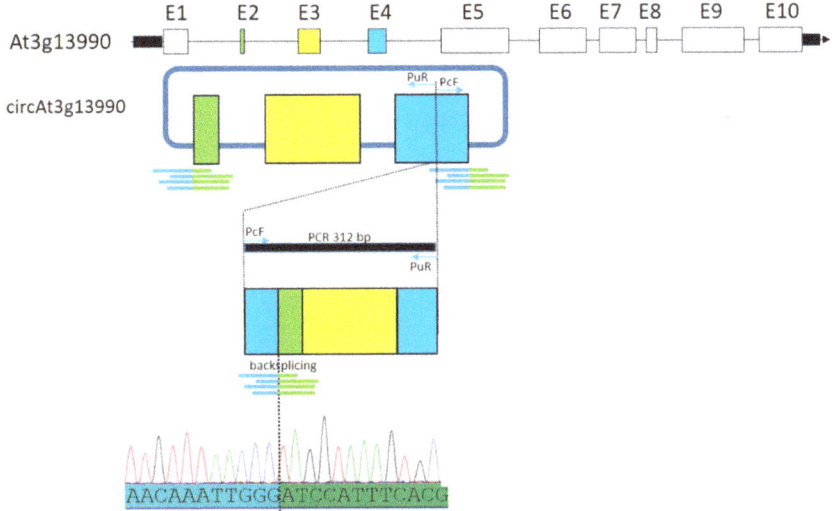

Figure 3. circRNA_At3g13990 back-splicing validation by sequencing. The parental gene structure is represented by exons (boxes), introns (black lines) and 5′ and 3′ untranslated regions (black rectangles). Filled boxes represent exons encompassing the circRNA. Sequencing reactions were performed using PcF and PuR primer. Lines indicated below the colored boxes represent reads matching the back-splicing junction. The nucleotide sequence flanking the back-splicing is represented as an electropherogram. Primer universal Reverse (PuR), Primer circular Forward (PcF) and base pairs (bp).

3. Discussion

At present, the role of circular RNA (circRNA) as one of the several classes of competing endogenous RNA (ceRNA) was only demonstrated in animals. They can act as sponges of miRNAs, modulating miRNA action upon target mRNAs. Nevertheless, circRNAs have been identified all across the eukaryotic tree of life [23]. Argonaute (AGO) is an important regulatory protein, with nuclease activity, that is involved in the pathway of RNA-induced silencing. AGO harbors a small RNA in its active site and places it in the correct sequence position in relation to the RNA target in the silencing complex (RISC). The interaction between miRNA-AGO and mRNA target forms a ternary complex and leads to transcripts regulation either by the mRNA cleavage or by translational repression [4]. Another possible molecular component in this ternary complex would be miRNA-AGO and circRNAs. Considering that circRNAs can act as miRNA sponges in mammals [12], publicly available sequencing data from AGO-IP RNAseq libraries were used to screen for circRNAs with this same function in *A. thaliana*. There are studies using the AGO-IP protocol [2,45]. However, because there are no other experiments available with AGO-IP followed by both small RNAseq and RNAseq, we used the data from Carbonell et al. [44], to develop our work.

There are several algorithms available to identify circRNAs. In this work, we used the CirComPara pipeline to detect, quantify and annotate circRNAs from RNA-seq data. This software comprises four different methods for back-splice identification. Each of them has its own features and requirements for the identification. This is the reason why we observed a such variability in the number of identified circRNAs. In order to increase the detection reliability of circRNAs, only those identified by at least 3 methods were selected. These circRNAs present a wide chromosomal distribution, since their loci are on chromosomes 1, 2, 3 and 5 of *A. thaliana*. In addition, their parental genes presented a considerable diversity of functions [46,47]. Using the PlantcircBase, which is a database for plant circular RNAs [48], only circRNA_At1g31810, circRNA_At2g35940, circRNA_At5g16880, circRNA_At3g13990 and circRNA_At5g27720 were previously identified. Besides, they are not conserved between others plant species. Until the present work, none of them had been validated. However, we show the circ_At3g13990 validation, by quantitative PCR and by sequencing. From the 12 circRNAs, only 5 have miRNA binding sites for miRNAs with read counts that are higher in AGO-IP libraries than in the empty vector library.

Interestingly, the majority of enriched miRNAs (6 out 10) have mismatches at the central region of hybridization with their target circRNAs and were predicted as having translation inhibition, which should avoid the degradation of the circRNAs, as observed in mammals' miRNAs sponges [49]. In this scenario, circRNAs would be able to capture miRNAs for longer periods and increase their sponge activity efficiency, avoiding the negative regulation of miRNAs on their target transcripts. At the same time, the miRNAs: mRNA targets found in AGO-IP libraries were predicted as being inhibited by cleavage (Table 4).

One circRNAs with miRNA binding sites (circ_At3g13990 and circ_At1g12080) was validated by both RT-PCR and sequencing. Except in flower, no amplification products were detected in the other tissues analyzed. This indicates that these circRNAs are tissue specific, a feature also observed in other works [27,50,51]. Besides that, circ_At1g12080 presented more than one amplification product in the electrophoresis analysis. This indicates that different circular isoforms can be produced from a given gene and that their expression can be specific to cell type, tissue, and developmental stage.

The circ_At3g13990 is a perfect case study. It was predicted by 3 methods; the amplified PCR product had the expected size of 312 bp and contains perfect back-splicing site confirmed by Sanger sequencing. RT-PCR performed in RNA samples treated with RNAse R produce the expected 312 bp amplification, with the same intensity as in untreated samples, thereby demonstrating the circular nature of this RNA molecule. The PCR product from the parental cDNA, which originated from RNA samples previously treated with RNase R, revealed a weak amplification. It could indicate that not all RNA was degraded. The circ_At3g13990 contains three predicted miRNA binding sites, but just one of them, miR4239-5p, was significantly more frequently found in AGO-IP libraries. It indicates that this circRNA may be acting as miRNA sponge, blocking the action of miRNA upon its target. Curiously, no mRNA target for miR4239-5p was identified among the reads enriched in the AGO-IP libraries. It suggests that the majority of miR4239-5p molecules are associated to the miRNA:AGO:circRNAs ternary complex. Thus, not allowing the formation of the alternative miRNA:AGO:mRNA complex that would downregulate the target gene expression. It is noteworthy that miR4239-5p has as predicted targets: the small RNA degrading nuclease 3-SDN3 (At5g67240), the UBP1-associated protein 2A (At3g56860) and the gamma tubulin complex protein (At3g43610). All these three gene present a higher expression in flowers and carpels compared to other tissues, according to the BAR eFP Browser from the TAIR database (www.arabidopsis.org).

Our data contributed to the knowledge about the role of circRNAs in plants, since no work until now had demonstrated the existence of a ternary complex formed by AGO:miRNA:circRNA. These findings allow us to propose that plants circRNAs are also able to act as miRNA sponges and modulate the mRNA target regulation by using miRNA.

4. Materials and Methods

4.1. mRNAseq and Small RNAs Libraries

The RNAseq and AGO-IP small RNAs libraries [44] were downloaded from Gene Expression Omnibus (GEO, accession number GSM989339—GSM989346 and GSM989350—GSM989352) of NCBI. Quality and the presence of adapters in these libraries was visualized using FastQC software (http://www.bioinformatics.babraham.ac.uk/projects/fastqc/). Next, quality trimming and adaptor removal in the small RNAs and RNAseq libraries were carried out using Cutadapt/Sickle (https://github.com/najoshi/sickle) and Trimmomatic [52], respectively.

4.2. circRNAs Identification in mRNAseq Libraries from AGO-IP

Clean data from the AGO-IP RNAseq libraries SRR546147, SRR546148, SRR546149 and SRR546150 were used to identify, quantify and annotate potential circRNAs using the CirComPara pipeline [53], which uses five different methods in parallel: FindCirc, TestRealign and CircExplorer2, that works with three different aligners (Segemhel, Star and Tophat). All methods realized back-splice identification in each library with a minimum of 2 reads. *A. thaliana* genome and annotation files obtained from Ensembl Plants (https://plants.ensembl.org/index.html) were used as references. Only circRNAs identified by at least 3 methods were selected for the subsequent analyses.

4.3. Analysis of Target mRNAs and miRNAs Counts in AGO-IP and Control Libraries

The psRNATarget tool [54] was used to identify potential miRNAs that could interact with the circRNAs identified in the AGO-IP libraries. All mature miRNAs of *A. thaliana* from miRBase release 21 [55] were used in this analysis. Those miRNAs with an Expectation value (number of mismatches allowed) of 5 or less were selected to subsequent analysis. The Bowtie algorithm [56] was used to align the small RNAs sequences from each library to the miRNA sequences of the selected miRNAs to obtain read count values. The default parameters were used for the alignment and no mismatch was allowed. The miRNAs read counts were normalized according to the size of the libraries. A miRNA was considered to be enriched in the AGO-IP libraries if the normalized read count values of the miRNA were higher in the two AGO-IP libraries compared to the control (input).

The data from the same libraries used for the identification of the circRNAs was used to evaluate if the mRNAs targets of the selected miRNAs were also enriched in the AGO-IP libraries. The putative target mRNAs were selected using the psRNATarget tool. The enriched miRNAs and the transcriptome from *A. thaliana* TAIR version 10 obtained from Phytozome database (https://phytozome.jgi.doe.gov/pz/portal.html) were selected for this analysis using a maximum expectation value of 3.

To obtain the read count value, the reads from each library were mapped against the *A. thaliana* transcriptome using the Bowtie2 algorithm [57] with the default parameters. The DESeq package from the R software [49] was used to identify the target mRNAs significantly more frequent (maximum adjusted *p*-value of 0.05) in the four AGO-IP libraries compared to the input total mRNA controls.

4.4. Plant Material and Growth Condition

A. thaliana plants of ecotype Columbia were used. After incubation in the dark at 4 °C for 3 days, seeds were cultivated in soil for six weeks, at a temperature of 22 °C and a photoperiod of 16 h of light. Samples of leaves, flowers, axis and siliques were collected and stored at liquid nitrogen for subsequent storage at −80 °C.

4.5. RNA Extraction, RNase R Treatment and cDNA Synthesis

The RNA was extracted using the Trizol (Invitrogen) reagent, according to the manufacturer's instructions. The RNA integrity was performed using 1% agarose gel electrophoresis, where it was visualized under UV light and a digital image generated by the Gel-Doc (Bio-Rad) system. Prior to cDNA synthesis, samples containing 1 g of total RNA were treated with 2 units of RNase R (Lucigen) for 60 min at 37 °C. For cDNA synthesis were used the reverse primer of each analyzed circRNA and the M-MLV Reverse Transcriptase (Promega), according to the manufacturer's instructions.

4.6. Primers Design

Primers were projected using the Primer3 tool [58]. To validate the circRNAs identified by bioinformatic, divergent primers were projected [59]. In order to amplify part of the circRNA we used the primer combination PcF/PuR and to amplify all circRNA sequence we used the primer combination PcF/PuRi. As control, a set of convergent primers were designed for the parental mRNA detection The Reverse universal primer was the same for both circRNA and mRNA detection (Supplementary Table S1).

4.7. circRNAs and Parental mRNAs Amplification and Sequencing

The expression of the five circRNAs was evaluated by RT-PCR, using divergent primers. Samples were analyzed in technical triplicates and biological quadruplicates. The Polymerase Chain Reactions (PCR) reactions were realized using the Platinum Taq DNA polymerase (Invitrogen) enzyme. All RT-PCR reactions were performed on the Applied Biosystems Veriti apparatus. PCR conditions were conducted in a volume of 20 µL containing 10 µL of the diluted cDNA (1:100), 0.4 mM dNTPs, 10× Buffer, 3 mM MgCl 2, 0.25 U Platinum Taq DNA polymerase (Invitrogen) and 0.1 µM of each oligonucleotide. PCR conditions were: an initial 2 min step at 95 °C followed by 40 cycles of 10 s denaturing at 95 °C, 15 s annealing at 60 °C and 15 s extension at 72 °C. Confirmation of the fragments was performed by 3% agarose gel electrophoresis. The circRNAs predicted by bioinformatics and confirmed by PCR were purified using the Wizard SV gel PCR clean-up system (Ludwig Biotecnologia) according to the manufacturer's recommendations. Sanger sequencing reactions were performed with purified PCR products at a final concentration of the reaction of 4.5 pmol/µL, using the PcF, PuR, or PuRi primers (Supplementary Table S1).

Supplementary Materials: The following are available online at http://www.mdpi.com/2223-7747/8/9/302/s1, Table S1: Primers used for PCR validation; Table S2: Description of total targets of miRNAs with sites in circRNAs; Table S3: Location of circRNAs identified in at least three methods.

Author Contributions: Conceptualization, E.F.C., G.C.F. and R.M.; Methodology, E.F.C.; Software, E.F.C., G.C.F. and F.G.; Validation, E.F.C. and G.C.F.; Formal Analysis, G.C.F.; Investigation, E.F.C., G.C.F. and R.M.; Data Curation, E.F.C., G.C.F., F.G. and R.M.; Writing—Original Draft Preparation, E.F.C.; Writing—Review & Editing, G.C.F. and R.M.; Supervision, R.M.; Project Administration, R.M.; Funding Acquisition, R.M.

Funding: This research was funded by Conselho Nacional de Pesquisa e Desenvolvimento Científico e Tecnológico and by Coordenação de Aperfeiçoamento de Pessoal de Nível Superior CAPES. The present study was also partially supported through a grant from INCT-Plant Stress Biotech.

Acknowledgments: RM is the recipient of a research fellowship 309030/2015-3 and EF is recipient of a PhD fellowship from Conselho Nacional de Pesquisa e Desenvolvimento Científico e Tecnológico. EF is recipient of M.Sc. fellowship and GC and FG of Post-Doctoral fellowships from Coordenação de Aperfeiçoamento de Pessoal de Nível Superior CAPES.

Conflicts of Interest: The funders had no role in the design of the study; in the collection, analyses, or interpretation of data; in the writing of the manuscript, or in the decision to publish the results. The authors declare no conflicts of interest.

References

1. Tay, Y.; Rinn, J.; Pandolfi, P.P. The multilayered complexity of ceRNA crosstalk and competition. *Nature* **2014**, *505*, 344–352. [CrossRef]
2. Chi, S.W.; Zang, J.B.; Mele, A.; Darnell, R.B. Argonaute HITS-CLIP decodes microRNA-mRNA interaction maps. *Nature* **2009**, *460*, 479–486. [CrossRef]
3. Licatalosi, D.D.; Mele, A.; Fak, J.J.; Ule, J.; Kayikci, M.; Chi, S.W.; Clark, T.A.; Schweitzer, A.C.; Blume, J.E.; Wang, X.; et al. HITS-CLIP yields genome-wide insights into brain alternative RNA processing. *Nature* **2008**, *456*, 464–469. [CrossRef]
4. Huntzinger, E.; Izaurralde, E. Gene silencing by microRNAs: Contributions of translational repression and mRNA decay. *Nat. Rev. Genet.* **2011**, *12*, 99–110. [CrossRef]
5. Pasquinelli, A.E. MicroRNAs and their targets: Recognition, regulation and an emerging reciprocal relationship. *Nat. Rev. Genet.* **2012**, *13*, 271–282. [CrossRef]
6. Poliseno, L.; Salmena, L.; Zhang, J.; Carver, B.; Haveman, W.J.; Pandolfi, P.P. A coding-independent function of gene and pseudogene mRNAs regulates tumour biology. *Nature* **2010**, *465*, 1033–1038. [CrossRef]
7. Salmena, L.; Poliseno, L.; Tay, Y.; Kats, L.; Pandolfi, P.P. A ceRNA Hypothesis: The Rosetta Stone of a Hidden RNA Language? *Cell* **2011**, *146*, 353–358. [CrossRef]
8. Seitz, H. Redefining MicroRNA Targets. *Curr. Biol.* **2009**, *19*, 870–873. [CrossRef]
9. Tay, Y.; Kats, L.; Salmena, L.; Weiss, D.; Tan, S.M.; Ala, U.; Karreth, F.; Poliseno, L.; Provero, P.; Di Cunto, F.; et al. Coding-Independent Regulation of the Tumor Suppressor PTEN by Competing Endogenous mRNAs. *Cell* **2011**, *147*, 344–357. [CrossRef]
10. Franco-Zorrilla, J.M.; Valli, A.; Todesco, M.; Mateos, I.; Puga, M.I.; Rubio-Somoza, I.; Leyva, A.; Weigel, D.; García, J.A.; Paz-Ares, J. Target mimicry provides a new mechanism for regulation of microRNA activity. *Nat. Genet.* **2007**, *39*, 1033–1037. [CrossRef]
11. Hansen, T.B.; Wiklund, E.D.; Bramsen, J.B.; Villadsen, S.B.; Statham, A.L.; Clark, S.J.; Kjems, J. miRNA-dependent gene silencing involving Ago2-mediated cleavage of a circular antisense RNA. *EMBO J.* **2011**, *30*, 4414–4422. [CrossRef]
12. Memczak, S.; Jens, M.; Elefsinioti, A.; Torti, F.; Krueger, J.; Rybak, A.; Maier, L.; Mackowiak, S.D.; Gregersen, L.H.; Munschauer, M.; et al. Circular RNAs are a large class of animal RNAs with regulatory potency. *Nature* **2013**, *495*, 333–338. [CrossRef]
13. Capel, B.; Swain, A.; Nicolis, S.; Hacker, A.; Walter, M.; Koopman, P.; Goodfellow, P.; Lovell-Badge, R. Circular transcripts of the testis-determining gene Sry in adult mouse testis. *Cell* **1993**, *73*, 1019–1030. [CrossRef]
14. Cocquerelle, C.; Daubersies, P.; Majérus, M.A.; Kerckaert, J.P.; Bailleul, B. Splicing with inverted order of exons occurs proximal to large introns. *EMBO J.* **1992**, *11*, 1095–1098. [CrossRef]
15. Nigro, J.M.; Cho, K.R.; Fearon, E.R.; Kern, S.E.; Ruppert, J.M.; Oliner, J.D.; Kinzler, K.W.; Vogelstein, B. Scrambled exons. *Cell* **1991**, *64*, 607–613. [CrossRef]
16. Kos, A.; Dijkema, R.; Arnberg, A.C.; van der Meide, P.H.; Schellekens, H. The hepatitis delta (delta) virus possesses a circular RNA. *Nature* **1986**, *323*, 558–560. [CrossRef]
17. Sanger, H.L.; Klotz, G.; Riesner, D.; Gross, H.J.; Kleinschmidt, A.K. Viroids are single-stranded covalently closed circular RNA molecules existing as highly base-paired rod-like structures. *Proc. Natl. Acad. Sci. USA* **1976**, *73*, 3852–3856. [CrossRef]
18. Salzman, J.; Chen, R.E.; Olsen, M.N.; Wang, P.L.; Brown, P.O. Cell-Type Specific Features of Circular RNA Expression. *PLoS Genet.* **2013**, *9*, e1003777. [CrossRef]

19. Ye, C.-Y.; Chen, L.; Liu, C.; Zhu, Q.-H.; Fan, L. Widespread noncoding circular RNAs in plants. *New Phytol.* **2015**, *208*, 88–95. [CrossRef]
20. Salzman, J.; Gawad, C.; Wang, P.L.; Lacayo, N.; Brown, P.O. Circular RNAs Are the Predominant Transcript Isoform from Hundreds of Human Genes in Diverse Cell Types. *PLoS ONE* **2012**, *7*, e30733. [CrossRef]
21. Zhang, Y.; Zhang, X.-O.; Chen, T.; Xiang, J.-F.; Yin, Q.-F.; Xing, Y.-H.; Zhu, S.; Yang, L.; Chen, L.-L. Circular Intronic Long Noncoding RNAs. *Mol. Cell* **2013**, *51*, 792–806. [CrossRef]
22. Zhang, X.-O.; Wang, H.-B.; Zhang, Y.; Lu, X.; Chen, L.-L.; Yang, L. Complementary Sequence-Mediated Exon Circularization. *Cell* **2014**, *159*, 134–147. [CrossRef]
23. Wang, P.L.; Bao, Y.; Yee, M.-C.; Barrett, S.P.; Hogan, G.J.; Olsen, M.N.; Dinneny, J.R.; Brown, P.O.; Salzman, J. Circular RNA Is Expressed across the Eukaryotic Tree of Life. *PLoS ONE* **2014**, *9*, e90859. [CrossRef]
24. Danan, M.; Schwartz, S.; Edelheit, S.; Sorek, R. Transcriptome-wide discovery of circular RNAs in Archaea. *Nucleic Acids Res.* **2012**, *40*, 3131–3142. [CrossRef]
25. Conn, V.M.; Hugouvieux, V.; Nayak, A.; Conos, S.A.; Capovilla, G.; Cildir, G.; Jourdain, A.; Tergaonkar, V.; Schmid, M.; Zubieta, C.; et al. A circRNA from SEPALLATA3 regulates splicing of its cognate mRNA through R-loop formation. Nature Plants, v. 3, n. 5, p. 17053, 18 abr. 2017. cognate mRNA through R-loop formation. *Nat. Plants* **2017**, *3*, 17053. [CrossRef]
26. Darbani, B.; Noeparvar, S.; Borg, S. Identification of Circular RNAs from the Parental Genes Involved in Multiple Aspects of Cellular Metabolism in Barley. *Front. Plant Sci.* **2016**, *7*, 776. [CrossRef]
27. Lu, T.; Cui, L.; Zhou, Y.; Zhu, C.; Fan, D.; Gong, H.; Zhao, Q.; Zhou, C.; Zhao, Y.; Lu, D.; et al. Transcriptome-wide investigation of circular RNAs in rice. *RNA* **2015**, *21*, 2076–2087. [CrossRef]
28. Wang, Y.; Yang, M.; Wei, S.; Qin, F.; Zhao, H.; Suo, B. Identification of Circular RNAs and Their Targets in Leaves of *Triticum aestivum* L. under Dehydration Stress. *Front. Plant Sci.* **2017**, *7*, 2024. [CrossRef]
29. Zuo, J.; Wang, Q.; Zhu, B.; Luo, Y.; Gao, L. Deciphering the roles of circRNAs on chilling injury in tomato. *Biochem. Biophys. Res. Commun.* **2016**, *479*, 132–138. [CrossRef]
30. Rybak-Wolf, A.; Stottmeister, C.; Glažar, P.; Jens, M.; Pino, N.; Giusti, S.; Hanan, M.; Behm, M.; Bartok, O.; Ashwal-Fluss, R.; et al. Circular RNAs in the Mammalian Brain Are Highly Abundant, Conserved, and Dynamically Expressed. *Mol. Cell* **2015**, *58*, 870–885. [CrossRef]
31. Westholm, J.O.; Miura, P.; Olson, S.; Shenker, S.; Joseph, B.; Sanfilippo, P.; Celniker, S.E.; Graveley, B.R.; Lai, E.C. Genome-wide Analysis of Drosophila Circular RNAs Reveals Their Structural and Sequence Properties and Age-Dependent Neural Accumulation. *Cell Rep.* **2014**, *9*, 1966–1980. [CrossRef]
32. Suzuki, H.; Tsukahara, T. A view of pre-mRNA splicing from RNase R resistant RNAs. *Int. J. Mol. Sci.* **2014**, *15*, 9331–9342. [CrossRef]
33. Burd, C.E.; Jeck, W.R.; Liu, Y.; Sanoff, H.K.; Wang, Z.; Sharpless, N.E. Expression of Linear and Novel Circular Forms of an INK4/ARF-Associated Non-Coding RNA Correlates with Atherosclerosis Risk. *PLoS Genet.* **2010**, *6*, e1001233. [CrossRef]
34. Jeck, W.R.; Sorrentino, J.A.; Wang, K.; Slevin, M.K.; Burd, C.E.; Liu, J.; Marzluff, W.F.; Sharpless, N.E. Circular RNAs are abundant, conserved, and associated with ALU repeats. *RNA* **2013**, *19*, 141–157. [CrossRef]
35. Talhouarne, G.J.S.; Gall, J.G. Lariat intronic RNAs in the cytoplasm of Xenopus tropicalis oocytes. *RNA* **2014**, *20*, 1476–1487. [CrossRef]
36. Li, Z.; Huang, C.; Bao, C.; Chen, L.; Lin, M.; Wang, X.; Zhong, G.; Yu, B.; Hu, W.; Dai, L.; et al. Exon-intron circular RNAs regulate transcription in the nucleus. *Nat. Struct. Mol. Biol.* **2015**, *22*, 256–264. [CrossRef]
37. Suzuki, H.; Zuo, Y.; Wang, J.; Zhang, M.Q.; Malhotra, A.; Mayeda, A. Characterization of RNase R-digested cellular RNA source that consists of lariat and circular RNAs from pre-mRNA splicing. *Nucleic Acids Res.* **2006**, *34*, e63. [CrossRef]
38. Ashwal-Fluss, R.; Meyer, M.; Pamudurti, N.R.; Ivanov, A.; Bartok, O.; Hanan, M.; Evantal, N.; Memczak, S.; Rajewsky, N.; Kadener, S. circRNA Biogenesis Competes with Pre-mRNA Splicing. *Mol. Cell* **2014**, *56*, 55–66. [CrossRef]
39. Chao, C.W.; Chan, D.C.; Kuo, A.; Leder, P. The mouse formin (Fmn) gene: Abundant circular RNA transcripts and gene-targeted deletion analysis. *Mol. Med.* **1998**, *4*, 614–628. [CrossRef]
40. Hentze, M.W.; Preiss, T. Circular RNAs: splicing's enigma variations. *EMBO J.* **2013**, *32*, 923–925. [CrossRef]
41. Romeo, T. Global regulation by the small RNA-binding protein CsrA and the non-coding RNA molecule CsrB. *Mol. Microbiol.* **1998**, *29*, 1321–1330. [CrossRef]

42. Thomas, L.F.; Sætrom, P. Circular RNAs are depleted of polymorphisms at microRNA binding sites. *Bioinformatics* **2014**, *30*, 2243–2246. [CrossRef]
43. Sablok, G.; Zhao, H.; Sun, X. Plant Circular RNAs (circRNAs): Transcriptional Regulation Beyond miRNAs in Plants. *Mol. Plant* **2016**, *9*, 192–194. [CrossRef]
44. Carbonell, A.; Fahlgren, N.; Garcia-Ruiz, H.; Gilbert, K.B.; Montgomery, T.A.; Nguyen, T.; Cuperus, J.T.; Carrington, J.C. Functional Analysis of Three Arabidopsis ARGONAUTES Using Slicer-Defective Mutants. *Plant Cell* **2012**, *24*, 3613–3629. [CrossRef]
45. Voinnet, O.; Ponce, M.R.; Vaucheret, H.; Baumberger, N.; Sarazin, A.; Clavel, M.; Micol, J.L.; Ziegler-Graff, V.; Genschik, P.; Derrien, B.; et al. A Suppressor Screen for AGO1 Degradation by the Viral F-Box P0 Protein Uncovers a Role for AGO DUF1785 in sRNA Duplex Unwinding. *Plant Cell* **2018**, *30*, 1353–1374.
46. Lasda, E.; Parker, R. Circular RNAs: Diversity of form and function. *RNA* **2014**, *20*, 1829–1842. [CrossRef]
47. Li, X.; Yang, L.; Chen, L.-L. The Biogenesis, Functions, and Challenges of Circular RNAs. *Mol. Cell* **2018**, *71*, 428–442. [CrossRef]
48. Chu, Q.; Zhang, X.; Zhu, X.; Liu, C.; Mao, L.; Ye, C.; Zhu, Q.; Fan, L. PlantcircBase: A database for plant circular RNAs. *Mol. Plant.* **2017**, *10*, 1126–1128. [CrossRef]
49. Anders, S.; Huber, W. Differential expression analysis for sequence count data. *Genome Biol.* **2010**, *11*, R106. [CrossRef]
50. Ye, J.; Wang, L.; Li, S.; Zhang, Q.; Zhang, Q.; Tang, W.; Wang, K.; Song, K.; Sablok, G.; Sun, X.; et al. AtCircDB: A tissue-specific database for Arabidopsis circular RNAs. *Brief. Bioinform.* **2017**, *20*, 58–65. [CrossRef]
51. Zhao, W.; Cheng, Y.; Zhang, C.; You, Q.; Shen, X.; Guo, W.; Jiao, Y. Genome-wide identification and characterization of circular RNAs by high throughput sequencing in soybean. *Sci. Rep.* **2017**, *7*, 5636. [CrossRef]
52. Bolger, A.M.; Lohse, M.; Usadel, B. Trimmomatic: A flexible trimmer for Illumina sequence data. *Bioinformatics* **2014**, *30*, 2114–2120. [CrossRef]
53. Gaffo, E.; Bonizzato, A.; Kronnie, G.; Bortoluzzi, S. CirComPara: A Multi-Method Comparative Bioinformatics Pipeline to Detect and Study circRNAs from RNA-seq Data. *Non Coding RNA* **2017**, *3*, 8. [CrossRef]
54. Dai, X.; Zhao, P.X. psRNATarget: A plant small RNA target analysis server. *Nucleic Acids Res.* **2011**, *39*, W155–W159. [CrossRef]
55. Kozomara, A.; Griffiths-Jones, S. miRBase: Annotating high confidence microRNAs using deep sequencing data. *Nucleic Acids Res.* **2014**, *42*, D68–D73. [CrossRef]
56. Langmead, B.; Trapnell, C.; Pop, M.; Salzberg, S.L. Ultrafast and memory-efficient alignment of short DNA sequences to the human genome. *Genome Biol.* **2009**, *10*, R25. [CrossRef]
57. Langmead, B.; Salzberg, S.L. Fast gapped-read alignment with Bowtie 2. *Nat. Methods* **2012**, *9*, 357–359. [CrossRef]
58. Rozen, S.; Skaletsky, H. Primer3 on the WWW for general users and for biologist programmers. *Methods Mol. Biol.* **2000**, *132*, 365–386.
59. Kulcheski, F.R.; Christoff, A.P.; Margis, R. Circular RNAs are miRNA sponges and can be used as a new class of biomarker. *J. Biotechnol.* **2016**, *238*, 42–51. [CrossRef]

© 2019 by the authors. Licensee MDPI, Basel, Switzerland. This article is an open access article distributed under the terms and conditions of the Creative Commons Attribution (CC BY) license (http://creativecommons.org/licenses/by/4.0/).

Article

Expression of miR159 Is Altered in Tomato Plants Undergoing Drought Stress

María José López-Galiano [1], Inmaculada García-Robles [1], Ana I. González-Hernández [2], Gemma Camañes [2], Begonya Vicedo [2], M. Dolores Real [1] and Carolina Rausell [1,*]

1. Department of Genetics, University of Valencia, Burjassot, 46100 Valencia, Spain
2. Plant Physiology Area, Biochemistry and Biotechnology Group, Department CAMN, University Jaume I, 12071 Castellón, Spain
* Correspondence: carolina.rausell@uv.es; Tel.: +34-96-354-3397

Received: 27 May 2019; Accepted: 27 June 2019; Published: 2 July 2019

Abstract: In a scenario of global climate change, water scarcity is a major threat for agriculture, severely limiting crop yields. Therefore, alternatives are urgently needed for improving plant adaptation to drought stress. Among them, gene expression reprogramming by microRNAs (miRNAs) might offer a biotechnologically sound strategy. Drought-responsive miRNAs have been reported in many plant species, and some of them are known to participate in complex regulatory networks via their regulation of transcription factors involved in water stress signaling. We explored the role of miR159 in the response of *Solanum lycopersicum* Mill. plants to drought stress by analyzing the expression of sly-miR159 and its target SlMYB transcription factor genes in tomato plants of cv. Ailsa Craig grown in deprived water conditions or in response to mechanical damage caused by the Colorado potato beetle, a devastating insect pest of Solanaceae plants. Results showed that sly-miR159 regulatory function in the tomato plants response to distinct stresses might be mediated by differential stress-specific MYB transcription factor targeting. sly-miR159 targeting of SlMYB33 transcription factor transcript correlated with accumulation of the osmoprotective compounds proline and putrescine, which promote drought tolerance. This highlights the potential role of sly-miR159 in tomato plants' adaptation to water deficit conditions.

Keywords: *Solanum lycopersicum*; drought; Colorado potato beetle; miR159; MYB transcription factors; *P5CS*; proline; putrescine

1. Introduction

Climate change due to increasing concentration of CO_2 in the atmosphere is leading to rising temperatures, altered rainfall patterns, and more frequent and severe drought episodes [1], which negatively impact crop production. Therefore, gaining knowledge about how plants regulate their adaptation to stress is critical to find ways to enhance plant performance in eventually drier environments.

To cope with drought, plants activate a complex cascade of events at the cellular level that include extensive metabolic and gene transcriptional reprogramming to protect cells from osmotic stress, and limit water loss. The response of plants to drought stress involves genes related to diverse functional categories such as genes encoding proteins participating in the direct protection of essential proteins and membranes (osmoprotectants, free radical scavengers, etc.), genes encoding membrane transporters and ion channels that promote water uptake, and genes encoding stress related regulatory proteins such as kinases and transcription factors belonging to the V-myb myeloblastosis viral oncogene homolog (MYB), basic-helix-loop-helix (bHLH), basic region/leucine zipper (bZIP), NAM, ATAF1/2, and CUC (NAC), and APETALA2/ethylene-responsive element binding protein (AP2/EREBP) families [2].

The phytohormone Abscisic acid (ABA) coordinates the plant's response to reduced water availability by modulating the expression of some of the drought responsive genes [3]. Interestingly, microRNAs (miRNAs) have been recently reported to mediate drought tolerance by post-transcriptionally regulating drought-responsive genes, some of which are known to be controlled by ABA signaling pathways [4]. An example of such intricate regulatory network is provided by miR159, which in *Arabidopsis* germinating seeds, has been reported to be induced by ABA and drought treatments, and promote transcript cleavage of the ABA positive regulators MYB33 and MYB101 transcription factors, thereby playing a key role in ABA response [5].

The miR159 family is highly conserved among monocot and dicot plants, but in plants undergoing drought, miR159's relative abundance varies in a tissue- and species-specific manner. For instance, miR159 was reported to be up-regulated by drought stress in *Arabidopsis* [6], and maize [7], but down-regulated in cotton [8], and potato [9], whereas in barley and alfalfa, miR159 was down-regulated in roots and up-regulated in leaves in response to drought stress [10,11]. Pegler et al. [12] proposed that the differential miRNA abundance across species following drought or salt stress exposure might be in part due to differential distribution of regulatory transcription factor binding sites within the putative promoter region of the miRNA gene, which encodes the highly conserved, stress-responsive miRNA.

To expand our knowledge on the miR159 regulatory network involved in tomato plants' response to drought stress, in the present work we analyzed the expression of miR159 and its predicted target genes in tomato plants of *Solanum lycopersicum* Mill. cv. Ailsa Craig undergoing drought stress, in which we previously reported that ABA hormone is accumulated after water deprivation [13].

2. Results and Discussion

2.1. Expression of miR159 in Tomato Plants Undergoing Drought Stress

To assess miR159 expression in tomato plants of *Solanum lycopersicum* Mill. cv. Ailsa Craig following a seven-day water deprivation, we analyzed sly-miR159 (GenBank: 102464332) transcript levels by RT-qPCR in control tomato plants and plants undergoing drought stress. Results showed significantly reduced expression of sly-miR159 in response to stress (Figure 1A). However, in recent high-throughput sequencing studies performed by Liu et al. [14,15], miR159 was not found among the miRNAs differentially expressed after 10 days of drought stress in a sensitive and a tolerant tomato cultivar. This apparent discrepancy with our results might be due to the differential experimental conditions or techniques used to measure miRNA expression, but is most probably due to the fact that the tomato cultivars were different, since it has been reported that miRNAs respond to environmental stresses in a genotype-dependent manner [16]. As in plants, most miRNAs negatively regulate their target genes, we hypothesized that sly-miR159 gene targets that are upregulated in tomato plants grown in water-limited conditions in our experimental conditions may play beneficial roles in the adaptive responses to drought stress.

In *Arabidopsis*, a clade of seven closely related *GAMYB*-like genes (*MYB33*, *MYB101*, *MYB65*, *MYB81*, *MYB97*, *MYB104*, and *MYB120*) share a conserved putative miR159-binding site [17]. The *GAMYB*-like genes encode a highly conserved family of R2R3-type MYB domain transcription factors that are regulated by Gibberellic acid (GA) and ABA and participate in the GA signaling pathway [18]. Recent studies in potato plants highlight the involvement of miR159 and its targets *GAMYB*-like genes in the response of this species to water stress [9]. Using psRNATarget software [19] we identified the following putative *GAMYB*-like transcription factor genes that are sly-miR159 targets in tomato: *SlMYB33* (Solyc01g009070.2.1), *SlMYB65* (Solyc06g073640.2.1), *SlMYB104* (Solyc11g072060.1.1), *SlMYB97* (Solyc10g019260.1.1), and *SlMYB120* (Solyc01g090530.1.1). Figure 1B shows the nucleotide sequence of the sly-miR159-binding sites in the tomato *SlMYB* transcripts identified, which strongly resemble those found in *AtMYB* transcripts targeted by miR159 in *Arabidopsis* [20].

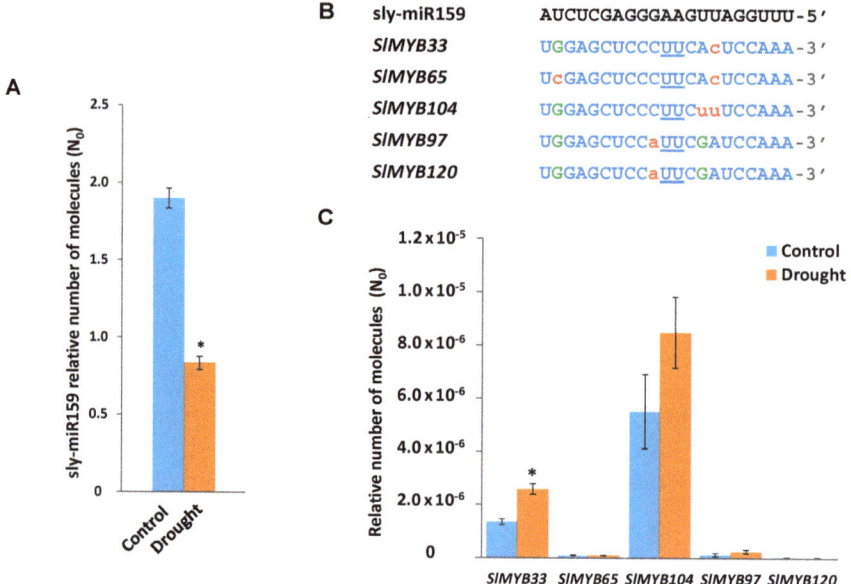

Figure 1. RT-qPCR analysis of sly-miR159 expression and its *MYB* predicted targets in tomato plants undergoing drought stress. (**A**) RT-qPCR analysis of sly-miR159 expression in control tomato plants and tomato plants following 7-day water deprivation. (**B**) Nucleotide sequence of sly-miR159-binding sites in tomato GAMYB-like transcripts. Nucleotides in the cleavage site are underlined, lower-case red letters indicate mismatches to sly-miR159, and G:U pairing is shown in uppercase green letters. (**C**) RT-qPCR analysis of *SlMYB33*, *SlMYB65*, *SlMYB104*, *SlMYB97*, and *SlMYB120* genes expression in control tomato plants and tomato plants following 7-day water deprivation. In panels (A) and (C), data shown are the mean of three independent experiments ± standard error (SE). Asterisk indicates that differences between means of control and undergoing drought stress tomato plants were statistically significant (Student's *t*-test, $p < 0.05$).

Li et al. [21] identified 127 *MYB* genes in the tomato genome and classified the corresponding proteins into 18 subgroups based on domain similarity and phylogenetic topology, and suggested that conserved motifs outside the MYB domain might reflect their functional conservation. SlMYB33, SlMYB65, and SlMYB104 proteins cluster in subgroup 12, in which the three of them are the only ones (out of the thirteen subgroup members) sharing the conserved motifs 14 and 15 outside the MYB domain. SlMYB97 and SlMYB120 proteins constitute subgroup 15, which is composed only by these two MYB proteins that have no conserved motifs outside the MYB domain.

We analyzed the expression of sly-miR159 *MYB* predicted targets in control tomato plants and plants undergoing drought stress by RT-qPCR (Figure 1C). Only *SlMYB33* gene showed statistically significant induction in water-stressed tomato plants, exhibiting an opposite pattern of expression relative to that of sly-miR159, which suggests that this *MYB* gene may be regulated by sly-miR159 in tomato plants in response to drought stress. In line with this hypothesis, in potato plants in which the *CBP80* gene encoding a protein involved in RNA processing was silenced, improved tolerance to water stress was correlated with decreased levels of miR159 and enhanced *MYB33* gene expression [22].

To further assess the involvement of sly-miR159 in the regulation of *SlMYB33* gene expression under drought stress, we aimed at analyzing *SlMYB33* cleavage fragments. We designed two pairs of primers to amplify *SlMYB33* mRNA fragments in small RNA samples isolated from total RNA of control tomato plants and tomato plants following a seven-day water deprivation (Materials and Methods, Section 4.3). Figure 2A shows the annealing positions of both pairs of PCR primers. The primer pair

O$_{Fw}$ and O$_{Rv}$ anneals to sequences within a *SlMYB33* mRNA region downstream of the predicted sly-miR159-binding site, yielding a 199 bp *SlMYB33* amplification product. The primer pair F$_{Fw}$ and F$_{Rv}$ anneals to sequences flanking the putative cleavage site in the predicted sly-miR159-binding region, yielding a 200 bp *SlMYB33* amplification product only when the SlMYB33 mRNA is not cleaved at the sly-miR159 cleavage site. Therefore, we hypothesized that if sly-miR159 is not involved in the regulation of *SlMYB33* gene expression of the same amplification patterns of control vs. drought, then small RNA samples with both primer pairs would be expected. Figure 2B shows the results obtained in the RT-PCR amplifications using the two pairs of primers. Lower amounts of amplification products were obtained using primers O$_{Fw}$ and O$_{Rv}$ in tomato plants grown under water scarcity compared to control plants. In contrast, higher amount of PCR amplified product was observed in drought-stressed tomato plants than in control tomato plants using primers F$_{Fw}$ and F$_{Rv}$. Collectively, these results support targeted cleavage of *SlMYB33* transcripts by sly-miR159 that might participate in the transcriptional regulation of the tomato plants' response to drought stress.

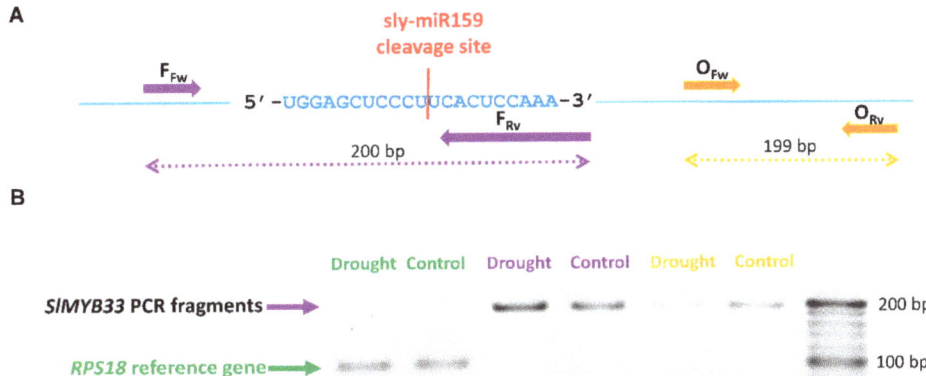

Figure 2. RT-PCR amplification of SlMYB33 mRNA fragments. (**A**) Nucleotide sequence of sly-miR159-binding sites in SlMYB33 transcripts. Bar in red depicts the putative cleavage site and arrows indicate the annealing positions of primer pair OFw and ORv, outside the sly-miR159-binding region, and primer pair FFw and FRv, flanking the putative cleavage site in sly-miR159-binding region. (**B**) RT-PCR analysis of SlMYB33 small RNA fragments in control tomato plants and tomato plants following 7-day water deprivation using primers OFw and ORv, or FFw and FRv. RPS18 gene expression was used as normalization control. For each sample, three biological replicates were pooled and analyzed.

Interestingly, Qin et al. [23] proposed that MYB33 transcription factor may enhance drought tolerance by means of promoting osmotic pressure balance reconstruction and reactive oxidative species (ROS) scavenging, since ectopic over-expression of wheat *MYB33* gene in *Arabidopsis* induced the expression of *AtP5CS* and *AtZAT12* genes involved in proline synthesis and ascorbate peroxidase synthesis, respectively. Accordingly, we observed an induction of *SlP5CS* gene expression and a remarkable increase in proline levels relative to other amino acids in tomato plants grown in water-shortage conditions compared to irrigated control plants (Figure 3A,B), suggesting that sly-miR159 might participate in the tomato plants' adaptive response to drought stress via induction of *SlMYB33* transcription factor gene expression. Nevertheless, further research is needed to demonstrate whether the sly-miR159-SlMYB33 pathway is necessary for drought tolerance in the tomato cultivar Ailsa Craig.

Tonon et al. [24] proposed a strong metabolic coordination between polyamines and proline pathways in response to osmotic stresses. Therefore, we analyzed polyamine levels in tomato plants undergoing drought stress and non-stressed control plants (Figure 3C), and results showed increased accumulation of putrescine, a polyamine reported to have a role in protecting plants during water-deficient conditions, as well as oxidative stress [25]. In wheat, Pál et al. [26] recently described that ABA pre-treatments induced the expression of *P5CS* gene and enhanced the accumulation of putrescine. Authors suggested that the connection between polyamine metabolism and ABA signaling may control the regulation and maintenance of polyamine and proline levels under osmotic stress conditions in wheat seedlings.

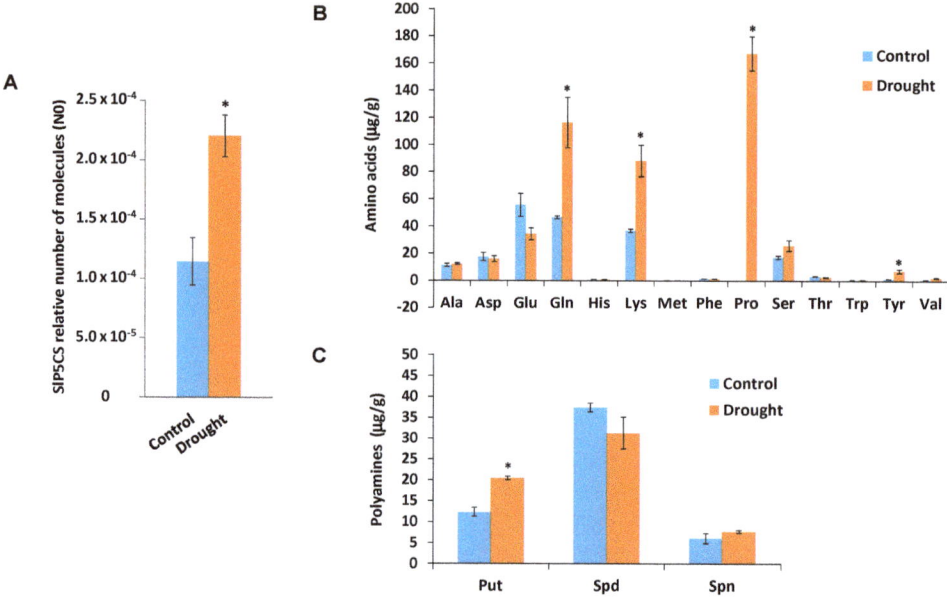

Figure 3. Analysis of *SlP5CS* gene expression, and amino acid and polyamines in tomato plants undergoing drought stress. (**A**) RT-qPCR analysis of *SlP5CS* expression in control tomato plants and tomato plants following 7-day water deprivation. (**B**) Amino acids levels upon drought treatment. Amino acids levels are expressed in µg/g DW. (**C**) Polyamines levels upon drought treatment. Polyamines levels are expressed in µg/g DW. Put (putrescine), Spd (Spermidine), Spn (Spermine). Tomato leaves were collected from plants that were properly irrigated (Control) or deprived of water 1 week (Drought). Data shown are the mean of three independent experiments ± standard error (SE). Asterisk indicates that differences between means of control and undergoing drought stress tomato plants were statistically significant (Student's *t*-test, $p < 0.05$).

2.2. Assessment of sly-miR159 Stress-Specific Targeting of SlMYB33

To ascertain whether *SlMYB33* targeting by sly-miR159 is stress-specific, we analyzed the expression of sly-miR159 and its predicted MYB target genes in tomato plants attacked by the coleopteran insect pest Colorado potato beetle (CPB), in which we previously reported that, as opposed to tomato plants undergoing drought stress, ABA was not accumulated [13]. In the present work, neither *SlP5CS* gene expression were induced, nor were increased proline and putrescine levels observed in infested tomato plants compared to tomato control plants (Figure 4), corroborating that the plants' response to this biotic stress is different from the plant response to water stress.

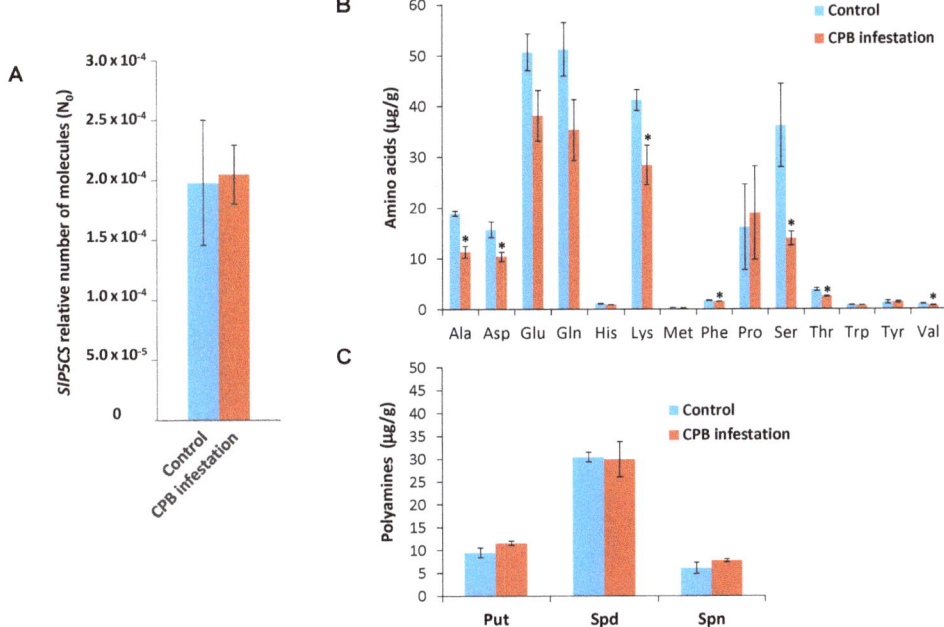

Figure 4. Analysis of *SlP5CS* gene expression, and amino acid and polyamines in tomato plants infested by Colorado potato beetle (CPB) larvae. (**A**) RT-qPCR analysis of *SlP5CS* expression in control tomato plants and tomato plants infested by CPB larvae. (**B**) Amino acids levels upon CPB infestation. Tomato leaves were collected from non-infested plants (Control) or plants infested by CPB. Amino acids levels are expressed in µg/g DW. (**C**) Polyamines levels upon CPB larvae infestation. Tomato leaves were collected from non-infested plants (Control) or plants infested by CPB. Polyamines levels are expressed in µg/g DW. Put (putrescine), Spd (Spermidine), Spn (Spermine). Data shown are the mean of three independent experiments ± standard error (SE). Asterisk indicates that differences between means of control and undergoing drought stress tomato plants were statistically significant (Student´s *t*-test, $p < 0.05$).

Intriguingly, as it was observed in plants deprived of water, in infested tomato plants, sly-miR159 was significantly down-regulated compared to non-infested control plants (Figure 5A). However, in plants attacked by CPB, among sly-miR159 putative *MYB* targets, only the *SlMYB104* transcript factor gene was significantly up-regulated (Figure 5B), suggesting that sly-miR159 might be regulating this specific MYB transcription factor in response to CPB damage.

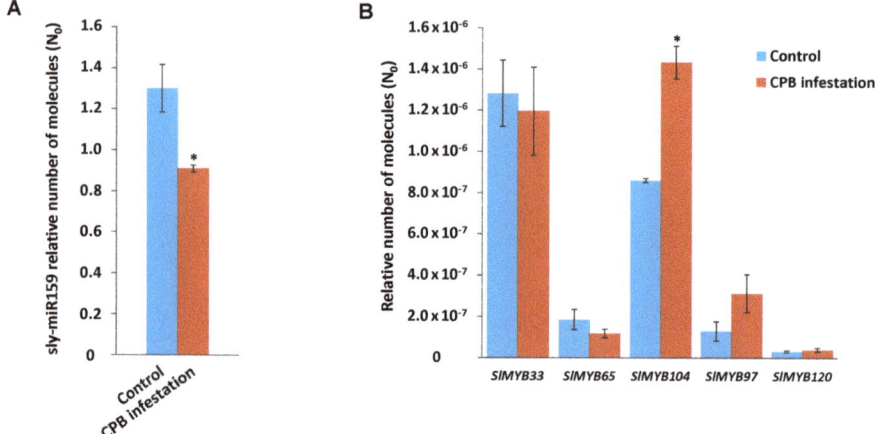

Figure 5. RT-qPCR analysis of sly-miR159 expression and its *MYB* predicted targets in tomato plants infested by CPB larvae. (**A**) RT-qPCR analysis of sly-miR159 expression in control tomato plants and tomato plants infested by CPB larvae. (**B**) RT-qPCR analysis of *SlMYB33*, *SlMYB65*, *SlMYB104*, *SlMYB97*, and *SlMYB120* genes expression in control tomato plants and tomato plants upon CPB larvae infestation. In panels (A) and (B), data shown are the mean of three independent experiments ± standard error (SE). Asterisk indicates that differences between means of control tomato plants and tomato plants infested by CPB larvae were statistically significant (Student´s *t*-test, $p < 0.05$).

In contrast, correlating with the lack of proline and putrescine accumulation, no variation was detected in *SlMYB33* transcription factor gene expression. This suggests that the specificity of the stress response regulated by sly-miR159 might, at least in part, rely on the distinct *MYB* transcription factor transcript that the sly-miR159 sRNA specifically regulates under each stress condition. It has been proposed that additional factors other than complementarity and cleavage, such as target accessibility and secondary structure, RNA binding proteins, and target site context may modulate silencing efficiency [27], which might lie at the root of the stress specific miR159 regulation of MYB transcription factors, and deserve further research.

3. Conclusions

Overall, the results obtained in this work show the potential involvement of sly-miR159 in the tomato plants' response to different stresses through stress-specific *MYB* transcription factor targeting. Under drought-stress, sly-miR159 targeting of *SlMYB33* correlates with induction of *SlP5CS* gene expression and accumulation of the osmoprotective compounds proline and putrescine, pointing to the possible participation of this miR in the regulation of drought stress tolerance. Understanding the regulatory network underlying drought stress response may provide new biotechnological approaches to generate plants better adapted to dry environments. Our results support that in addition to using *SlMYB33* transcription factor as a biotechnological target for metabolic engineering by ectopic expression, *SlMYB33* gene expression reprogramming by sly-miR159 might develop into a useful system to improve plant drought tolerance in tomato plants.

4. Materials and Methods

4.1. Plants

Thirty-day-old tomato plants of *Solanum lycopersicum* Mill. cv. Ailsa Craig (four-week-old) were grown from germinated seeds in a growth chamber under the following environmental conditions: 16/8 h light/night cycle, 26/18 °C day/night temperature cycle, and 60% relative humidity (RH). Seeds were irrigated twice a week with distilled water during the first week, and with Hoagland solution thereafter [28].

For drought stress experiments, thirty-day-old tomato plants were deprived of water for 7 days, and leaf tissue from 3rd and 4th leaves was collected, frozen in liquid nitrogen, and stored at −80 °C. Leaf tissue from 3rd and 4th leaves of irrigated plants was also collected as control.

For Colorado potato beetle (CPB) infestation, 15 CPB larvae of different developmental stages were placed on the 3rd and 4th leaves of thirty-day-old tomato plants. When necessary, non-cooperative larvae (molting or not eating) were removed and substituted. Leaf tissue left after 3 h of CPB feeding and that of the non-infested control plants were harvested, frozen in liquid nitrogen, and stored at −80 °C.

4.2. Total RNA Isolation and RT-qPCR Analysis

Total RNA was isolated from leaves of control tomato plants and plants undergoing drought stress or CPB infestation using RiboPure Kit (Ambion, Cat. No. AM1924), following the manufacturer's protocol. TURBO DNA-free kit (Ambion, Cat. No. AM1907) was used to remove contaminating genomic DNA from RNA preparations and RNA quality was evaluated by 1% agarose gel electrophoresis and quantified spectrophotometrically (NanoDrop 2000, Thermo Scientific, Waltham, MA, USA).

RT-qPCR amplification was performed using SYBR Premix Ex Taq II (Takara).

For sly-miR159 amplification 1 μg of RNA was polyadenylated in a final volume of 10 μL, including 1 μL of 10x poly(A) polymerase buffer, 1 mM of ATP, and 1 unit of poly(A) polymerase (New England Biolabs, Ipswich, MA, USA), and incubated at 37 °C for 15 min and then at 65 °C for 20 min. Polyadenylated RNA was reverse transcribed to complementary DNA (cDNA) using the Universal RT-primer (Integrated DNA Technologies, Coralville, IA, USA) described in Balcells et al. [29] (5′-CAGGTCCAGTTTTTTTTTTTTTTTVN-3′, where V is A, C, and G, and N is A, C, G, and T). Reverse transcription reaction was performed using PrimeScript™ RT reagent Kit (Takara) in a final volume of 10 μL, including 2 μL of 5X PrimeScript™ Buffer, 0.5 μL of PrimeScript™ RT Enzyme Mix I, and 1 μM of Universal RT-primer, and it was incubated at 37 °C for 15 min followed by enzyme inactivation at 85 °C for 5 s. Forward and reverse primers for miRNA RT-qPCR amplification were designed according to Balcells et al. [29] (Table 1).

Table 1. Primers used to analyze by RT-qPCR sly-miR159, *SlMYB*, and *SlP5CS* gene expression in tomato plants.

Gene	Forward Primer (5′-3′)	Reverse Primer (5′-3′)	Product Size (bp)
sly-miR159	CGCAGTTTGGATTGAAGGGAG	CAGGTCCAGTTTTTTTTTTTTTTTAGAG	50
SlMYB33	TATGGGCATCCAGTCTCTCC	TGGGACTGGAAAAGATCGTC	199
SlMYB65	TCTGCTGCATCGGTGTTTAG	TCTGGCCTGGGACAGATAAG	164
SlMYB104	TTTCGGAATTGTTTGGAAGC	TGAAGAAGTTGCCGACAATG	110
SlMYB97	CATGTCCCCTTGGAAGATTTAG	CTAGTGGCAAAGCAAAGTCATC	181
SlMYB120	CACATTCCAGTCCAAACCAAC	CCTAGGTCGGAAGCACTGAG	116
SlP5CS	TGCTCAACAGGCCGGATATG	AAAGTGTGACCAAGGGGCTC	126
U6 snRNA	GGGGACATCCGATAAAATTGGAAC	TGGACCATTTCTCGATTTGTGC	88
RPS18	GGGCATTCGTATTTCATAGTCAGAG	CGGTTCTTGATTAATGAAAACATCCT	105

For *SlMYB33*, *SlMYB65*, *SlMYB104*, *SlMYB97*, *SlMYB120*, and *SlP5CS* transcript amplification, the PrimeScript™ RT reagent kit (Takara) was used for cDNA synthesis according to the manufacturer's protocol using 50 ng/μL oligo(dT) (Promega), and 2.5 μM random hexamers (Applied Biosystems). Ten ng cDNA, and gene specific forward (F) and reverse (R) primers (Table 1), designed with PRIMER3PLUS software [30], were used.

A StepOnePlus Real-Time PCR system (Applied Biosystems) was used, under the conditions recommended by the manufacturer, and the cycling parameters were: Initial polymerase activation step at 95 °C for 30 s, 40 cycles of denaturation at 95 °C for 5 s, annealing, and elongation at 60 °C for 30 s. For each sample, three biological replicates (with 3 technical replicates each) were analyzed. Relative-fold calculations were made using *RPS18* (ribosomal protein S18, GeneBank: 3950409) gene to normalize gene expression, and *U6* snRNA gene (GenBank: X51447.1) to normalize sly-miR159 expression (Table 1). LingReg software [31] was employed for the analysis of RT-qPCR experiments and data were analyzed by Student's *t*-test for statistically significant differences ($p < 0.05$).

Each biological sample from the 3rd and 4th leaves of plants undergoing drought stress and their corresponding controls consisted of a pool of total RNA from 25 plants. Biological samples in CPB infestation experiments and their corresponding controls also consisted of a pool of total RNA from 25 plants.

4.3. Small RNA Isolation and RT-PCR Analysis

The small RNA fraction in total RNA samples of control tomato plants and tomato plants following 7-day water deprivation was isolated using Nucleospin® miRNA (Macherey-Nagel, Bethlehem, PA, USA) following the manufacturer's instructions.

For *SlMYB33* small mRNA amplification, the PrimeScript™ RT reagent kit (Takara, Shiga, Japan) was used for cDNA synthesis according to the manufacturer's protocol using 50 ng/μL oligo(dT) (Promega, Madison, WI, USA) and 2.5 μM random hexamers (Applied Biosystems, Waltham, MA, USA), 10 ng cDNA, and gene specific forward (F) and reverse (R) primers (Table 2), designed with PRIMER3PLUS software [30]. *RPS18* (ribosomal protein S18, GeneBank: 3950409) was used as a reference gene.

Table 2. Primers used to analyze *SlMYB33* small transcript fragments by RT-PCR in tomato plants annealing to a region outside the predicted sly-miR159 binding site in *SlMYB33* mRNA (O_{Fw}, O_{Rv}) or flanking the putative cleavage site within the predicted sly-miR159 binding site in *SlMYB33* mRNA (F_{Fw}, F_{Rv}).

Primer Pair	Forward Primer (5′-3′)	Reverse Primer (5′-3′)	Product Size (bp)
O_{Fw}, O_{Rv}	TATGGGCATCCAGTCTCTCC	TGGGACTGGAAAAGATCGTC	199
F_{Fw}, F_{Rv}	ATGACGGTTCTTTGCTTGCT	CTGTCTGGTTTTGGAGTGAAGG	200
$RPS18_{FW}$, $RPS18_{RV}$	GGGCATTCGTATTTCATAGTCAGAG	CGGTTCTTGATTAATGAAAACATCCT	105

The cycling parameters were as follows: Initial polymerase activation step at 95 °C for 30 s, 40 cycles of denaturation at 95 °C for 5 s, annealing, and elongation at 60 °C for 30 s. For each sample, three biological replicates were pooled and analyzed. Five microliters of the reaction volume were separated in a 3% agarose gel.

4.4. Amino Acids and Polyamines Quantification

Leaves were recollected after stress condition and frozen in liquid N_2, ground, and lyophilized.

For amino acids analysis, dry tissue (0.1 g) was homogenized with 800 μL of extraction solution: 400 μL of distilled water, 200 μL of chloroform, and 200 μL of methanol per sample. Moreover, a mixture of internal standards was added prior to extraction (100 ng of Phe $^{13}C_9^{15}N$ and 100 ng of Thr $^{13}C_4^{15}N$). Samples were filtered, and a final concentration of 1 mM perfluoroheptanoic acid as ion-pairing reagent was added to each sample. A 20 μL aliquot was injected into a high-performance liquid chromatography system (HPLC) with an XSelect HSS C18 column (5 μm 2.1 × 100 mm) which was interfaced with a triple quadrupole mass spectrometer (TQD, Waters, Manchester, UK).

Polyamine analysis was conducted according to the method described by Sánchez-López et al. [32], using as internal standards a mixture of [$^{13}C_4$]-putrescine and 1,7-diamineheptane. To analyze each condition, ten independent biological replicates per sample were generated and three independent experiments were conducted.

Author Contributions: Conceptualization, I.G.-R., M.D.R. and C.R.; Funding acquisition, I.G.-R., G.C., B.V., M.D.R. and C.R.; investigation, M.J.L.-G., I.G.-R., A.I.G.-H. and G.C.; writing—original draft, M.D.R. and C.R.; writing—review and editing, M.J.L.-G., I.G.-R., G.C. and B.V.

Funding: This work was supported by "MINECO" and "FEDER" (AGL2013-49023-C3-1,2,3-R; AGL2017-85987-C3-1,2-R). M.J. López-Galiano was awarded with a University of Valencia Ph. D. fellowship.

Acknowledgments: We thank the Genomics, Proteomics and Greenhouse Facilities from the SCSIE of the University of Valencia and the SCIC from the University Jaume I.

Conflicts of Interest: The authors declare no conflict of interest. The funders had no role in the design of the study; in the collection, analyses, or interpretation of data; in the writing of the manuscript, or in the decision to publish the results.

References

1. Swann, A.L.S. Plants and drought in a changing climate. *Curr. Clim. Chang. Rep.* **2018**, *4*, 192–201. [CrossRef]
2. Hossain, M.A.; Wani, S.H.; Bhattacharjee, S.; Burritt, D.J.; Tran, L.-S.P. *Drought Stress Tolerance in Plants*, 1st ed.; Springer International Publishing: Basel, Switzerland, 2016; Volume 2, ISBN 978-3-319-32421-0.
3. Cutler, S.R.; Rodriguez, P.R.; Finkelstein, R.R.; Abrams, S.R. Abscisic acid: Emergence of a core signaling network. *Annu. Rev. Plant Biol.* **2010**, *61*, 651–679. [CrossRef] [PubMed]
4. Ding, Y.; Tao, Y.; Zhu, C. Emerging roles of microRNAs in the mediation of drought stress response in plants. *J. Exp. Bot.* **2013**, *64*, 3077–3086. [CrossRef] [PubMed]
5. Reyes, J.L.; Chua, N.-H. ABA induction of miR159 controls transcript levels of two MYB factors during *Arabidopsis* seed germination. *Plant J.* **2007**, *49*, 592–606. [CrossRef] [PubMed]
6. Liu, H.-H.; Tian, X.; Li, Y.-J.; Wu, C.-A.; Zheng, C.-C. Microarray-based analysis of stress-regulated microRNAs in *Arabidopsis thaliana*. *RNA* **2008**, *14*, 836–843. [CrossRef]
7. Wei, L.; Zhang, D.; Xiang, F.; Zhang, Z. Differentially expressed miRNAs potentially involved in the regulation of defense mechanism to drought stress in maize seedlings. *Int. J. Plant Sci.* **2009**, *170*, 979–989. [CrossRef]
8. Xie, F.; Wang, Q.; Sun, R.; Zhang, B. Deep sequencing reveals important roles of microRNAs in response to drought and salinity stress in cotton. *J. Exp. Bot.* **2015**, *66*, 789–804. [CrossRef]
9. Yang, J.; Zhang, N.; Mi, X.; Wu, L.; Ma, R.; Zhu, X.; Yao, L.; Jin, X.; Si, H.; Wang, D. Identification of miR159s and their target genes and expression analysis under drought stress in potato. *Comput. Biol. Chem.* **2014**, *53*, 204–213. [CrossRef]
10. Hackenberg, M.; Gustafson, P.; Langridge, P.; Shi, B.-J. Differential expression of microRNAs and other small RNAs in barley between water and drought conditions. *Plant Biotechnol. J.* **2015**, *13*, 2–13. [CrossRef]
11. Li, Y.; Wan, L.; Bi, S.; Wan, X.; Li, Z.; Cao, J.; Tong, Z.; Xu, H.; He, F.; Li, X. Identification of drought-responsive microRNAs from roots and leaves of alfalfa by high-throughput sequencing. *Genes* **2017**, *8*, 119. [CrossRef]
12. Pegler, J.L.; Grof, C.P.L.; Eamens, A.L. Profiling of the differential abundance of drought and salt stress-responsive microRNAs across grass crop and genetic model plant species. *Agronomy* **2018**, *8*, 118. [CrossRef]
13. López-Galiano, M.J.; González-Hernández, A.I.; Crespo-Salvador, O.; Rausell, C.; Real, M.D.; Escamilla, M.; Camañes, G.; García-Agustín, P.; González-Bosch, C.; García-Robles, I. Epigenetic regulation of the expression of WRKY75 transcription factor in response to biotic and abiotic stresses in Solanaceae plants. *Plant Cell Rep.* **2018**, *37*, 167–176. [CrossRef] [PubMed]
14. Liu, M.; Yu, H.; Zhao, G.; Huang, Q.; Lu, Y.; Ouyang, B. Profiling of drought-responsive microRNA and mRNA in tomato using high-throughput sequencing. *BMC Genom.* **2017**, *18*, 481. [CrossRef]
15. Liu, M.; Yu, H.; Zhao, G.; Huang, Q.; Lu, Y.; Ouyang, B. Identification of drought-responsive microRNAs in tomato using high-throughput sequencing. *Funct. Integr. Genom.* **2018**, *18*, 67–78. [CrossRef] [PubMed]
16. Zhang, B. MicroRNA: A new target for improving plant tolerance to abiotic stress. *J. Exp. Bot.* **2015**, *66*, 1749–1761. [CrossRef]
17. Allen, R.S.; Li, J.; Stahle, M.I.; Dubroué, A.; Gubler, F.; Millar, A.A. Genetic analysis reveals functional redundancy and the major target genes of the *Arabidopsis* miR159 family. *Proc. Natl. Acad. Sci. USA* **2007**, *104*, 16371–16376. [CrossRef]
18. Woodger, F.J.; Millar, A.; Murray, F.; Jacobsen, J.V.; Gubler, F. The role of GAMYB transcription factors in GA-regulated gene expression. *J. Plant Growth Regul.* **2003**, *22*, 176–184. [CrossRef]

19. Dai, X.; Zhao, P.X. psRNATarget: A plant small RNA target analysis server. *Nucleic Acids Res.* **2011**, *39*, W155–W159. [CrossRef]
20. Zheng, Z.; Reichel, M.; Deveson, I.; Wong, G.; Li, J.; Millar, A.A. Target RNA secondary structure is a major determinant of miR159 efficacy. *Plant Physiol.* **2017**, *174*, 1764–1778. [CrossRef]
21. Li, Z.; Peng, R.; Tian, Y.; Han, H.; Xu, J.; Yao, Q. Genome-wide identification and analysis of the MYB transcription factor superfamily in *Solanum lycopersicum*. *Plant Cell Physiol.* **2016**, *57*, 1657–1677. [CrossRef]
22. Pieczynski, M.; Marczewski, W.; Hennig, J.; Dolata, J.; Bielewicz, D.; Piontek, P.; Wyrzykowska, A.; Krusiewicz, D.; Strzelczyk-Zyta, D.; Konopka-Postupolska, D.; et al. Down-regulation of CBP80 gene expression as a strategy to engineer a drought-tolerant potato. *Plant Biotechnol. J.* **2013**, *11*, 459–469. [CrossRef] [PubMed]
23. Qin, Y.; Wang, M.; Tian, Y.; He, W.; Han, L.; Xia, G. Over-expression of TaMYB33 encoding a novel wheat MYB transcription factor increases salt and drought tolerance in *Arabidopsis*. *Mol. Biol. Rep.* **2012**, *39*, 7183–7192. [CrossRef] [PubMed]
24. Tonon, G.; Kevers, C.; Faivre-Rampant, O.; Graziani, M.; Gaspar, T. Effect of NaCl and mannitol iso-osmotic stresses on proline and free polyamine levels in embryogenic *Fraxinus angustifolia* callus. *J. Plant Physiol.* **2004**, *161*, 701–708. [CrossRef] [PubMed]
25. Alcázar, R.; Cuevas, J.C.; Patron, M.; Altabella, T.; Tiburcio, A.F. Abscisic acid modulates polyamine metabolism under water stress in *Arabidopsis thaliana*. *Physiol. Plant* **2006**, *128*, 448–455. [CrossRef]
26. Pál, M.; Tajti, J.; Szalai, G.; Peeva, V.; Végh, B.; Janda, T. Interaction of polyamines, abscisic acid and proline under osmotic stress in the leaves of wheat plants. *Sci. Rep.* **2018**, *8*, 12839. [CrossRef] [PubMed]
27. Li, J.; Reichel, M.; Li, Y.; Millar, A.A. The functional scope of plant microRNA-mediated silencing. *Trends Plant Sci.* **2014**, *19*, 750–756. [CrossRef] [PubMed]
28. Hoagland, D.R.; Arnon, D.I. The water-culture method for growing plants without soil. *Circ. Calif. Agric. Exp. Sta.* **1950**, *347*, 1–32.
29. Balcells, I.; Cirera, S.; Busk, P.K. Specific and sensitive quantitative RT-PCR of miRNAs with DNA primers. *BMC Biotechnol.* **2011**, *11*, 70. [CrossRef] [PubMed]
30. Untergasser, A.; Nijveen, H.; Rao, X.; Bisseling, T.; Geurts, R.; Leunissen, J.A.M. Primer3Plus, an enhanced web interface to Primer3. *Nucleic Acids Res.* **2007**, *35*, W71–W74. [CrossRef]
31. Ruijter, J.M.; Ramakers, C.; Hoogaars, W.M.H.; Karlen, Y.; Bakker, O.; van den Hoff, M.J.B.; Moorman, A.F.M. Amplification efficiency: Linking baseline and bias in the analysis of quantitative PCR data. *Nucleic Acids Res.* **2009**, *37*, e45. [CrossRef]
32. Sánchez-López, J.; Camañes, G.; Flors, V.; Vicent, C.; Pastor, V.; Vicedo, B.; Cerezo, M.; García-Agustín, P. Underivatized polyamine analysis in plant samples by ion pair LC coupled with electrospray tandem mass spectrometry. *Plant Physiol. Biochem.* **2009**, *47*, 592–598. [CrossRef]

© 2019 by the authors. Licensee MDPI, Basel, Switzerland. This article is an open access article distributed under the terms and conditions of the Creative Commons Attribution (CC BY) license (http://creativecommons.org/licenses/by/4.0/).

Article

Conserved Cu-MicroRNAs in *Arabidopsis thaliana* Function in Copper Economy under Deficiency

Muhammad Shahbaz and Marinus Pilon *

Biology Department, Colorado State University, Fort Collins, CO 80523-1878, USA; ms@sppg.ca
* Correspondence: pilon@colostate.edu; Tel.: 1-970-495-4390

Received: 1 May 2019; Accepted: 24 May 2019; Published: 29 May 2019

Abstract: Copper (Cu) is a micronutrient for plants. Three small RNAs, which are up-regulated by Cu deficiency and target transcripts for Cu proteins, are among the most conserved microRNAs in plants. It was hypothesized that these Cu-microRNAs help save Cu for the most essential Cu-proteins under deficiency. Testing this hypothesis has been a challenge due to the redundancy of the Cu microRNAs and the properties of the regulatory circuits that control Cu homeostasis. In order to investigate the role of Cu-microRNAs in Cu homeostasis during vegetative growth, we used a tandem target mimicry strategy to simultaneously inhibit the function of three conserved Cu-microRNAs in *Arabidopsis thaliana*. When compared to wild-type, transgenic lines that express the tandem target mimicry construct showed reduced Cu-microRNA accumulation and increased accumulation of transcripts that encode Cu proteins. As a result, these mimicry lines showed impaired photosynthesis and growth compared to wild type on low Cu, which could be ascribed to a defect in accumulation of plastocyanin, a Cu-containing photosynthetic electron carrier, which is itself not a Cu-microRNA target. These data provide experimental support for a Cu economy model where the Cu-microRNAs together function to allow maturation of essential Cu proteins under impending deficiency.

Keywords: plastocyanin; photosynthesis; copper deficiency; Cu-microRNA; copper protein; target mimicry

1. Introduction

Copper deficiency in plants leads to defects in photosynthesis, chlorosis, reduced respiration, and wilting of leaves. Copper is not a "mobile" element. This means that under impending deficiency, Cu is not efficiently transported from older tissues to newly developing leaves, which become chlorotic [1]. In green tissue, the majority of Cu is found in the chloroplast [2,3]. Plastocyanin (PC) is a blue Cu protein that mediates electron transfer from the cytochrome-b_6f complex to photosystem I (PSI) in the thylakoid lumen of oxygenic photosynthetic organisms [4]. *Arabidopsis* has two genes for PC, *PC1 (PETE1)* and *PC2 (PETE2)*. The PC2 protein is more abundant and responsive to Cu [5,6]. PC2 accumulation is a consequence of protein stability due to cofactor presence and is not due to transcript abundance changes [6–8]. In higher plants, PC is the only protein that can accept electrons from the cytochrome-b_6f complex and *Arabidopsis* mutants with insertions in both PC genes are seedling-lethal on soil [9,10]. Cu is also a cofactor of Cu/Zn superoxide dismutase (CSD) proteins that function in the metabolism of reactive oxygen radicals. In *Arabidopsis*, CSD2 is active in the plastids, and CSD1 is active in the cytosol [11]. Both CSD1 and CSD2 receive the Cu cofactor from a copper chaperone called copper chaperone for superoxide dismutase, CCS [12,13]. In mitochondria, copper is required for the function of cytochrome-c oxidase (COX), the proton-pumping terminal oxidase in the respiratory electron transport chain in the inner membrane [14]. Three Cu atoms are bound by the core COX subunits I, II, and III, which are encoded in the mitochondrial genome. The ethylene receptors in the endomembrane system are copper-binding proteins [15,16]. All remaining Cu proteins are most likely apoplastic. These include the plant-specific blue copper proteins called phytocyanins, which includes plantacyanin, apoplastic ascorbate oxidases, amine oxidases, and laccases [17–24]. The laccase family has 17 members in *Arabidopsis* [22], and for several of these, a role in lignification and secondary growth of the vasculature has now been established [25,26].

MicroRNAs that are regulated by Cu availability and that target mRNAs encoding for Cu proteins are called Cu-microRNAs [27]. The four Cu-microRNAs of *Arabidopsis*—*miR397, miR398, miR408,* and *miR857*—were first discovered in deep sequencing projects [28] and regulation of CSD1 and CSD2 by *miR398* was shown by Sunkar et al. [29]. It was later found that *miR398*, and thus CSD1, CSD2, and CCS, are regulated primarily by Cu levels [13,30]. In addition, *miR397, miR408, and miR857* regulate the abundance of other Cu proteins in *Arabidopsis*, specifically laccases and the secreted protein plantacyanin in response to Cu availability [31]. The Cu-microRNAs are in turn regulated via a Cu-responsive transcription factor called SPL7 (squamosa promotor binding protein-like7), which also regulates Cu assimilation [8,32–34]. Three of the four *Arabidopsis* Cu microRNAs (*miR397, miR398,* and *miR408*) are among the most highly conserved microRNAs in plants [35]. This conservation suggests an important function. Because of the strong link with Cu, it was hypothesized that the Cu-microRNAs function in the Cu economy. According to this idea, in plants, which have symplasmic connections that allow both the sharing of nutrients and communication via small RNAs, the regulation via microRNAs provides a mechanism to save Cu and to allow essential Cu protein maturation, such as PC and COX, in actively growing cells of a tissue during impending deficiency [27]. The hypothesis rests on three legs. Two well-supported tenets are the very tight bonding of Cu atoms to its ligands [36] and the presence of symplasmic connections between cells in plants. The third leg of the hypothesis is that effective signaling in and between cells via microRNAs works to signal Cu status and that it can indeed help to tune Cu protein expression. This third leg needs further experimental support. The possible role of Cu-microRNAs in the Cu economy is based mostly on correlative evidence with some more direct support in *Arabidopsis* and in poplar [30,31,33,34,37,38]. In fact, the idea that Cu delivery to PC is a priority is not supported by the observation that PC2 protein levels are strongly affected by Cu deficiency even if mRNA levels were not affected [6]. Do Cu microRNAs actually make a difference for Cu economy? To test this, we aimed to inactivate the conserved Cu-microRNAs (*miR397, miR398,* and *miR408*) and to analyze, especially under impending deficiency, the effects on growth and Cu allocation to abundant and essential Cu proteins, such as PC1 and PC2, which are not down-regulated via a microRNA. Because Cu deficiency affects flower and pollen development with several compounding effects, we limited our study to the vegetative shoot before flowering [31,39].

The Cu-microRNAs are under control of SPL7 [8,32]. Because in a *spl7*-loss-of-function mutant, Cu uptake is also defective, a good test of Cu-microRNA function requires that its regulation is uncoupled from SPL7. We have tested single microRNA loss-of-function mutants for *miR398a/c* and for *miR408* and found no discernable phenotypes and certainly no defect in PC maturation [3,13]. Similarly, *miR398* overexpression caused no severe phenotypes on soil or on agar media [13,40]. However, Zhang et al. have reported a small effect on PC accumulation due to altered miR408 expression in a hy5/spl7 background [34]. Due to properties of the SPL7-mediated system, changes in single Cu-microRNAs can be predicted to show attenuated effects on Cu homeostasis because the addition or elimination of Cu-binding targets will affect the Cu pool sensed by SPL7, which will reset the expression of Cu-microRNAs and Cu uptake systems [27]. Due to the "dampening" properties of the system, we predict that disruption of just one Cu-microRNA will have very small effects. Thus, perhaps all or the majority of the Cu-microRNAs must be perturbed simultaneously before strong effects on Cu homeostasis can be expected. We focused here on the conserved Cu-microRNAs. Each of *miR397* and *miR398* have multiple loci and thus combinations (crosses) of knock-out (KO) lines to inactivate all Cu-microRNAs are virtually impossible. We therefore used a target mimicry construct to simultaneously inhibit all the conserved Cu-microRNAs [41]. Target mimicry relies on the production of a "designed" RNA transcript, which is modified from INDUCED BY PHOSPHATE STARVATION1 (*IPS1*), a non-coding RNA in *Arabidopsis* that functions to sequester miR399, a microRNA that functions in the regulation of phosphate homeostasis [41]. The native *IPS1* transcript contains a microRNA target site for *miR399* that cannot be cleaved and that should sequester and inactivate the microRNA. The original mimicry target site in *miR399* can be modified to inhibit other microRNAs [41].

2. Results

In order to simultaneously inhibit all three Cu-microRNAs, we designed a tandem target mimicry strategy (see Supplementary Figure S1). In this approach we modified the native IPS1 sequence [41] by replacing the miR399 target sequence with three, in tandem, target sites for miR397, miR398, and miR408, respectively. The construct was placed under the control of a constitutive 35S-CaMV promoter. The three conserved Cu-microRNAs should be sequestered by the modified IPS1-Cu-microRNA tandem mimicry construct, which is not a target for microRNA-directed cleavage due to inserted mismatches at the predicted target cleavage sites.

Three transgenic *Arabidopsis* lines with high expression of the tandem mimicry construct were selected and compared to wild-type plants. To control Cu status, we germinated plants on Cu replete agar media for 10 days before transplanting the seedlings to hydroponics. We first verified that the expression in the shoot of the tandem mimicry construct was not affected by Cu-feeding status after 3.5 weeks in hydroponics (5 weeks after the start of germination) (Figure 1). We used three Cu regimes; no Cu addition for mild deficiency after 3–4 weeks in this growth condition; 5 nM $CuSO_4$ for Cu-sufficient; and 50 nM $CuSO_4$ for Cu-replete, a condition where all Cu proteins are expected to acquire this cofactor [31]. As expected, the target mimicry construct expression was not affected by Cu availability (Figure 1).

The tandem target mimicry construct was designed to affect the function of the three conserved Cu-microRNAs and prevent strong down-regulation of the transcripts for Cu proteins normally targeting the Cu-microRNAs, especially in low Cu conditions. We investigated, using quantitative reverse transcription polymerase chain reaction (qRT-PCR), the effect of the tandem target mimicry construct on the accumulation of the three conserved Cu-microRNAs at the three Cu concentrations (Figure 1). For both miR397 and miR398, three loci are present, whereas for miR408, a single locus exists. For miR397 and miR398, we utilized primers optimized to amplify the more abundant and Cu induced isoforms [8]. As expected, all three conserved Cu-microRNAs were strongly regulated by Cu availability in the wild-type, with highest expression seen when Cu was omitted from the growth medium (Figure 1). Conversely, in plants grown on 50 nM $CuSO_4$, the expression of all three Cu-microRNAs was strongly repressed. Expression of the tandem target mimicry construct caused a notable reduction of all three Cu microRNAs. The effect on Cu-microRNA expression was more pronounced for miR398 and miR408 compared to miR397 (Figure 1). The effect of target tandem mimicry was most pronounced at 0 nM and 5 nM $CuSO_4$, where in the wild-type, the Cu-microRNAs accumulated. Higher expression of the mimicry construct in lines M5 and M17 correlated with a stronger effect on Cu-microRNA accumulation.

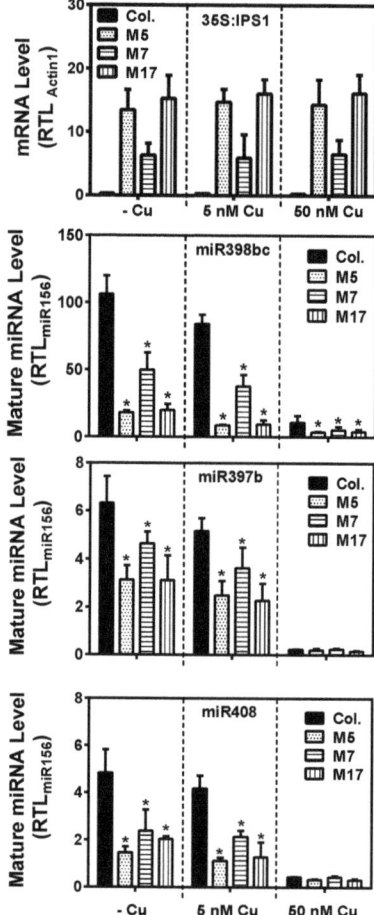

Figure 1. Expression of the IPS1 tandem target mimicry construct and three conserved Cu-microRNAs in wild type (Col) and three transgenic lines (M5, M7, and M17). 35S:IPS1 transgene relative transcript levels (RTL) were determined using qRT-PCR. Values are normalized relative to actin 1 expression. Mature microRNA levels were measured by qRT-PCR using primers designed for miR398bc, miR397b, and miR408. Relative microRNA levels were normalized to miR156 expression. All data and are given as averages ± SD ($n = 3$). * indicate significant differences from the control ($p < 0.05$, Student's t-test).

We next analyzed the effect of tandem target mimicry on the transcript levels of selected confirmed Cu-microRNA targets (Figure 2). As expected, a strong effect of tandem target mimicry was seen for all target transcripts, this effect was most pronounced under Cu deficiency (- Cu). The deregulation by tandem target mimicry on Cu deficiency was strong for all the tested targets of the three conserved microRNAs (Figure 2). We conclude that accumulation of all the three conserved Cu-microRNAs was disrupted in the shoot by tandem target mimicry, and in turn this caused aberrant accumulation of transcripts encoding Cu proteins under Cu deficiency.

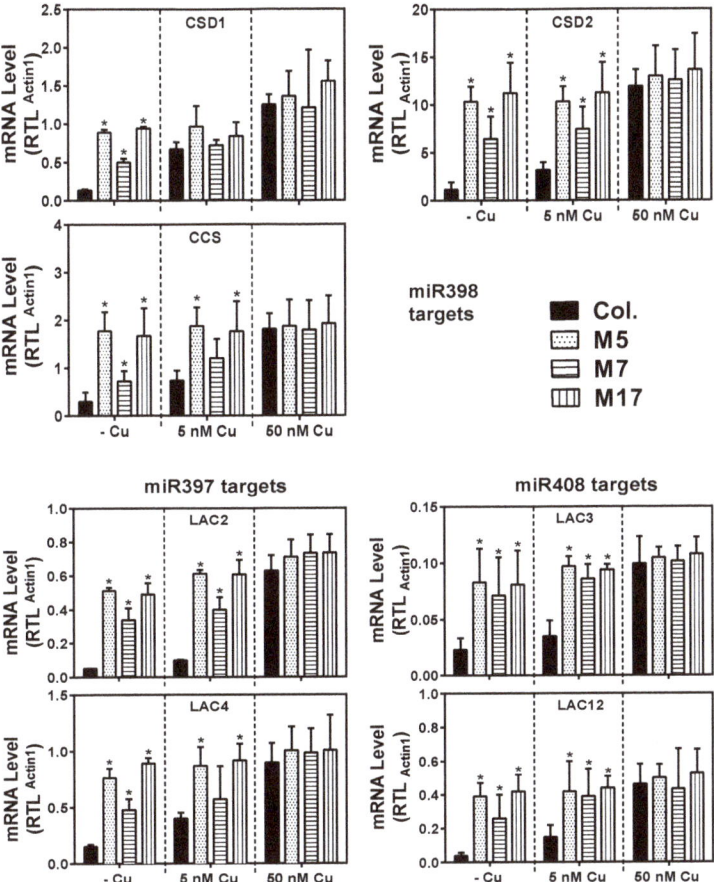

Figure 2. Expression analysis of selected Cu-microRNA target transcripts using qRT-PCR. mRNA-relative transcript levels (RTL) were normalized relative to actin 1 expression and given as averages ± SD ($n = 3$). * indicate significant differences from the control ($p < 0.05$, Student's t-test).

We next analyzed vegetative growth. Because the Cu-microRNAs are highly expressed in plants grown on low Cu and repressed in Cu-replete conditions, phenotypes due to target mimicry were mainly expected on low Cu conditions. Because we were especially interested in the effects on photosynthesis during vegetative growth, we compared plants at 5 weeks of age, which in our short-day growth condition is about a week before bolting, with the onset of symptoms of deficiency showing in the wild-type, without irreversible secondary effects. Because there is Cu in the seeds and in the germination medium, Cu depletion treatment in hydroponics takes several weeks to show visible deficiency symptoms. The plants of all genotypes showed overall comparable morphology on all three conditions after 3.5 weeks in hydroponics (Figure 3A). However, when the plant biomass was measured, a clear effect was seen for Cu omission compared to 5 nM and 50 nM $CuSO_4$ (Figure 3B). Interestingly, in the deficient condition, a significant further reduction in biomass was seen for the three tandem target mimicry lines. We measured the elemental composition of the shoots. As expected, $CuSO_4$ in the medium strongly affected Cu concentrations in the shoots. There was, however, no difference between the transgenics and wild-type, indicating that tandem target mimicry did not affect Cu levels (Figure 3C). Other elements did not show line-specific differences (See Table S1). In summary, tandem target mimicry for Cu-microRNAs resulted in a mild but significant effect on biomass accumulation, but only under Cu deficiency.

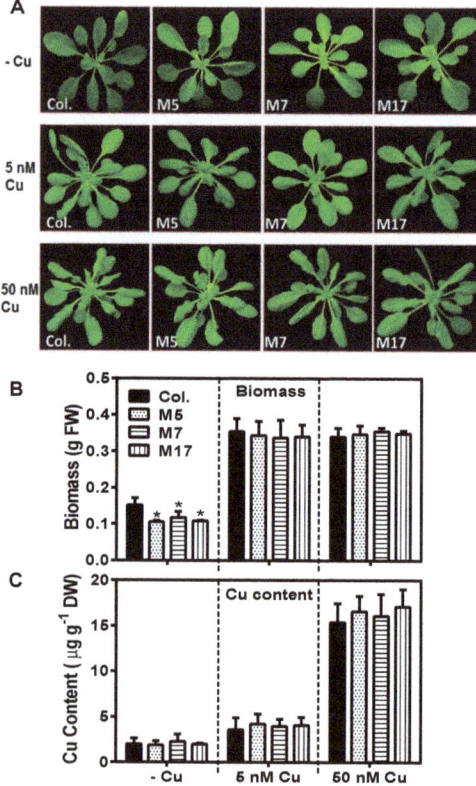

Figure 3. Effect of Cu-microRNA target mimicry on plant growth in hydroponics. (**A**) Representative images of 5-week-old WT (Col.); M5, M7, and M17 are All3-Cu miRNA mimicry lines. Plants were grown on 1/10th strength Hoagland solution containing 0, 5, or 50 nM $CuSO_4$. (**B**) Fresh weight and (**C**) Cu content ($\mu g\ g^{-1}$ DW) of 5-week grown plants. Values are given as averages ± SD ($n = 6$). * indicate significant differences ($p < 0.05$, Student's t-test).

We hypothesized that tandem target mimicry, because of the misregulated and higher expression of Cu-microRNA targets, affected growth by limiting the pool of Cu available for plastocyanin, which is not a microRNA target. A lack of plastocyanin function should result in a decreased photosynthetic electron transport and a more reduced plastoquinone pool. Measurement of chlorophyll fluorescence parameters (Figure 4) indicated that Cu depletion in the wild-type caused a mild decrease in the flux through PSII (ΦPSII), which is an estimate of photosynthetic electron transport activity. However, there was a significantly larger decrease in ΦPSII for the three tandem target mimicry lines compared to the wild-type, but only under Cu deficiency. Similarly, the parameter 1-qP, which indicates the redox state of the plastoquinone pool, was affected by Cu depletion with a strong effect seen in tandem target mimicry. These results are strongly indicative of a defect in plastocyanin function in plants grown on low Cu, which is exacerbated by tandem target mimicry.

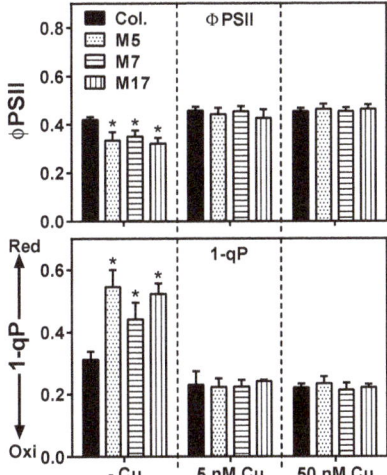

Figure 4. Photosynthetic electron transport parameters as measured by chlorophyll fluorescence. FluorCam measurement done on dark-adopted rosette leaves from WT (Col.) and transgenic lines (M5, M7, M17) growing in hydroponic culture at 0, 5, or 50 nM $CuSO_4$, where values are given as averages ± SD ($n = 6$). Top panel: ΦPSII (estimate of electron transport). Bottom panel: 1-qP (indicative of the redox state of the plastoquinone pool). * indicate significant differences between lines within a growth condition ($p < 0.05$, Student's t-test).

We next analyzed Cu protein accumulation using immunoblotting levels (Figure 5A). In *Arabidopsis*, two plastocyanin isoforms are expressed, PC1 and PC2 [10]. The more abundant PC2 isoform is known to be strongly affected at the protein level by Cu depletion, most likely due to a post-translational process as neither PC transcript is a target of a microRNA [6]. Indeed, Cu depletion in the wild-type resulted in a marked decrease in PC2 accumulation, while PC1 was relatively less affected (Figure 5A). Remarkably, there was a much stronger reduction in both PC isoforms in the three tandem target mimicry lines on low Cu. The severity of the effect on PC correlated strongly with the tandem target mimicry expression level (Figures 1 and 5A). At 5 and 50 nM $CuSO_4$, there was however no noticeable difference between the wild-type and transgenics. Cytochrome-c oxidase (COX) is a major Cu protein in the mitochondria where it functions as the terminal oxidase. We used antibodies specific to the Cu-binding COX core subunit II (COXII) as a proxy for COX accumulation (Figure 5A). COXII was affected by Cu depletion in the wild-type, albeit that the effect seemed not as strong as for PC in the chloroplast. There was a mild but noticeable larger effect of Cu depletion on COXII accumulation in the three target mimicry lines (Figure 5A).

We also verified accumulation of the miR398 targets: CSD1, CSD2, and CCS. As expected, accumulation of these proteins was strongly affected by Cu levels (Figure 5A). However, tandem target mimicry attenuated the effects of lowering Cu on the accumulation of these proteins. For CSD1 and CSD2 protein accumulation, the strongest effect of target mimicry was seen at 5 nM $CuSO_4$, whereas for CCS protein accumulation, a strong effect was also seen at 0 nM $CuSO_4$ (Figure 5A). The superoxide dismutase isozyme activity was analyzed using native gel assays (Figure 5B). As expected, the decrease in CSD1 and CSD2 protein levels resulted in a reduced CSD activity. FeSOD expression and activity, which is known to be highly Cu responsive and regulated via SPL7 [7,8,32], was not affected by tandem-target mimicry. In conclusion, lines that express the tandem target mimicry construct show a stronger reduction in plastocyanin and COXII levels on low Cu, which is accompanied by an elevated accumulation and activity of the miR398 targets.

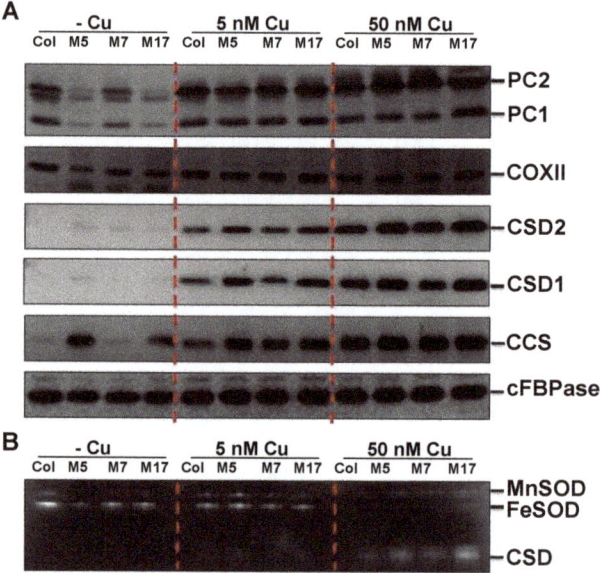

Figure 5. Comparison of Cu protein accumulation in wild-type (Col) and three transgenics (M5, M7, M17). (**A**) Western-blot analysis using the following antibodies: PC, plastocyanin two isoforms (PC2 and PC1 are detected); COXII, cytochrome-c oxidase subunit II; CSD2 and CSD1 (chloroplast and cytosolic isoforms of Cu/Z superoxide dismutase; and CCS, copper chaperone for Cu/ZnSOD. cFBPase was used as a loading control. Total soluble proteins (30 μg) were fractioned using SDS-PAGE (15% gel) and blotted onto nitrocellulose membranes. (**B**) For SOD isoform activity, total soluble proteins (30 μg) were fractioned on a non-denaturing 15% acryl amide gel and stained for SOD activity. The indicated detected isoforms were identified based on known relative mobility.

3. Discussion

The conservation of the Cu microRNA sequences, their target transcripts encoding Cu proteins, and their regulation by Cu via SPL7 suggests an important function related to copper directly. Neither plastocyanin nor cytochrome-c oxidase core subunits are targets of a microRNA in plants, presumably because these proteins are indispensible. Could a role of the Cu-microRNAs be to maintain a pool of Cu accessible for plastocyanin and cytochrome-c oxidase? While such a role in the Cu economy has been proposed before, the direct evidence has been limited. We aimed to test a role of conserved Cu-microRNAs in the Cu economy in *Arabidopsis* under conditions that require photoautotrophic growth. In order to characterize a function of the conserved Cu microRNAs, we employed a tandem target mimicry strategy. Expression of target mimicry constructs caused a decrease in the microRNAs that were targeted (Figure 1). Such a decrease has been reported before for other microRNAs and indicates that target mimicry not only leads to target sequestration, but also causes increased microRNA instability. The Cu-microRNA deregulation was well correlated with the extent of mimicry construct expression. The tandem target mimicry approach clearly caused a deregulation of a large set of transcripts but (as expected) only on low Cu, a condition where Cu-microRNAs are highly expressed in the wild-type (Figures 1 and 2). As a consequence of tandem target mimicry, the plants accumulated lower amounts of the photosynthetic electron carrier plastocyanin, one of the most abundant Cu proteins in plants. The exacerbated reduction in plastocyanin content in tandem target mimicry lines was strongly correlated with a defect in photosynthetic electron transfer and reduced biomass on low Cu (Figures 3–5). These data provide strong and direct support for a role of the Cu-microRNAs in the Cu economy in *Arabidopsis*, requiring optimization of vegetative growth under low-Cu conditions.

While the effect on growth was relatively small and required low Cu to be noted, we think it is likely a strong enough phenotype to explain the conservation of the Cu-microRNAs. In this context it should be noted that the non-conserved and fourth Cu-microRNA (miR857) was not targeted here [31]. Perhaps stronger phenotypes could be observed if this fourth target was also de-regulated.

Because of the presence of Cu in seeds and in the germination medium, which is needed to ensure comparable development, there is a significant lag time of about two and a half to three weeks before symptoms of Cu deficiency manifest themselves [31]. We wanted to avoid secondary effects, and therefore grew all plants for less than four weeks in hydroponics without Cu. We did this to avoid the compounding effects of bolting and flowering, which are strongly affected by Cu deficiency treatment, but only after strong and irreversible symptoms of Cu deficiency, such as leaf curling and browning, are evident in the vegetative shoot [31]. Therefore, in this study, we focused on vegetative growth and on shoots because this is where the Cu demand for plastocyanin is high. It seems likely, though, that the role of Cu-microRNAs, which are expressed in roots, stems, leaves, and flowers, is not limited to the vegetative shoot and indeed Cu feeding status is known to affect the timing of flowering, with deficient plants showing a delay [31]. Furthermore, Cu is also important for pollen development, presumably to allow for sufficient mitochondrial cytochrome-c oxidase activity [39]. We did not note any defect in fertility or seed set for our three target mimicry lines compared to wild-type. This was probably due to the sufficiency of Cu in the soil. It is possible that if the plants are to be subjected to Cu deficiency after the onset of flowering that a defect in pollen maturation would cause a decrease in fertility. However, such an experiment is difficult to control, and effects of both Cu and development would have to be considered simultaneously. Nevertheless, accumulation of mitochondrial COXII protein, one of three mitochondrial-encoded Cu-binding core subunits of cytochrome-c oxidase in the respiratory electron transport chain [14], was clearly negatively affected by tandem target mimicry in the shoots. This observation indicates that optimal cytochrome-c oxidase activity under Cu deficiency requires functional Cu-microRNAs. A role in regulating grain yield was also reported for miR408 in rice [42]. In this same study, overexpression of the rice miR408 target UCL8 (uclacyanin 8, a phytocyanin), showed a negative effect on plastocyanin and chloroplastic Cu accumulation in leaves [42].

In response to Cu availability, the Cu-microRNAs are regulated directly by SPL7, which is the plant homolog of the chlamydomonas (CRR1) copper response regulator [43]. SPL7 and CRR1 regulation requires *cis* elements called Cu-response elements (CuRe) with a GTAC core motif [43]. Thus, a strong mechanistic link exists for Cu and Cu-microRNA regulation. Another identified transacting factor that directly mediates miR408 Cu-microRNA expression via promoter area binding is HY5 (elongated hypocotyl 5). Co-regulation was shown for *miR408* by both SPL7 and HY5 [33,34]. This regulation makes sense since HY5 mediates gene expression in the light required for photosynthetic growth, which has a high demand for Cu. Plants that lack SPL7 or HY5 function had only a small defect in PC [34]. Overexpression of miR408 in the *spl7* and *hy5* mutants could, to some extent, alleviate this defect in PC maturation, albeit that plants were grown in vitro on agar media to see this effect [34]. Interestingly, miR408 overexpression could also rescue some developmental defects in *hy5/spl7* lines [34]. Therefore, a feedback loop was proposed as a mechanism to allow better PC maturation in miR408-overproducing lines, but it can also be argued that an improved Cu economy in such lines allows for better Cu cofactor availability to PC [34]. Consistent with this idea are the observations reported in several plant species where miR408 expression causes higher chloroplast Cu content and improved photosynthesis, growth, and seed yield in diverse plant species [42,44,45]. On the other hand, Carrió-Seguí et al. have reported that both overexpression and loss of miR408 caused a diminished plant performance, especially in low Fe conditions [46].

Besides Cu availability, other environmental conditions have been reported to affect Cu-microRNA expression [27]. This was convincingly shown in *Arabidopsis* for cold [47–49], Fe availability [46,50], or stresses that are predicted to induce reactive oxygen species accumulation such as high light, the herbicide methyl-viologen, and excessive toxic metal [29]. In our setup, we controlled growth conditions and attempted to avoid any additional stressors. It seems likely that Cu-microRNA expression is further tuned by environmental conditions apart from Cu and light. However, unlike for Cu and light, where *cis*-regions and interacting *trans*-acting factors (SPL7 and HY5) are identified, it is presently unclear at a mechanistic level how these conditions could be linked to gene expression of Cu-microRNAs. Therefore, the effects of other stresses on Cu microRNAs could be direct or indirect. For Fe, for instance, it is well-reported in the literature that lower Cu levels increases Fe uptake and vice versa. A mechanism for this interaction can be proposed based on the observation that both Fe and Cu uptake at the root require a reductase activity of the ferric reductase oxidase (FRO) family. A cross reactivity for the low Cu-induced FRO4/FRO5 with Fe, or for the low-Fe induced FRO2 with Cu, would lead to increased Cu uptake under Fe deficiency, thus resetting the SPL7 regulation of Cu microRNAs, making it appear as if Fe regulates Cu-microRNAs [8,51]. Therefore, to more directly link Cu-microRNAs to other abiotic stresses, it will be important to uncover how potential *cis*-regions in the promoters and transacting factors interact to mediate stress response.

4. Materials and Methods

4.1. Plant Material and Growth Conditions

Lines were propagated on PRO-MIX HP soil that was fertilized with Miracle-Gro Liquid All Purpose Plant Food (Scotts Company, Marysville, OH, USA). Wild type (WT; Col.) and All3-mimicry transgenic lines seeds were surface-sterilized and germinated on agar plates containing one-half strength Murashige and Skoog (MS) medium [52] supplemented with 1% sucrose. For hydroponics, 7–10 days old seedlings were placed on a one-tenth-strength Hoagland's solution prepared with deionized water [53]. The nutrient solution was aerated and was replaced each week. For copper depletion (deficiency), Cu was omitted from the nutrient solution, while $CuSO_4$ was added to 5 nM for low but sufficient Cu (sufficient for high PC activity, but with low CSD activity in the wild-type) and to 50 nM for the Cu-replete conditions (where all Cu proteins accumulate to maximum levels in the wild-type) [31]. To minimize Cu contamination, all containers and buckets used for Cu-deprivation had never been in contact with Cu. The plants were grown in a light intensity of 150 µmol m^{-2} s^{-1}, with a 10-h-light/14-h-dark cycle and the temperature was maintained at 25 °C ± 2 °C in a climate-controlled room. After 4-weeks, plant material was harvested and stored frozen at −80 °C before analysis.

4.2. Transgenic Lines

Artificial target mimicry constructs were generated by modifying the sequence of the IPS1 gene [41]. The sequence of the construct, which was ordered as a synthesized piece of DNA (GenScript USA Inc. Piscataway, NJ, USA), is given in the Supplementary Materials. The All3-Cu miRNA target mimic construct was placed behind the CaMV 35S promoter in the pGWB41 vector, conferring resistance to kanamycin and hygromycin [54]. The construct was introduced into *A. thaliana* (accession Col-0) plants via a *Agrobacterium tumefaciens*-mediated transformation followed by selection on kanamycin containing agar media [55].

4.3. Elemental Analysis

For mineral content analysis, the plant tissue was placed in a drying oven at 55 °C for 48–72 h. One hundred milligrams of the dried material was digested in 1 mL of trace element grade nitric acid and heated at 60 °C for 2 h, followed by 130 °C for 6 h. The digests were subsequently diluted to 10 mL with double-distilled water before analysis using inductively coupled plasma – atomic emission spectrometry (ICP-AES) as described [56].

4.4. Chlorophyll Fluorescence Measurements

For chlorophyll fluorescence assays, whole rosettes of intact plants were used and dark-adapted for 30 minutes prior to analysis. Chlorophyll fluorescence imaging was done using a FluoroCam 701 MF

(Photon Systems International, Brno, Czech Republic) at an actinic light intensity of 150 micro-Einsteins as described [13]. The parameters ΦPSII (flux through PSII, an estimate of photosynthetic electron the transport rate) and 1-qP (an estimate of the redox state of plastoquinone pool) were calculated as described [57].

4.5. Protein Accumulation

Soluble proteins were extracted as described [38]. Protein concentration was determined using the Pierce BCA protein assay kit (Thermo Scientific, Waltham, MA, USA) using bovine serum albumin as a standard. For western blotting, 20 µg of total protein was separated using 15% SDS-PAGE and then transferred onto a nitrocellulose membrane. Antibodies used for immunodetection of PC, CSD1, CSD2, and CCS have been described [13,37]. Antisera for COXII and cFBPase were obtained from Agrisera (Vannas, Sweden). All protein detection experiments were done at least in biological triplicate with comparable results, and representative gels are shown.

4.6. Quantitative Reverse Transcription-PCR

RNA from plants sampled in biological triplicate was isolated using the Trizol reagent following the manufacturer's recommended protocol (Life Technologies, Carlsbad, CA, USA). After determination of total RNA concentrations, equal amounts per sample were reverse-transcribed using a First Strand cDNA Synthesis Kit (Life Technologies) and random hexamer primers. Quantitative RT-PCR was performed using the Light Cycler SYBR Green l master mix (Life Technologies) using gene-specific primer pairs as described previously for *Populus trichocarpa* using actin 1 as a housekeeping reference gene [37]. Samples without a template were used as negative controls. All primers are listed in Supplementary Table S2. qRT-PCR results and quality controls were analyzed using Light-Cycler 480 data-analysis software (version 1.5.1, Roche, Basel, Switzerland). The $\Delta\Delta Ct$ method was used to calculated relative transcript expression levels.

4.7. Mature miRNA Stem-Loop qRT-PCR

For the quantification of mature microRNAs, a stem-loop pulsed RT was used [37]. The RNA was extracted using the Trizol method as described above; however, ethanol washes were avoided and nucleic acid precipitation steps were carried out after addition of 1/10th volume of sodium acetate (3 M; pH 5.2), followed by an equal volume of isopropanol [37]. The stem-loop pulsed RT and miRNA qRT-PCR were performed as described previously [37,58]. miRNA397b, 398b/c, and 408 abundance were analyzed using gene-specific primers, as described previously for *P. trichocarpa* [37]; see Supplementary Table S2). Relative mature miRNA abundance was standardized using miR156 expression [37]. Each sample was analyzed in biological triplicate.

4.8. Statistical Analysis

For statistical analyses the JMP software package (version 9.0.2; SAS Institute, Cary, NC, USA) was used. All results represent the averages and SD from at least three independent biological replicates. A Student's *t*-test was used to calculate significant differences ($p < 0.05$).

5. Conclusions

Simultaneous inhibition of the function of three conserved Cu-microRNAs caused a mild but significant growth phenotype on low-Cu media, which can be ascribed to a lack of function of plastocyanin, which is essential for photosynthesis. These observations provide support for a function of Cu-microRNAs is the Cu economy.

Supplementary Materials: The following are available online at http://www.mdpi.com/2223-7747/8/6/141/s1. Figure S1: Conceptual model for tandem target mimicry, Table S1: Mineral composition of lines under three Cu regimes in mg/kg dw, Table S2: List of the primers used for qRT-PCR and mature miRNA stem-loop qRT-PCR.

Author Contributions: Authors' contributions: M.S and M.P. conceived the research plans; M.S. did the experiments; M.S. and M.P. analyzed the data; M.S. drafted the first manuscript; M.P. supervised the overall project and finalized the writing.

Acknowledgments: This project was supported by the Agriculture and Food Research Initiative competitive grant 2012-67-13-19416 of the USDA National Institute of Food and Agriculture to M.P.

Conflicts of Interest: The authors declare no conflict of interest

Abbreviations

CSD, Cu/Zn-superoxide dismutase; COX, cytochrome-c oxidase; PSII, flux through PSII; FeSOD, iron superoxide dismutase; 1-qP, redox state of plastoquinone pool; LAC, laccase; PC, Plastocyanin.

References

1. Marschner, H. *Mineral Nutrition of Higher Plants*; Academic Press: London. UK, 1995.
2. Shikanai, T.; Muller-Moule, P.; Munekage, Y.; Niyogi, K.K.; Pilon, M. PAA1, a P-type ATPase of arabidopsis, functions in copper transport in chloroplasts. *Plant Cell* **2003**, *15*, 1333–1346. [CrossRef]
3. Tapken, W.; Ravet, K.; Pilon, M. Plastocyanin controls the stabilization of the thylakoid Cu-transporting P-type ATPase PAA2/HMA8 in response to low copper in Arabidopsis. *J. Biol. Chem.* **2012**, *287*, 18544–18550. [CrossRef] [PubMed]
4. Raven, J.A.; Evans, M.C.; Korb, R.E. The role of trace metals in photosynthetic electron transport in O_2-evolving organisms. *Photosynth. Res.* **1999**, *60*, 111–149. [CrossRef]
5. Pesaresi, P.; Scharfenberg, M.; Weigel, M.; Granlund, I.; Schroeder, W.P.; Finazzi, G.; Rappaport, F.; Masiero, S.; Furini, A.; Jahns, P.; et al. Mutants, Overexpressors, and Interactors of *Arabidopsis* Plastocyanin Isoforms: Revised Roles of Plastocyanin in Photosynthetic Electron Flow and Thylakoid Redox State. *Mol. Plant* **2009**, *2*, 236–248. [CrossRef] [PubMed]
6. Abdel-Ghany, S.E. Contribution of plastocyanin isoforms to photosynthesis and copper homeostasis in *Arabidopsis thaliana* grown at different copper regimes. *Planta* **2009**, *229*, 767–779. [CrossRef]
7. Abdel-Ghany, S.E.; Muller-Moule, P.; Niyogi, K.K.; Pilon, M.; Shikanai, T. Two P-type ATPases are required for copper delivery in *Arabidopsis thaliana* chloroplasts. *Plant Cell* **2005**, *17*, 1233–1251. [CrossRef]
8. Bernal, M.; Casero, D.; Singh, V.; Wilson, G.T.; Grande, A.; Yang, H.; Dodani, S.C.; Pellegrini, M.; Huijser, P.; Connolly, E.L.; et al. Transcriptome sequencing identifies SPL7-regulated copper acquisition genes FRO4/FRO5 and the copper dependence of iron homeostasis in Arabidopsis. *Plant Cell* **2012**, *24*, 738–761. [CrossRef] [PubMed]
9. Molina-Heredia, F.P.; Wastl, J.; Navarro, J.A.; Bendall, D.S.; Hervás, M.; Howe, C.J.; De La Rosa, M.A. Photosynthesis: a new function for an old cytochrome? *Nature* **2003**, *424*, 33–34. [CrossRef]
10. Weigel, M.; Varotto, C.; Pesaresi, P.; Finazzi, G.; Rappaport, F.; Salamini, F.; Leister, D. Plastocyanin Is Indispensable for Photosynthetic Electron Flow in *Arabidopsis thaliana*. *J. Biol. Chem.* **2003**, *278*, 31286–31289. [CrossRef] [PubMed]
11. Kliebenstein, D.J.; Monde, R.-A.; Last, R.L. Superoxide dismutase in Arabidopsis: An eclecticenzyme family with disparate regulation and protein localization. *Plant Physiol.* **1998**, *118*, 637–650. [CrossRef]
12. Chu, C.C.; Lee, W.C.; Guo, W.Y.; Pan, S.M.; Chen, L.J.; Li, H.M.; Jinn, T.L. A copper chaperone for superoxide dismutase that confers three types of copper/zinc superoxide dismutase activity in Arabidopsis. *Plant Physiol.* **2005**, *139*, 425–436. [CrossRef]
13. Cohu, C.M.; Abdel-Ghany, S.E.; Gogolin Reynolds, K.A.; Onofrio, A.M.; Bodecker, J.R.; Kimbrel, J.A.; Niyogi, K.K.; Pilon, M. Copper delivery by the copper chaperone for chloroplast and cytosolic copper/zinc-superoxide dismutases: regulation and unexpected phenotypes in an Arabidopsis mutant. *Mol. Plant* **2009**, *2*, 1336–1350. [CrossRef]
14. Carr, H.S.; Winge, D.R. Assembly of Cytochrome *c* Oxidase within the Mitochondrion. *Acc. Chem. Res.* **2003**, *36*, 309–316. [CrossRef]
15. Rodriguez, F.I.; Esch, J.J.; Hall, A.E.; Binder, B.M.; Schaller, G.E.; Bleecker, A.B. A copper cofactor for the ethylene receptor ETR1 from Arabidopsis. *Science* **1999**, *283*, 996–998. [CrossRef]
16. Alonso, J.M.; Stepanova, A.N. The Ethylene Signaling Pathway. *Science* **2004**, *306*, 1513–1515. [CrossRef]
17. Dong, J.; Kim, S.T.; Lord, E.M. Plantacyanin plays a role in reproduction in Arabidopsis. *Plant Physiol.* **2005**, *138*, 778–789. [CrossRef]

18. Pignocchi, C.; Fletcher, J.M.; Wilkinson, J.E.; Barnes, J.D.; Foyer, C.H. The function of ascorbate oxidase in tobacco. *Plant Physiol.* **2003**, *132*, 1631–1641. [CrossRef]
19. Yamamoto, A.; Bhuiyan, M.N.; Waditee, R.; Tanaka, Y.; Esaka, M.; Oba, K.; Jagendorf, A.T.; Takabe, T. Suppressed expression of the apoplastic ascorbate oxidase gene increases salt tolerance in tobacco and Arabidopsis plants. *J. Exp. Bot.* **2005**, *56*, 1785–1796. [CrossRef]
20. Frébort, I.; Sebela, M.; Svendsen, I.; Hirota, S.; Endo, M.; Yamauchi, O.; Bellelli, A.; Lemr, K.; Pec, P. Molecular mode of interaction of plant amine oxidase with the mechanism-based inhibitor 2-butyne-1,4-diamine. *Eur. J. Biochem.* **2000**, *267*, 1423–1433.
21. Groß, F.; Rudolf, E.E.; Thiele, B.; Durner, J.; Astier, J. Copper amine oxidase 8 regulates arginine-dependent nitric oxide production in Arabidopsis thaliana. *J. Exp. Bot.* **2017**, *68*, 2149–2162. [CrossRef]
22. McCaig, B.C.; Meagher, R.B.; Dean, J.F. Gene structure and molecular analysis of the laccase-like multicopper oxidase (LMCO) gene family in Arabidopsis thaliana. *Planta* **2005**, *221*, 619–636. [CrossRef]
23. Turlapati, P.V.; Kim, K.W.; Davin, L.B.; Lewis, N.G. The laccase multigene family in Arabidopsis thaliana: towards addressing the mystery of their gene function(s). *Planta* **2001**, *233*, 439–470. [CrossRef] [PubMed]
24. Cai, X.; Davis, E.J.; Ballif, J.; Liang, M.; Bushman, E.; Haroldsen, V.; Torabinejad, J.; Wu, Y. Mutant identification and characterization of the laccase gene family in Arabidopsis. *J. Exp. Bot.* **2006**, *57*, 2563–2569. [CrossRef] [PubMed]
25. Berthet, S.; Demont-Caulet, N.; Pollet, B.; Bidzinski, P.; Cézard, L.; Le Bris, P.; Borrega, N.; Hervé, J.; Blondet, E.; Balzergue, S.; et al. Disruption of LACCASE4 and 17 Results in Tissue-Specific Alterations to Lignification of Arabidopsis thaliana Stems. *Plant Cell* **2011**, *23*, 1124–1137. [CrossRef]
26. Zhao, Y.; Lin, S.; Qiu, Z.; Cao, D.; Wen, J.; Deng, X.; Wang, X.; Lin, J.; Li, X. MicroRNA857 is Involved in the Regulation of Secondary Growth of Vascular Tissues in Arabidopsis. *Plant Physiol.* **2015**, *169*, 2539–2552. [CrossRef] [PubMed]
27. Pilon, M. The copper microRNAs. *New Phytol.* **2017**, *213*, 1030–1035. [CrossRef]
28. Jones-Rhoades, M.W.; Bartel, D.P.; Bartel, B. MicroRNAs and their regulatory roles in plants. *Ann. Rev. Plant Biol.* **2006**, *57*, 19–53. [CrossRef]
29. Sunkar, R.; Kapoor, A.; Zhu, J.-K. Posttranscriptional induction of two Cu/Zn superoxide dismutase genes in Arabidopsis is mediated by downregulation of miR398 and important for oxidative stress tolerance. *Plant Cell* **2006**, *18*, 2051–2065. [CrossRef]
30. Yamasaki, H.; Abdel-Ghany, S.E.; Cohu, C.M.; Kobayashi, Y.; Shikanai, T.; Pilon, M. Regulation of Copper Homeostasis by Micro-RNA in Arabidopsis. *J. Biol. Chem.* **2007**, *282*, 16369–16378. [CrossRef]
31. Abdel-Ghany, S.E.; Pilon, M. MicroRNA-mediated Systemic Down-regulation of Copper Protein Expression in Response to Low Copper Availability in Arabidopsis. *J. Biol. Chem.* **2008**, *283*, 15932–15945. [CrossRef]
32. Yamasaki, H.; Hayashi, M.; Fukazawa, M.; Kobayashi, Y.; Shikanai, T. SQUAMOSA Promoter Binding Protein-Like7 Is a Central Regulator for Copper Homeostasis in Arabidopsis. *Plant Cell* **2009**, *21*, 347–361. [CrossRef] [PubMed]
33. Zhang, H.; Li, L. SQUAMOSA promoter binding protein-like7 regulated microRNA408 is required for vegetative development in Arabidopsis. *Plant J.* **2013**, *74*, 98–109. [CrossRef] [PubMed]
34. Zhang, H.; Zhao, X.; Li, J.; Cai, H.; Deng, X.W.; Li, L. MicroRNA408 is critical for the HY5-SPL7 gene network that mediates the coordinated respose to light and copper. *Plant Cell* **2014**, *26*, 4933–4953. [CrossRef]
35. Cuperus, J.T.; Fahlgren, N.; Carrington, J.C. Evolution and functional diversification of MIRNA genes. *Plant Cell* **2011**, *23*, 431–442. [CrossRef] [PubMed]
36. Waldron, K.J.; Rutherford, J.C.; Ford, D.; Robinson, N.J. Metalloproteins and metal sensing. *Nature* **2009**, *460*, 823–830. [CrossRef]
37. Ravet, K.; Danford, F.L.; Dihle, A.; Pittarello, M.; Pilon, M. Spatial-temporal analysis of Cu homeostasis in Populus trichocarpa reveals an integrated molecular remodeling for a preferential allocation of copper to plastocyanin in the chloroplasts of developing leaves. *Plant Physiol.* **2011**, *157*, 1300–1312. [CrossRef]
38. Shahbaz, M.; Ravet, K.; Peers, G.; Pilon, M. Prioritization of copper for the use in photosynthetic electron transport in developing leaves of hybrid poplar. *Front. Plant Sci.* **2015**, *6*, 407. [CrossRef]
39. Yan, J.; Chia, J.C.; Sheng, H.; Jung, H.I.; Zavodna, T.O.; Zhang, L.; Huang, R.; Jiao, C.; Craft, E.J.; Fei, Z.; et al. Arabidopsis Pollen Fertility Requires the Transcription Factors CITF1 and SPL7 That Regulate Copper Delivery to Anthers and Jasmonic Acid Synthesis. *Plant Cell* **2017**, *29*, 3012–3029. [CrossRef]

40. Dugas, D.V.; Bartel, B. Sucrose induction of Arabidopsis miR398 represses two Cu/Zn superoxide dismutases. *Plant Mol. Biol.* **2008**, *67*, 403–417. [CrossRef]
41. Franco-Zorrilla, J.M.; Valli, A.; Todesco, M.; Mateos, I.; Puga, M.I.; Rubio-Somoza, I.; Leyva, A.; Weigel, D.; García, J.A.; Paz-Ares, J. Target mimicry provides a new mechanism for regulation of microRNA activity. *Nat. Genet.* **2007**, *39*, 1033–1037. [CrossRef]
42. Zhang, J.P.; Yu, Y.; Feng, Y.Z.; Zhou, Y.F.; Zhang, F.; Yang, Y.W.; Lei, M.Q.; Zhang, Y.C.; Chen, Y.Q. MiR408 Regulates Grain Yield and Photosynthesis via a Phytocyanin Protein. *Plant Physiol.* **2017**, *175*, 1175–1185. [CrossRef] [PubMed]
43. Kropat, J.; Tottey, S.; Birkenbihl, R.P.; Depege, N.; Huijser, P.; Merchant, S. A regulator of nutritional copper signaling in Chlamydomonas is an SBP domain protein that recognizes the GTAC core of copper response element. *Proc. Natl. Acad. Sci. USA* **2005**, *102*, 18730–18735. [CrossRef] [PubMed]
44. Pan, J.; Huang, D.; Guo, Z.; Kuang, Z.; Zhang, H.; Xie, X.; Ma, Z.; Gao, S.; Lerdau, M.T.; Chu, C.; et al. Overexpression of microRNA408 enhances photosynthesis, growth, and seed yield in diverse plants. *J. Integr. Plant Biol.* **2018**, *60*, 323–340. [CrossRef] [PubMed]
45. Song, Z.; Zhang, L.; Wang, Y.; Li, H.; Li, S.; Zhao, H.; Zhang, H. Constitutive Expression of miR408 Improves Biomass and Seed Yield in Arabidopsis. *Front. Plant Sci.* **2018**, *8*, 2114. [CrossRef] [PubMed]
46. Carrió-Seguí, À.; Ruiz-Rivero, O.; Villamayor-Belinchón, L.; Puig, S.; Perea-García, A.; Peñarrubia, L. The Altered Expression of microRNA408 Influences the Arabidopsis Response to Iron Deficiency. *Front. Plant Sci.* **2019**, *10*, 324. [CrossRef]
47. Sunkar, R.; Zhu, J.K. Novel and stress-regulated micro-RNAs and other small RNAs from Arabidopsis. *Plant Cell* **2004**, *16*, 2001–2019. [CrossRef]
48. Chen, Y.; Jiang, J.; Song, A.; Chen, S.; Shan, H.; Luo, H.; Gu, C.; Sun, J.; Zhu, L.; Fang, W.; et al. Ambient temperature enhanced freezing tolerance of Chrysanthemum dichrum CdICE1 Arabidopsis via miR398. *BMC Biol.* **2013**, *11*, 121. [CrossRef]
49. Guan, Q.; Lu, X.; Zeng, H.; Zhang, Y.; Zhu, J. Heat stress induction of miR398 triggers a regulatory loop that is critical for thermotolerance in Arabidopsis. *Plant J.* **2013**, *74*, 840–851. [CrossRef]
50. Waters, B.M.; McInturf, S.A.; Stein, R.J. Rosette iron deficiency transcript and microRNA profiling reveals links between copper and iron homeostasis in Arabidopsis thaliana. *J. Exp. Bot.* **2012**, *63*, 5903–5918. [CrossRef]
51. Robinson, N.J.; Procter, C.M.; Connolly, E.L.; Guerinot, M.L. A ferric-chelate reductase for iron uptake from soils. *Nature* **1999**, *397*, 694–697. [CrossRef]
52. Murashige, T.; Skoog, F. A revised medium for rapid growth and bio assays with tobacco tissue cultures. *Physiol. Plant.* **1962**, *15*, 473–497. [CrossRef]
53. Epstein, E.; Bloom, A. *Mineral Nutrition in Plants: Principles and Perspectives*, 2nd ed.; Sinauer Associates: Sunderland, MA, USA, 2005.
54. Nakagawa, T.; Kurose, T.; Hino, T.; Tanaka, K.; Kawamukai, M.; Niwa, Y.; Toyooka, K.; Matsuoka, K.; Jinbo, T.; Kimura, T. Development of series of gateway binary vectors, pGWBs, for realizing efficient construction of fusion genes for plant transformation. *J. Biosci. Bioeng.* **2007**, *104*, 34–41. [CrossRef] [PubMed]
55. Clough, S.J.; Bent, A.F. Floral dip: a simplified method for Agrobacterium-mediated transformation of Arabidopsis thaliana. *Plant J.* **1998**, *16*, 735–743. [CrossRef] [PubMed]
56. Pilon-Smits, E.A.; Hwang, S.; Lytle, C.M.; Zhu, Y.; Tai, J.C.; Bravo, R.C.; Chen, Y.; Leustek, T.; Terry, N. Overexpression of ATP sulfurylase in Indian mustard leads to increased selenite uptake, reduction and tolerance. *Plant Physiol.* **1999**, *119*, 123–132. [CrossRef]
57. Maxwell, K.; Johnson, G.N. Chlorophyll fluorescence—a practical guide. *J. Exp. Bot.* **2000**, *51*, 659–668. [CrossRef]
58. Varkonyi-Gasic, E.; Hellens, R.P. Quantitative stem-loop RT-PCR for detection of microRNAs. *Methods Mol. Biol.* **2007**, *744*, 145–157.

© 2019 by the authors. Licensee MDPI, Basel, Switzerland. This article is an open access article distributed under the terms and conditions of the Creative Commons Attribution (CC BY) license (http://creativecommons.org/licenses/by/4.0/).

Article

DRB1, DRB2 and DRB4 Are Required for Appropriate Regulation of the microRNA399/*PHOSPHATE2* Expression Module in *Arabidopsis thaliana*

Joseph L. Pegler, Jackson M. J. Oultram, Christopher P. L. Grof and Andrew L. Eamens *

Centre for Plant Science, School of Environmental and Life Sciences, Faculty of Science, University of Newcastle, Callaghan 2308, New South Wales, Australia; Joseph.Pegler@uon.edu.au (J.L.P.); Jackson.Oultram@uon.edu.au (J.M.J.O.); chris.grof@newcastle.edu.au (C.P.L.G.)
* Correspondence: andy.eamens@newcastle.edu.au; Tel.: +61-249-217-784

Received: 15 April 2019; Accepted: 9 May 2019; Published: 13 May 2019

Abstract: Adequate phosphorous (P) is essential to plant cells to ensure normal plant growth and development. Therefore, plants employ elegant mechanisms to regulate P abundance across their developmentally distinct tissues. One such mechanism is PHOSPHATE2 (PHO2)-directed ubiquitin-mediated degradation of a cohort of phosphate (PO_4) transporters. *PHO2* is itself under tight regulation by the PO_4 responsive microRNA (miRNA), miR399. The DOUBLE-STRANDED RNA BINDING (DRB) proteins, DRB1, DRB2 and DRB4, have each been assigned a specific functional role in the *Arabidopsis thaliana* (*Arabidopsis*) miRNA pathway. Here, we assessed the requirement of DRB1, DRB2 and DRB4 to regulate the miR399/*PHO2* expression module under PO_4 starvations conditions. Via the phenotypic and molecular assessment of the knockout mutant plant lines, *drb1*, *drb2* and *drb4*, we show here that; (1) DRB1 and DRB2 are required to maintain P homeostasis in *Arabidopsis* shoot and root tissues; (2) DRB1 is the primary DRB required for miR399 production; (3) DRB2 and DRB4 play secondary roles in regulating miR399 production, and; (4) miR399 appears to direct expression regulation of the *PHO2* transcript via both an mRNA cleavage and translational repression mode of RNA silencing. Together, the hierarchical contribution of DRB1, DRB2 and DRB4 demonstrated here to be required for the appropriate regulation of the miR399/*PHO2* expression module identifies the extreme importance of P homeostasis maintenance in *Arabidopsis* to ensure that numerous vital cellular processes are maintained across *Arabidopsis* tissues under a changing cellular environment.

Keywords: *Arabidopsis thaliana*; phosphorous (P); phosphate (PO_4) stress; microRNA (miRNA); miR399; *PHOSPHATE2* (*PHO2*); DOUBLE-STRANDED RNA BINDING (DRB) proteins DRB1; DRB2; DRB4; miR399-directed *PHO2* expression regulation; RT-qPCR

1. Introduction

Phosphorous (P) is one of the most limiting factors for plant growth worldwide [1–3], with large quantities of P an essential requirement for numerous processes vital to the plant cell, including energy trafficking, signaling cascades, enzymatic reactions and nucleic acid and phospholipid synthesis [3,4]. Inorganic phosphate (Pi), in the form of PO_4, is the predominant form of P taken up by a plant from the soil, however, soil PO_4 primarily exists in organic or insoluble forms that are largely inaccessible to plant root uptake mechanisms [1]. Therefore, due to limited soil PO_4 availability, combined with the importance of an adequate concentration of P in plant cells to ensure normal growth and development, plants employ elegant mechanisms to spatially regulate P abundance across their developmentally distinct tissues [5,6]. Phosphorous homeostasis is therefore tightly controlled and involves both the remobilization of internal P stores and the increased acquisition of external PO_4 [5,7]. For example,

P limitation triggers the release of organic acids from the plant root system into the soil rhizosphere to chelate with metal ions to promote soluble PO_4 uptake to maintain or increase intracellular P concentration [1,8]. In addition, the P stored in the older leaves of a plant when the plant experiences P stress is remobilized; this allows for (1) continued growth of actively expanding tissues, and (2) the promotion of new growth. Enhanced P trafficking is achieved via promoting the expression of genes encoding PO_4 transporter proteins, and in turn, elevated PO_4 transporter protein abundance generally ensures that the cellular P concentration is maintained irrespective of external PO_4 levels [1,7].

In *Arabidopsis thaliana* (*Arabidopsis*), the first protein identified to be required for the maintenance of P homeostasis under PO_4 limiting conditions was PHOSPHATE1 (PHO1) [9]. The gene encoding PHO1 (*PHO1*; *AT1G14040*) was identified by [9] via their characterization of *pho1* plants, an *Arabidopsis* mutant line demonstrated to over-accumulate P in root tissues due to defective P translocation to the shoot. Although the *Arabidopsis* PHO1 protein, and the PHO1 proteins of other plant species characterized to date, do not closely resemble other PO_4 transporter proteins, PHO1 is indeed central to P movement in plants. The PHO1 protein is essential for PO_4 efflux into the root vascular cylinder; the first step in P transportation to the upper aerial tissues [10,11]. PHOSPHATE2 (PHO2) was the second protein demonstrated essential for the maintenance of P homeostasis with the *pho2* mutant shown to accumulate P to toxic levels in shoot tissues [12,13]. The *PHO2* gene (*AT2G33770*) has since been shown to encode a ubiquitin conjugating enzyme24 (UBC24), with the PHO2 UBC24 proposed to direct ubiquitin-mediated degradation of PO_4 transporters, PHOSPHATE TRANSPORTER1;4 (PHT1;4), PHT1;8 and PHT1;9 [14]. Further, *PHO2* is almost ubiquitously expressed in *Arabidopsis* shoot and root tissues [15], with the loss of PHO2-directed suppression of PHT1;4, PHT1;8 and PHT1;9 abundance in *pho2* plants leading to the enhanced translocation of P from the roots to the shoot tissue [14]. In addition to PHO1 and PHO2, traditional mutagenesis-based approaches have further identified other proteins essential to P homeostasis maintenance, including PHOSPHATE STARVATION RESPONSE1 (PHR1), a MYB domain transcription factor that regulates the expression of numerous P responsive genes [16,17].

More contemporary research, however, has concentrated on documenting the regulatory role directed at the posttranscriptional level by small regulatory RNAs (sRNA), specifically the microRNA (miRNA) class of sRNA, in order to maintain P homeostasis [18,19]. The advent of high throughput sequencing technologies has made sRNA profiling across plant species, and under different growth regimes, including exposure of a plant to abiotic and biotic stress, a routine experimental procedure in modern research [14,20,21]. Such profiling has identified a common suite of conserved miRNAs (miRNAs identified across multiple, evolutionary unrelated plant species) that accumulate differentially when mineral nutrients are lacking, including P, nitrogen (N), copper and sulphur [20,21]. Responsiveness of a single miRNA to multiple mineral nutrient stresses is not surprising considering the considerable overlap in the complex regulation of metal ion transport and/or uptake in plants [14,22,23]. In *Arabidopsis* for example, P and N uptake mechanisms are reciprocally linked to one another, therefore; a miRNA with enhanced accumulation during periods of P stress will usually be reduced in abundance during N starvation [19,24,25].

The miRNA, miR399, has been conclusively linked with the maintenance of P homeostasis and the regulation of PO_4 uptake in *Arabidopsis* [18,19]. In *Arabidopsis*, the miR399 sRNA is processed from six precursor transcripts, namely *PRE-MIR399A* to *PRE-MIR399F*, transcribed from five genomic loci (*MIR399A-MIR399D* and *MIR399E/F*). The miR399 sRNA is unique amongst *Arabidopsis* miRNAs in that it acts as a mobile systemic signal upon PO_4 stress [21,26]. More specifically, when P becomes limited in *Arabidopsis* shoots, *MIR399* gene expression is stimulated by PHR1 [27], and following processing of the now abundant miR399 precursor transcripts by the protein machinery of the *Arabidopsis* miRNA pathway, the mature miR399 sRNA is transported to the roots. Here, miR399 is actively loaded by the miRNA-induced silencing complex (miRISC) to direct miRISC-mediated cleavage of *PHO2*, the target transcript of miR399 [7,21,27]. Reduced PHO2 protein abundance, due to elevated miRISC-mediated cleavage of the *PHO2* transcript, in turn removes the PHO2-mediated suppression of PO_4 transporters, PHT1;4, PHT1;8 and PHT1;9, to ultimately promote root-to-shoot P transport in an attempt to maintain shoot P homeostasis in P limited conditions [28–31]. Additional regulatory complexity to the miR399/*PHO2* expression module is offered by the non-protein-coding RNA, *INDUCED BY PHOSPHATE STARVATION1* (*IPS1*) [32]. Once transcribed, *IPS1* acts as an endogenous target mimic (eTM) of miR399 activity [33]. Specifically, the miR399 target site harbored by *IPS1* contains a three nucleotide mismatch bulge across miR399 nucleotide positions 10 and 11: the position at which the catalytic core of miRISC, ARGONAUTE1 (AGO1), catalyzes the cleavage of miRNA target transcripts [34]. The bulge that forms at this position once miR399-directed AGO1 binds *IPS1*, renders *IPS1* resistant to AGO1-catalyzed cleavage, thereby effectively sequestering away miR399 activity [33].

Three of the five members of the *Arabidopsis* DOUBLE-STRANDED RNA BINDING (DRB) protein family, including DRB1, DRB2 and DRB4, have been assigned functional roles in the *Arabidopsis* miRNA pathway [35–39]. Both DRB1 and DRB4 form functional partnerships with DICER-LIKE (DCL) proteins, RNase III-like endonucleases that cleave molecules of double-stranded RNA (dsRNA). More specifically, the DRB1/DCL1 partnership processes stem-loop structured molecules of imperfectly dsRNA that form post miRNA precursor transcript folding [35–37], and the DRB4/DCL4 partnership is central for the processing of a small subset of miRNA precursor transcripts that fold to form stem-loop structures with high levels of base-pairing due to the almost perfect complementarity of the nucleotide sequences of the stem-loop arms [39]. More recently, DRB2 has also been assigned a functional role in the *Arabidopsis* miRNA pathway due to its demonstrated antagonism and/or synergism with the roles of both DRB1 and DRB4 in sRNA production [37,40]. Here, we therefore assessed the requirement of DRB1, DRB2 and DRB4 in the regulation of the miR399/*PHO2* expression module, both under non-stressed growth conditions and when wild-type *Arabidopsis* plants (ecotype Columbia-0 (Col-0)) and the *drb1*, *drb2* and *drb4* mutant lines are exposed to PO_4 starvation. More specifically, we aimed to determine; (1) the contribution of DRB1, DRB2 and/or DRB4 to miR399 production; (2) the mode of silencing directed by miR399 to regulate *PHO2* expression, and; (3) whether either DRB1, DRB2 or DRB4 are required for P homeostasis maintenance. Phenotypic and molecular assessment of Col-0, *drb1*, *drb2* and *drb4* plants post exposure to a 7-day period of PO_4 starvation, revealed that DRB1 and DRB2 are required for P homeostasis maintenance. Further, DRB1 was established as the primary DRB protein required to regulate miR399 production. However, DRB2 and DRB4 were demonstrated to play a secondary role in miR399 production regulation. Furthermore, miR399 appears to regulate the expression of its targeted transcript, *PHO2*, via both the canonical mechanism of plant miRNA-directed target gene expression repression, target mRNA cleavage, and via the alternative mode of target gene expression regulation, translational repression. Taken together, the hierarchical contribution of DRB1, DRB2 and DRB4 to the regulation of the miR399/*PHO2* expression module in *Arabidopsis* shoots and roots identifies the extreme importance of maintaining P homeostasis to ensure that numerous vital cellular processes are maintained across *Arabidopsis* tissue types and under a changing cellular environment.

2. Results

2.1. The Phenotypic and Physiological Response to PO_4 Stress in the Shoot Tissues of Arabidopsis Plant Lines Defective in DRB Protein Activity

To determine the consequence of loss of DRB activity on P homeostasis maintenance in 15-day old *Arabidopsis* plants post a 7-day period of PO_4 starvation, a series of phenotypic and physiological parameters were assessed in Col-0, *drb1*, *drb2* and *drb4* shoots. The severe developmental phenotype of the *drb1* mutant has been reported previously [36,41,42]. Figure 1A clearly reveals the reduced size of the *drb1* mutant at 15 days of age, compared to Col-0 plants, when both *Arabidopsis* lines are cultivated on standard growth media (P^+ media). The retarded development of the *drb1* mutant is further evidenced in Figure 1B where the fresh weight of 8-day old Col-0 and *drb1* seedlings is presented. Specifically, prior to seedling transfer to either P^+ or P^- media, the fresh weight of an 8-day old *drb1* seedling (13.5 ± 1.0 mg) is 53.4% less than that a Col-0 seedling (29.0 ± 3.5 mg). Compared to *drb1*, the *drb2* and *drb4* mutants display mild developmental phenotypes [37,42] as evidenced by those displayed by 15-day old *drb2* and *drb4* plants cultivated on P^+ growth media (Figure 1A), and by the fresh weights of 8-day old *drb2* (26.8 ± 4.2 mg) and *drb4* (22.9 ± 1.4 mg) seedlings. Although the *drb1* mutant displayed the most severe phenotype, *drb1* development appeared to be the least affected by the 7-day PO$_4$ stress treatment. The fresh weight of P$^-$ *drb1* plants (35.5 ± 1.0 mg) was only reduced by 21.6% compared to P$^+$ *drb1* plants (45.3 mg ± 1.5 mg) (Figure 1C). The development of Col-0, *drb2* and *drb4* plants was negatively impacted to a similar degree by the 7-day PO$_4$ stress treatment, with their fresh weights reduced by 36.6%, 39.1% and 36.3%, respectively (Figure 1C). Determination of rosette area revealed largely similar trends across the *drb* mutant lines analyzed, that is, *drb1* rosette area was reduced by 29.3%, while the rosette development of P$^-$ *drb2* and P$^-$ *drb4* plants was reduced by 48.0% and 38.7%, respectively (Figure 1D). Interestingly, the observed reductions to the rosette area of P$^-$ *drb1*, P$^-$ *drb2* and P$^-$ *drb4* plants was considerably less than the 60.1% reduction to the rosette area of P$^-$ Col-0 plants (11.2 ± 1.7 mm^2) compared to P$^+$ Col-0 plants (28.1 ± 5.5 mm^2) (Figure 1D).

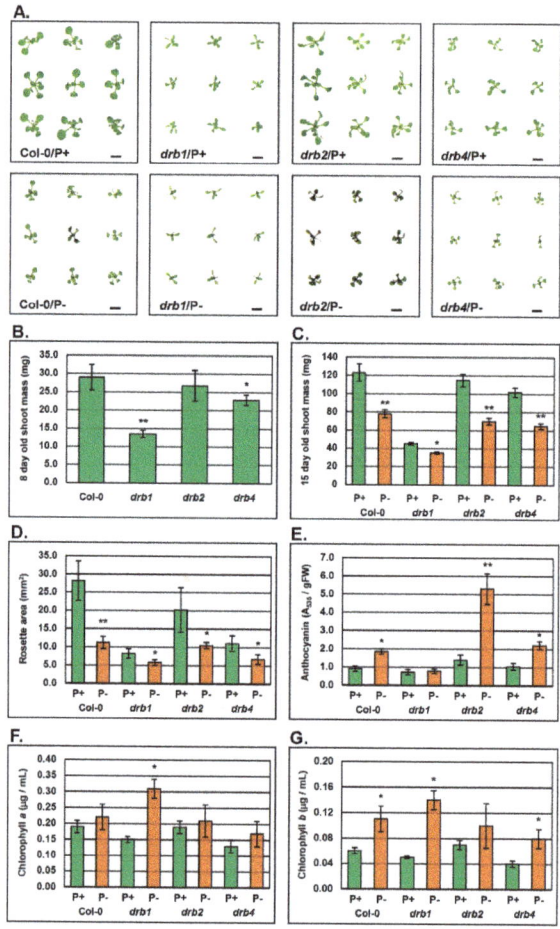

Figure 1. The aerial tissue phenotypes displayed by 15-day old *Arabidopsis* plant lines Col-0, *drb1*, *drb2* and *drb4* post exposure to a 7-day period of PO$_4$ starvation. (**A**) The aerial tissue phenotypes expressed by non-stressed (top row of panels) and PO$_4$-stressed (bottom row of panels) 15-day old Col-0, *drb1*, *drb2* and *drb4* plants. Scale bar = 1cm. (**B**) Quantification of the shoot mass of 8-day old Col-0, *drb1*, *drb2* and *drb4* seedlings germinated and cultivated under standard growth conditions. (**C**) The shoot mass of non-stressed and PO$_4$-stressed 15-day old Col-0, *drb1*, *drb2* and *drb4* plants. (**D**) The rosette area of

non-stressed and PO$_4$-stressed 15-day old *Arabidopsis* lines, Col-0, *drb1*, *drb2* and *drb4*. (**E**) Anthocyanin accumulation in the shoot tissues of 15-day old Col-0, *drb1*, *drb2* and *drb4* plants cultivated under standard growth conditions, or for 7-days under PO$_4$ starvation. (**F** and **G**) Chlorophyll *a* (**F**) and chlorophyll *b* (**G**) abundance in the aerial tissues of non-stressed and PO$_4$-stressed Col-0, *drb1*, *drb2* and *drb4* plants. (**B-G**) Error bars represent the standard deviation of four biological replicates and each biological replicate consisted of a pool of twelve individual plants. The presence of an asterisk above a column represents a statistically significant difference either between non-stressed Col-0 plants and each assessed *drb* mutant post cultivation under either a non-stressed or stressed growth regime (**B**) or between the non-stressed and PO$_4$-stressed sample of each plant line (**C-G**) (*p*-value: * < 0.05; ** < 0.005; *** < 0.001).

Anthocyanin, chlorophyll *a* and chlorophyll *b* content of Col-0, *drb1*, *drb2* and *drb4* shoots was also determined. Phosphate starvation has been previously shown to elevate the levels of PRODUCTION OF ANTHOCYANIN PIGMENT1 (PAP1/MYB75), PAP2 (MYB90) and MYB113, three MYB domain transcription factors that in turn stimulate the expression of a cohort of genes required for anthocyanin production in vegetative tissues [19,43]. These reports, in combination with the readily observable pigmentation that accumulated in the rosette leaves of P$^-$ Col-0, P$^-$ *drb2* and P$^-$ *drb4* plants (Figure 1A), identified anthocyanin as an ideal metric to further assess the response of each *drb* mutant to PO$_4$ starvation. The anthocyanin content of non-stressed Col-0, *drb1*, *drb2* and *drb4* shoots was similar (Figure 1E). However, when PO$_4$ is limited, an approximate 2.0-fold increase in anthocyanin accumulation was detected for P$^-$ Col-0 shoots. Further promotion of anthocyanin accumulation was determined for PO$_4$-stressed *drb2* and *drb4* plants, with anthocyanin content elevated 3.7- and 2.8-fold in P$^-$ *drb2* and P$^-$ *drb4* plants, respectively (Figure 1E). As readily observable in Figure 1A, anthocyanin accumulation was not promoted in the shoot tissue of P$^-$ *drb1* plants. However, spectrophotometry revealed abundance changes for both chlorophyll *a* and chlorophyll *b* in the shoot tissue of P$^-$ *drb1* plants. Specifically, chlorophyll *a* (Figure 1F) and chlorophyll *b* (Figure 1G) abundance was elevated by 2.1- and 2.8-fold in P$^-$ *drb1* shoots, compared to P$^+$ *drb1* shoots. In PO$_4$-stressed Col-0, *drb2* and *drb4* shoots, the chlorophyll *a* level remained largely unchanged compared to the non-stressed counterpart of each plant line (Figure 1F). Chlorophyll *b* accumulation however, was determined to be promoted in Col-0 and *drb4* shoots, by 1.8- and 2.0-fold, by the 7-day PO$_4$ starvation period (Figure 1G).

2.2. Molecular Profiling of the miR399/PHO2 Expression Module in the Shoot Tissues of Arabidopsis Plant Lines Defective in DRB Protein Activity

The results presented in Figure 1 strongly indicated that each *drb* mutant was responding differently to the applied stress and when this finding is considered together with the documented roles of DRB1, DRB2 and DRB4 in the *Arabidopsis* miRNA pathway [35–39], including the demonstrated antagonism between DRB1 and DRB2 [37] and between DRB2 and DRB4 [40] in miRNA production, the miR399/*PHO2* expression module was next profiled via a RT-qPCR-based approach. RT-qPCR profiling was conducted in an attempt to determine if the observed differences in the response of each *drb* mutant line to PO$_4$ stress was a result of dysfunction of the miR399/*PHO2* expression module.

In *Arabidopsis* shoots, *PHR1* promotes *MIR399* gene expression when PO$_4$ supplies become limited, resulting in elevated miR399 abundance [27]. Therefore, RT-qPCR was first used to assess *PHR1* expression in control and PO$_4$-stressed Col-0, *drb1*, *drb2* and *drb4* shoots (Figure 2A). *PHR1* expression was only mildly elevated by 1.5-, 1.6- and 1.7-fold in P$^+$ *drb1*, P$^+$ *drb2* and P$^+$ *drb4* shoots respectively, compared to its levels in non-stressed Col-0 shoots (Figure 2A). RT-qPCR further revealed that PO$_4$ stress only induced mild elevations to *PHR1* expression in P$^-$ Col-0 (1.00 to 1.22 relative expression) and P$^-$ *drb2* shoots (1.62 to 1.74 relative expression) (Figure 2A). This result was not unexpected in view of the previous report of only mild *PHR1* expression induction in PO$_4$-stressed *Arabidopsis* [17]. Interestingly, *PHR1* expression was reduced by 19.6% and 31.2% in P$^-$ *drb1* and P$^-$ *drb4* shoots, respectively (Figure 2A), and not mildly elevated as expected.

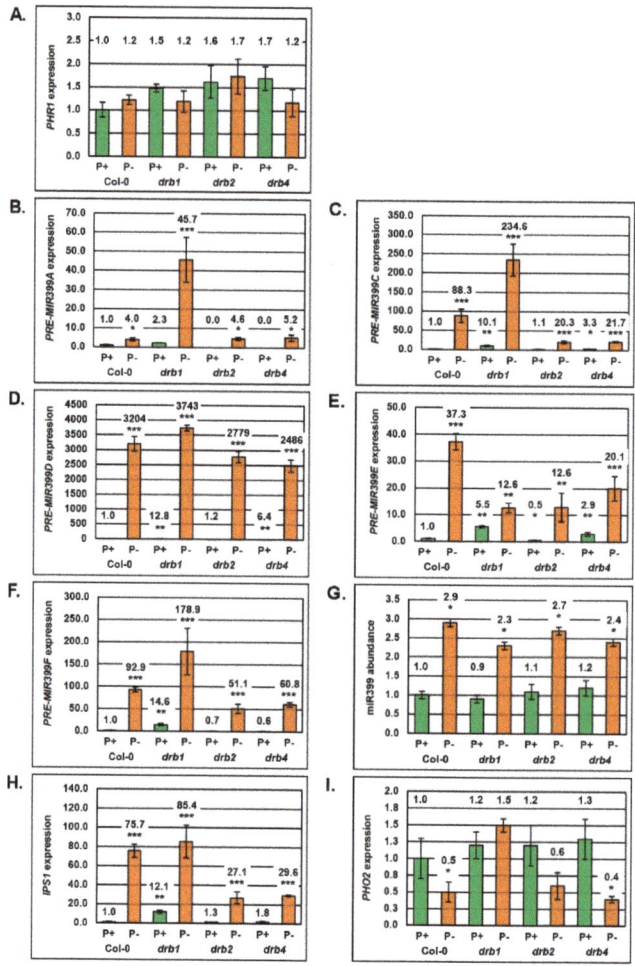

Figure 2. Molecular profiling of the miR399/PHO2 expression module in the aerial tissues of non-stressed and PO$_4$-stressed Col-0, *drb1*, *drb2* and *drb4* plants. (**A**) RT-qPCR assessment of the expression of the PO$_4$ responsive transcription factor PHR1 in the aerial tissues of non-stressed and PO$_4$-stressed Col-0, *drb1*, *drb2* and *drb4* plants. (**B** to **F**) RT-qPCR profiling of miR399 precursor transcript abundance in the aerial tissues of non-stressed and PO$_4$-stressed Col-0, *drb1*, *drb2* and *drb4* plants, including precursors *PRE-MIR399A* (**B**), *PRE-MIR399C* (**C**), *PRE-MIR399D* (**D**), *PRE-MIR399E* (**E**) and *PRE-MIR399F* (**F**). (**G**) Quantification of miR399 abundance in the aerial tissues of non-stressed and PO$_4$-stressed Col-0, *drb1*, *drb2* and *drb4* plants. (**H**) Assessment of the expression of the non-cleavable decoy of miR399 activity, *IPS1*, via RT-qPCR in the aerial tissues of non-stressed and PO$_4$-stressed *Arabidopsis* lines, Col-0, *drb1*, *drb2* and *drb4*. (**I**) RT-qPCR analysis of the expression of the miR399 target gene, *PHO2*, in the aerial tissues of non-stressed and PO$_4$-stressed *Arabidopsis* lines, Col-0, *drb1*, *drb2* and *drb4*. (**A–I**) Error bars represent the standard deviation of four biological replicates and each biological replicate consisted of a pool of twelve individual plants. Due to the vastly different levels of each assessed transcript, the relative expression value for each plant line/growth regime is provided above the corresponding column. The presence of an asterisk above a column represents a statistically significant difference between non-stressed Col-0 plants and each of the assessed *drb* mutant lines, post cultivation under either a standard or stressed growth regime (p-value: * < 0.05; ** < 0.005; *** < 0.001).

The miR399 sRNA is processed from six structurally distinct precursor transcripts (*PRE-MIR399A* to *PRE-MIR399F*), transcribed from five genomic loci (*MIR399A* to *MIR399D* and *MIR399E/F*) in *Arabidopsis*. RT-qPCR only failed to detect *PRE-MIR399B* expression in Col-0 shoots. RT-qPCR did however clearly reveal that PO_4 stress induced the expression of the five detectable miR399 precursor transcripts by 4.0-, 88.3-, 3204-, 37.3- and 92.9-fold in the shoots of P^- Col-0 plants (Figure 2B–F). Of the three members of the *Arabidopsis* DRB protein family analyzed here, Figure 2B–F clearly show that DRB1 is the primary DRB protein required to regulate miR399 production in *Arabidopsis* shoots with the abundance of *PRE-MIR399A*, *PRE-MIR399C*, *PRE-MIR399D*, *PRE-MIR399E* and *PRE-MIR399F* elevated by 2.3-, 10.1-, 12.8-, 5.5- and 14.6-fold, respectively, in P^+ *drb1* shoots. The primary role of DRB1 in regulating miR399 production in *Arabidopsis* shoots was further highlighted for *PRE-MIR399A*, *PRE-MIR399C*, *PRE-MIR399D* and *PRE-MIR399F* via additional elevations to their respective expression levels, specifically 45.7-, 234.6- 3743- and 178.9-fold increases to transcript abundance in P- *drb1* shoots (Figure 2B–D,F).

Failure to detect the *PRE-MIR399A* precursor by RT-qPCR in P^+ *drb2* shoots, and a similar degree of over-accumulation of this precursor in P^- Col-0 (4.0-fold) and P^- *drb2* shoots (4.6-fold), indicated that DRB2 is not required to regulate miR399 production from this precursor (Figure 2B). Wild-type-like accumulation of *PRE-MIR399C* (1.1-fold) and *PRE-MIR399D* (1.2-fold) in P^+ *drb2* shoots, and a lower degree of over-accumulation of these two precursors in P^- *drb2* shoots, compared to P^- Col-0 shoots, indicated that DRB2 plays a secondary role in regulating miR399 production from these two precursors (Figure 2C,D). A similar level of expression of *PRE-MIE399E* in PO_4-stressed *drb1* and *drb2* shoots suggested that both DRB1 and DRB2 are required for miR399 production from this precursor (Figure 2E). However, lower transcript abundance (0.5 relative expression) in P^+ *drb2* shoots, compared to relative expression levels of 1.0 and 5.5 in P^+ Col-0 and P^+ *drb1* shoots, respectively (Figure 2E), again indicated that under standard growth conditions, DRB2 plays a secondary role in regulating miR399 production from the *PRE-MIR399E* precursor. The abundance of the *PRE-MIR399F* transcript is also reduced in P^+ *drb2* shoots compared to its levels in P^+ Col-0 shoots, and further, the degree of over-accumulation of *PRE-MIR399F* is less in P^- *drb2* shoots compared to its levels in P^- Col-0 shoots (Figure 2F). When these expression trends are considered together with those documented for P^+ and P^- *drb1* shoots, they again indicate a secondary role for DRB2 in regulating miR399 production from this precursor.

As demonstrated for P^+ *drb2* shoots, the *PRE-MIR399A* transcript remained below the detection sensitivity of RT-qPCR in P^+ *drb4* shoots (Figure 2B). RT-qPCR did however, reveal *PRE-MIR399A* expression to be elevated by 5.2-fold in P^- *drb4* shoots, a similar degree of transcript elevation to that observed in P^- Col-0 shoots (4.0-fold increase) (Figure 2B). This indicates that DRB4 is not involved in regulating miR399 production from this precursor. Comparison of the RT-qPCR generated expression trends for *PRE-MIR399C*, *PRE-MIR399D* and *PRE-MIR399E* in P^+ and P^- *drb4* shoots, to those of P^+ Col-0, P^- Col-0, P^+ *drb1* and P^- *drb1* shoots, revealed a secondary role for DRB4 in regulating miR399 production from these three precursor transcripts (Figure 2C–E). DRB4 also appears to play a role in regulating miR399 production from the *PRE-MIR399F* transcript, with *PRE-MIR399F* abundance reduced by 40% in P^+ *drb4* shoots (Figure 2F). RT-qPCR also revealed that the expression of this precursor transcript was elevated to a relative expression level of 60.8 in PO_4-stressed *drb4* shoots; a lower degree of relative expression than observed in either P^- Col-0 (92.9 relative expression) or P^- *drb1* (178.9 relative expression) shoots (Figure 2F). This finding suggests that in the absence of DRB4 activity, miR399 is more efficiently processed from the *PRE-MIR399F* precursor transcript.

RT-qPCR was next applied to quantify miR399 abundance in the shoot material of non-stressed or PO_4-stressed Col-0, *drb1*, *drb2* and *drb4* plants. This analysis revealed that in spite of the considerable variation in precursor transcript abundance in the shoot tissues of P^+ Col-0, P^+ *drb1*, P^+ *drb2* and P^+ *drb4* plants, miR399 levels remained largely unchanged (Figure 2G). This was an especially surprising finding for control *drb1* plants, with the *PRE-MIR399A*, *PRE-MIR399C*, *PRE-MIR399D*, *PRE-MIR399E* and *PRE-MIR399F* transcripts demonstrated to over-accumulate by 4.0-, 10.1-, 12.8-, 5.5- and 14.6-fold in P^+ *drb1* shoots, compared to their respective levels in P^+ Col-0 shoots. However, miR399 abundance was only reduced by 10% in P$^+$ *drb1* shoots. Similarly, although the expression level of the five miR399 precursors varied considerably in P$^+$ *drb2* and P$^+$ *drb4* shoots, miR399 abundance was only elevated by 10% and 20%, respectively (Figure 2G). Enhanced miR399 accumulation in P$^+$ *drb2* and P$^+$ *drb4* shoots did however further identify that both of these DRB proteins are required to correctly regulate miR399 abundance in *Arabidopsis* shoots. The degree of alteration to miR399 abundance was demonstrated to be higher in the shoot tissues of the four assessed plant lines when these lines were cultivated on PO$_4$ deplete media. Specifically, RT-qPCR revealed 2.9-, 2.6-, 2.5- and 2.0-fold enhancement to miR399 abundance in PO$_4$-stressed Col-0, *drb1*, *drb2* and *drb4* shoots, respectively (Figure 2G).

The mild alteration to miR399 abundance quantified by RT-qPCR in non-stressed and PO$_4$-stressed shoots (Figure 2G) led us to next assess the expression of *IPS1*, the eTM of miR399 [32–34]. Due to *IPS1* being a PO$_4$ stress-induced gene, it was unsurprising to only observe mild (P$^+$ *drb2* and P$^+$ *drb4* shoots) to moderate differences (P$^+$ *drb1* shoots) in *IPS1* transcript abundance in the shoot tissue of non-stressed Col-0, *drb1*, *drb2* and *drb4* plants (Figure 2H). Further, and as expected, RT-qPCR showed that PO$_4$ stress induced the expression of *IPS1*, with *IPS1* transcript abundance elevated by 75.7-, 7.1-, 20.8- and 16.4-fold in the shoot tissues of PO$_4$ stressed Col-0, *drb1*, *drb2* and *drb4* plants, respectively (compared to the non-stressed counterpart of each plant line).

Next, the expression of the target gene of miR399, *PHO2*, was determined by RT-qPCR to largely remain at wild-type levels (P$^+$ Col-0 shoots) in the shoot tissues of P$^+$ *drb1*, P$^+$ *drb2* and P$^+$ *drb4* plants (Figure 2I). This was an unsurprising result considering that RT-qPCR also revealed only mild changes to miR399 abundance across the three *drb* mutant lines assessed when each plant line was cultivated on standard *Arabidopsis* culture media (Figure 2G). RT-qPCR also revealed that elevated miR399 abundance in P$^-$ Col-0, P$^-$ *drb2* and P$^-$ *drb4* plants, promoted miR399-directed expression repression of *PHO2*, with the abundance of the *PHO2* transcript reduced by 50%, 40% and 60% in the shoot tissues of these three plant lines, respectively (Figure 2I). In P$^-$ *drb1* shoots however, the level of the *PHO2* transcript was increased by 50% (Figure 2I). Elevated *PHO2* expression in P$^-$ *drb1* shoots, a tissue where miR399 abundance was also demonstrated to be elevated, indicated that in the absence of DRB1 activity, miR399-directed mRNA cleavage-mediated regulation of *PHO2* expression is lost.

2.3. The Phenotypic and Physiological Response to PO$_4$ Stress of the Root System of Arabidopsis Plant Lines Defective in DRB Protein Activity

The unique phenotypic (Figure 1) and molecular (Figure 2) response displayed by *drb1*, *drb2* and *drb4* shoots to PO$_4$ starvation led us to next repeat these assessments on the root system of each mutant background. As reported for the aerial tissue phenotypes expressed by the *drb1*, *drb2* and *drb4* mutants (Figure 1), Figure 3A again clearly displays the severe developmental phenotype expressed by the *drb1* mutant as well as the comparatively mild phenotypes that result from the loss of either DRB2 or DRB4 activity in *drb2* and *drb4* plants, respectively. The severity of the developmental phenotypes expressed by the three *drb* mutants assessed in this study is further evidenced when the fresh weight of the root system of 8-day old seedlings cultivated on standard growth media was determined. Specifically, the fresh weight of the root system of 8-day old *drb2* and *drb4* seedlings, 7.95 ± 0.20 mg and 8.00 ± 0.15 mg respectively, was equivalent to the fresh weight of the root system of Col-0 plants, 8.25 ± 0.45 mg (Figure 3B). However, the fresh weight of the root system of 8-day old *drb1* plants, 4.25 ± 0.15 mg, was approximately 50% less than that of an 8-day old Col-0 seedling (Figure 3B).

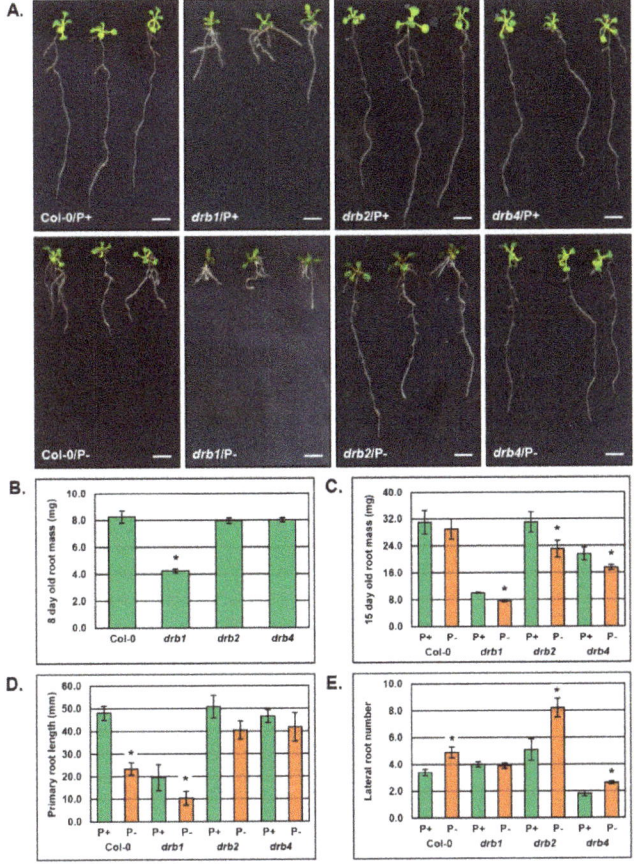

Figure 3. The root system phenotypes displayed by 15-day old *Arabidopsis* plant lines Col-0, *drb1*, *drb2* and *drb4* post exposure to a 7-day period of PO_4 starvation. (**A**) The root system phenotypes expressed by non-stressed (top row of panels) and PO_4-stressed (bottom row of panels) 15-day old Col-0, *drb1*, *drb2* and *drb4* plants. Scale bar = 1cm. (**B**) Quantification of the root mass of 8-day old Col-0, *drb1*, *drb2* and *drb4* seedlings cultivated under standard growth conditions. (**C**) The root mass of non-stressed and PO_4-stressed 15-day old Col-0, *drb1*, *drb2* and *drb4* plants. (**D**) The primary root length of non-stressed and PO_4-stressed 15-day old *Arabidopsis* lines, Col-0, *drb1*, *drb2* and *drb4*. (**E**) The number of lateral roots formed from the primary root of 15-day old Col-0, *drb1*, *drb2* and *drb4* plants cultivated under standard growth conditions, or post the 7-day PO_4 starvation period. (**B–E**) Error bars represent the standard deviation of four biological replicates and each biological replicate consisted of a pool of twelve individual plants. The presence of an asterisk above a column represents a statistically significant difference either between non-stressed Col-0 plants and each assessed *drb* mutant post cultivation under either a non-stressed or stressed growth regime (**B**) or between the non-stressed and PO_4-stressed sample of each plant line (**C-E**) (*p*-value: * < 0.05; ** < 0.005; *** < 0.001).

Figure 3C shows that at the completion of the 7-day PO_4 starvation period, the fresh weight of 15-day old P⁻ Col-0 roots (29.0 ± 3.0 mg) was only reduced by 2.0 mg compared to P⁺ Col-0 roots (31.0 ± 3.5 mg), a mild 6.5% reduction. The fresh weight of the root system of PO_4 stressed *drb1*, *drb2* and *drb4* plants all showed a much greater reduction when compared to their non-stressed counterparts (Figure 3C). That is, the fresh weight of the root system of 15-day old P⁻ *drb1* (7.5 ± 0.15 mg), P⁻ *drb2* (23.0 ± 2.5 mg) and P⁻ *drb4* plants (17.5 ± 0.75 mg) was reduced by 25.0%, 25.8% and 18.6%, respectively (Figure 3C).

Inhibition of primary root length is one of the main phenotypic responses of *Arabidopsis* to PO_4 stress [2,44], and accordingly, Figure 3A,D clearly show that the primary root length of 15-day old P^- Col-0 plants (23.4 ± 2.8 mm) was significantly reduced by 51.2% compared to non-stressed P^+ Col-0 plants (48.1 ± 3.1 mm) (Figure 3D). Although primary root length is already severely inhibited due to detrimental consequences of the loss of DRB1 activity on *Arabidopsis* development, the 7-day stress treatment caused a 46.7% reduction to the primary root length of P^- *drb1* plants (10.4 ± 3.1 mm) compared to P^+ *drb1* plants (19.5 ± 5.9 mm) (Figure 3D). Interestingly, PO_4 stress impacted primary root development to a much lower degree in both the *drb2* and *drb4* mutant backgrounds. Namely, primary root length was reduced by 20.3% and 10.3% in P^- *drb2* (40.5 ± 4.0 mm) and P^- *drb4* (41.8 ± 6.2 mm) plants respectively, compared to the primary root length of P^+ *drb2* (50.8 ± 5.0 mm) and P^+ *drb4* (46.6 ± 2.9 mm) plants (Figure 3D).

In parallel with inhibition to primary root length, promotion of lateral root development is a commonly reported phenotypic response of *Arabidopsis* plants exposed to PO_4 stress [2,44]. It was therefore unsurprising to document a 44% increase in the number of lateral roots that formed on 15-day old P^- Col-0 plants (4.9 ± 0.4) compared to P^+ Col-0 plants (3.4 ± 0.3) (Figure 3E). Interestingly, this phenotypic response to PO_4 stress appeared completely defective in the *drb1* mutant background with both P^+ *drb1* (4.0 ± 0.2) and P^- *drb1* (3.9 ± 0.2) plants forming approximately the same number of lateral roots. Unlike the *drb1* mutant, lateral root development was promoted by ~61% in the *drb2* mutant background with P^- *drb2* plants forming 8.2 ± 0.7 lateral roots compared to P^+ *drb2* plants which formed 5.1 ± 0.8 lateral roots. Lateral root formation was also induced by PO_4 stress in the *drb4* mutant with the number of lateral roots increased by 44% in P^- *drb4* plants (2.6 ± 0.1) compared to their number in P^+ *drb4* plants (1.8 ± 0.2).

2.4. Molecular Profiling of the miR399/PHO2 Expression Module in the Root System of Arabidopsis Plant Lines Defective in DRB Protein Activity

Due to its demonstrated role in inducing *MIR399* gene expression in PO_4 depleted conditions [27], RT-qPCR was initially used to profile *PHR1* expression in PO_4-stressed Col-0, *drb1*, *drb2* and *drb4* roots (Figure 4A). This analysis revealed that compared to the root system of each plant line's non-stressed counterpart, *PHR1* expression remained remarkably constant in P^- Col-0, P^- *drb1*, P^- *drb2* and P^- *drb4* roots (Figure 4A). Although RT-qPCR revealed that *PHR1* expression remained constant in the roots of control and PO_4-stressed plants, RT-qPCR was next applied to profile the expression of the six *MIR399* precursor transcripts in the roots of P^+ and P^- plants. Of the six miR399 precursors, RT-qPCR only allowed for expression quantification of three miR399 precursors, namely *PRE-MIR399A*, *PRE-MIR399C* and *PRE-MIR399D* in *Arabidopsis* roots (Figure 4B–D). In P^- Col-0 roots, RT-qPCR clearly revealed that PO_4 stress induced the expression of the miR399 precursors, *PRE-MIR399A*, *PRE-MIR399C* and *PRE-MIR399D*, by 4.0-, 40.6- and 1546-fold, respectively (Figure 4B–D). When compared to P^+ Col-0 roots, the moderate 2.3- and 3.6-fold elevation in the abundance of *PRE-MIR399A* and *PRE-MIR399C* in P^+ *drb1* roots, identified DRB1 as the primary DRB required for miR399 production regulation from these two precursor transcripts in the roots of wild-type *Arabidopsis* plants (Figure 4B,C). The primary role of DRB1 in *PRE-MIR399A* and *PRE-MIR399C* processing in non-stressed Col-0 roots is further evidenced by the wild-type equivalent accumulation of these two precursors in P^+ *drb2* and P^+ *drb4* roots, and by the highest degree of *PRE-MIR399A* and *PRE-MIR399C* precursor transcript over-accumulation in P^- *drb1* roots (Figure 4B,C). Considering this result, it was therefore of considerable interest to observe the greatest degree of *PRE-MIR399D* over-accumulation, an 8.2-fold increase, in P^+ *drb4* roots and not in P^+ *drb1* roots (4.3-fold increase) (Figure 4D). This finding suggests that in non-stressed wild-type *Arabidopsis* roots, DRB4 is the primary DRB responsible for regulating miR399 production from this precursor transcript. In addition, and under PO_4 stress conditions, the *PRE-MIR399D* transcript increased in its abundance to relative expression values of 829, 849 and 1271 in *drb1*, *drb2* and *drb4* roots, respectively (Figure 4D). Although these determined increases in precursor transcript abundance are all highly significant, they are not as significant as the 1546 relative expression value obtained for the

PRE-MIR399D transcript in P⁻ Col-0 roots. A lower degree of precursor transcript over-accumulation in each assessed *drb* mutant background, compared to the expression induction observed in wild-type roots, indicated that all three DRB proteins potentially play a role in fine-tuning the regulation of miR399 production from the *PRE-MIR399D* precursor in PO_4-stressed *Arabidopsis* roots (Figure 4D).

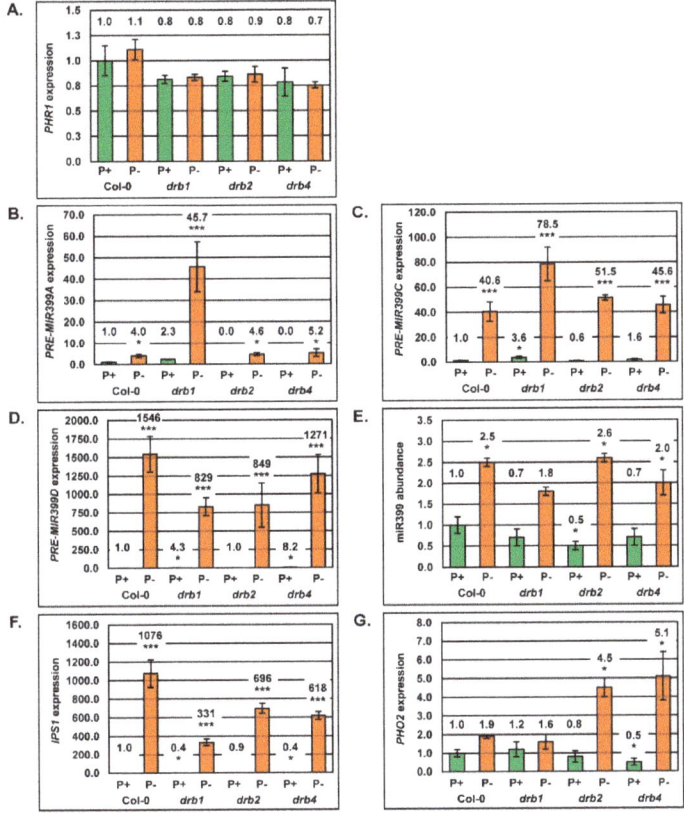

Figure 4. Molecular profiling of the miR399/*PHO2* expression module in the root system of non-stressed and PO_4-stressed Col-0, *drb1*, *drb2* and *drb4* plants. (**A**) RT-qPCR assessment of the expression of the PO_4 responsive transcription factor *PHR1* in the roots of non-stressed and PO_4-stressed Col-0, *drb1*, *drb2* and *drb4* plants. (**B–D**) RT-qPCR profiling of miR399 precursor transcript abundance in the root system of non-stressed and PO_4-stressed Col-0, *drb1*, *drb2* and *drb4* plants, including precursors *PRE-MIR399A* (**B**), *PRE-MIR399C* (**C**) and *PRE-MIR399D* (**D**). (**E**) Quantification of miR399 abundance in the roots of non-stressed and PO_4-stressed Col-0, *drb1*, *drb2* and *drb4* plants. (**F**) Assessment of *IPS1* transcript abundance in the roots of non-stressed and PO_4-stressed *Arabidopsis* lines, Col-0, *drb1*, *drb2* and *drb4*. (**G**) RT-qPCR analysis of *PHO2* expression, the target gene of miR399, in the root system of non-stressed and PO_4-stressed *Arabidopsis* lines, Col-0, *drb1*, *drb2* and *drb4*. (**A–G**) Error bars represent the standard deviation of four biological replicates and each biological replicate consisted of a pool of twelve individual plants. Due to the vastly different level of each assessed transcript, the relative expression value for each plant line/growth regime is provided above the corresponding column. The presence of an asterisk above a column represents a statistically significant difference between non-stressed Col-0 plants and each of the assessed *drb* mutant lines, post cultivation under either a standard or stressed growth regime (*p*-value: * < 0.05; ** < 0.005; *** < 0.001).

Post-establishment of highly variable expression profiles for *PRE-MIR399A*, *PRE-MIR399C* and *PRE-MIR399D* in non-stressed *drb1*, *drb2* and *drb4* roots (Figure 4B–D), miR399 abundance reductions of 30%, 50% and 30% in P$^+$ *drb1*, P$^+$ *drb2* and P$^+$ *drb4* roots, respectively, was expected (Figure 4E). Quantification of miR399 abundance, 2.5-, 1.8-, 2.6- and 2.0-fold elevations, respectively, in the root tissues of PO$_4$-stressed Col-0, *drb1*, *drb2* and *drb4* plants, revealed that the considerable induction to *PRE-MIR399A*, *PRE-MIR399C* and *PRE-MIR399D* expression (Figure 4B–D), did not however, result in an overly altered miR399 accumulation profile (Figure 4E).

Failure to establish a strong correlation between precursor transcript expression and miR399 abundance in either control or PO$_4$-stressed Col-0, *drb1*, *drb2* and *drb4* roots, led us to next assess *IPS1* expression in this tissue (Figure 4F). *IPS1* transcript abundance remained relatively unchanged in the root tissues of non-stressed Col-0 and *drb2* plants (Figure 4F). Interestingly, *IPS1* expression was reduced by 60% in P$^+$ *drb1* and P$^+$ *drb4* roots (Figure 4F). Significant induction of *IPS1* expression was observed in PO$_4$-stressed *drb1*, *drb2* and *drb4* roots, 331-, 696- and 618-fold elevations, respectively. Interestingly, RT-qPCR demonstrated that *IPS1* expression was promoted to its greatest degree, 1076-fold, in PO$_4$-stressed Col-0 roots (Figure 4F).

The expression of the miR399 target gene, *PHO2*, was next quantified by RT-qPCR in non-stressed and PO$_4$-stressed Col-0, *drb1*, *drb2* and *drb4* roots (Figure 4G). In P$^+$ *drb1* and P$^+$ *drb2* roots, RT-qPCR revealed *PHO2* expression to be elevated and reduced by 20%, respectively, and in P$^+$ *drb4* roots, *PHO2* expression was reduced by 30%. Elevated *PHO2* expression in P$^+$ *drb1* roots was expected considering the slight reduction to miR399 abundance observed in this tissue (Figure 4E). However, the reduced *PHO2* transcript levels in P$^+$ *drb2* and P$^+$ *drb4* roots was a surprise finding considering that miR399 abundance was also reduced in these two mutant lines by 50% and 30%, respectively (Figure 4E). *PHO2* expression was demonstrated by RT-qPCR to be elevated by 1.9-, 1.6-, 4.5- and 5.1-fold in PO$_4$-stressed Col-0, *drb1*, *drb2* and *drb4* roots, respectively (Figure 4G). This finding also formed an unexpected result considering that PO$_4$ starvation induced the accumulation of the miR399 sRNA in all four assessed plant lines (Figure 4E).

2.5. Correct Inorganic Phosphate Partitioning Between the Shoot and Root Tissue of Arabidopsis Requires DRB1 and DRB2

The molecular profiling of alterations to the miR399/*PHO2* expression module in the shoot and root tissue of *Arabidopsis* Col-0, *drb1*, *drb2* and *drb4* plants under PO$_4$ stress, in combination with each plant line displaying a unique phenotypic response to this stress, led us to next assess Pi partitioning in the aerial tissue and root system of P$^+$ and P$^-$ Col-0, *drb1*, *drb2* and *drb4* plants. In the shoot tissues of 15-day old plants cultivated in PO$_4$ replete conditions, Pi content was only altered in the *drb2* mutant background, with the Pi content of P$^+$ *drb2* shoots (13.8 µmol/gFW) reduced by 27.4% compared to the Pi content of P$^+$ Col-0 shoots (19.0 µmol/gFW) (Figure 5A). When cultivated in PO$_4$-stress conditions however, only the Pi content of P$^-$ *drb1* shoots (1.15 µmol/gFW) differed to that of P$^-$ Col-0 shoots (1.75 µmol/gFW); a 34.3% reduction (Figure 5A). In non-stressed roots, the Pi content of P$^+$ *drb1* (11.4 µmol/gFW) and P$^+$ *drb2* (9.8 ± 0.8 µmol/gFW) roots was determined to be elevated by 58.3% and 37.5% respectively, compared to P$^+$ Col-0 roots (7.2 µmol/gFW) (Figure 5B). As demonstrated for non-stressed *drb1* and *drb2* roots, the Pi content of P$^-$ *drb1* (1.84 µmol/gFW) and P$^-$ *drb2* (0.65 µmol/gFW) roots also differed to that of PO$_4$-stressed Col-0 roots (1.25 µmol/gFW), elevated and reduced by 47.2% and 48%, respectively (Figure 5B).

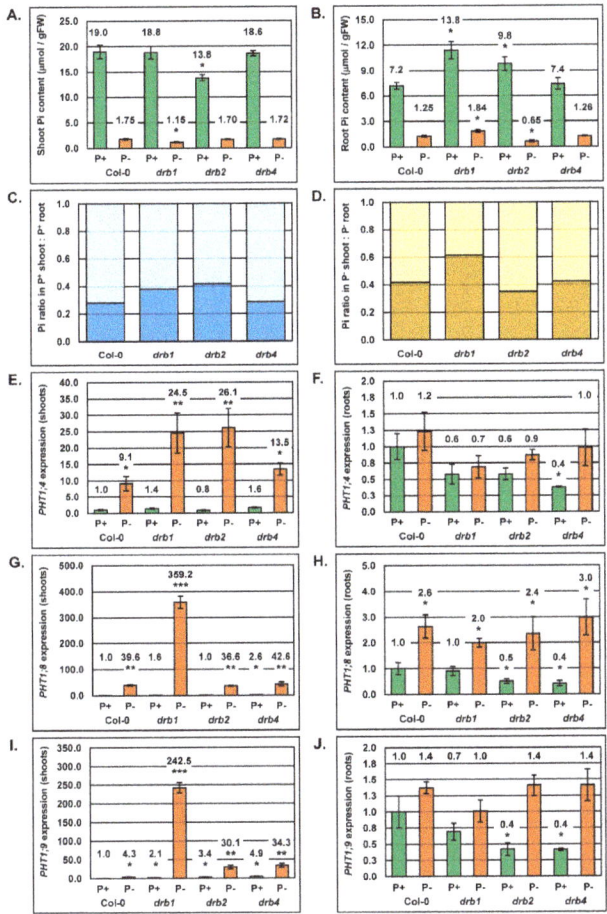

Figure 5. Pi content and PO$_4$ transporter gene expression in the shoot and root tissue of 15-day old *Arabidopsis* plant lines Col-0, *drb1*, *drb2* and *drb4* cultivated under either a standard growth regime or post-exposure to a 7-day period of PO$_4$ starvation. (**A,B**) Comparison of the Pi content of the shoots (A) and roots (B) of 15-day old non-stressed and PO$_4$-stressed Col-0, *drb1*, *drb2* and *drb4* plants. (**C**) Pi content shoot (light blue) to root (dark blue) ratio of 15-day old Col-0, *drb1*, *drb2* and *drb4* plants cultivated under standard growth conditions. (**D**) Pi content shoot (light gold) to root (dark gold) ratio of 15-day old Col-0, *drb1*, *drb2* and *drb4* plants post 7-days of PO$_4$ starvation. (**E,F**) Quantification of *PHT1;4* expression in the shoot (E) and root (F) tissues of 15-day old Col-0, *drb1*, *drb2* and *drb4* plants cultivated under standard growth conditions or post a 7-day period of PO$_4$ starvation. (**G,H**) RT-qPCR assessment of *PHT1;8* transcript abundance in the shoots (G) and roots (H) of 15-day old Col-0, *drb1*, *drb2* and *drb4* plants cultivated under either standard or PO$_4$ stress conditions. (**I,J**) *PHT1;9* expression in the shoot (I) and root (J) material of non-stressed or PO$_4$-stressed Col-0, *drb1*, *drb2* and *drb4* plants at 15 days of age. (A,B,E–J) Error bars represent the standard deviation of four biological replicates and each biological replicate consisted of a pool of twelve individual plants. Due to the vastly different levels of each assessed transcript, the relative expression value for each plant line/growth regime is provided above the corresponding column. The presence of an asterisk above a column represents a statistically significant difference between the non-stressed and PO$_4$-stressed sample of each plant line (A,B) or between non-stressed Col-0 plants and each *drb* mutant line, post cultivation under either a standard or stressed growth regime (E–J) (p-value: * < 0.05; ** < 0.005; *** < 0.001).

The reduced Pi content of P⁺ *drb2* shoots (Figure 5A), together with the elevated Pi contents of P⁺ *drb1* and P⁺ *drb2* roots (Figure 5B), suggested that Pi partitioning was potentially defective in these two mutant backgrounds. We therefore next determined the Pi content ratio of the shoot and root of non-stressed and PO_4-stressed Col-0, *drb1*, *drb2* and *drb4* plants. Figure 5C clearly shows that Pi partitioning between the shoot and root tissue of P⁺ *drb1* and P⁺ *drb2* plants is defective, even when these two mutant lines are cultivated on standard *Arabidopsis* growth media. Under PO_4 stress conditions, defective Pi partitioning is even more readily evident in the *drb1* mutant background which showed a 0.38:0.62 shoot to root Pi content ratio, compared to the shoot to root Pi content ratio of 0.58:0.42 for P⁻ Col-0 plants. Although not as striking as determined for P⁺ *drb2* plants, the altered shoot to root Pi content ratio (0.65:0.35) of PO_4-stressed *drb2* plants again indicated that Pi partitioning is defective in this mutant background (Figure 5D).

Altered shoot to root Pi content ratios in *drb1* and *drb2* plants strongly suggested that Pi partitioning is defective in these two mutant backgrounds. Considering that PO_4 transporters, PHT1;4, PHT1;8 and PHT1;9, are known targets of PHO2-mediated ubiquitination [7,14], together with our demonstration in Figures 2 and 4 that the miR399/PHO2 expression module is altered to differing degrees in the shoot and root tissues of each of the three assessed *drb* mutants, RT-qPCR was next applied to profile *PHT1;4*, *PHT1;8* and *PHT1;9* expression in non-stressed and PO_4-stressed Col-0, *drb1*, *drb2* and *drb4* plants. RT-qPCR revealed that PO_4 starvation promoted *PHT1;4*, *PHT1;8* and *PHT1;9* expression by 9.1-, 39.6- and 4.3-fold in Col-0 shoots (Figure 5E,G,I), and by 1.2-, 2.6- and 1.4-fold in Col-0 roots, respectively (Figure 5F,H,J). In non-stressed *drb1* shoots, the abundance of the *PHT1;4*, *PHT1;8* and *PHT1;9* transcripts were only mildly altered compared to their respective expression levels in P⁺ Col-0 shoots, returning 1.4-, 1.6- and 2.1-fold changes in expression. A similar mild degree of expression alteration was observed for P⁺ *drb1* roots. Specifically, compared to P⁺ Col-0 roots, the *PHT1;4*, *PHT1;8* and *PHT1;9* transcripts returned fold changes in abundance of 0.6, 1.0 and 0.7, respectively. The expression of these three PO_4 transporters was significantly induced by the 7-day stress period, returning abundance fold changes of 24.5 (*PHT1;4*), 359.2 (*PHT1;8*) and 242.5 (*PHT1;9*), respectively (Figure 5E,G,I), in P⁻ *drb1* shoots. In spite of the significant induction of *PHT1* gene expression in P⁻ *drb1* shoots, *PHT1;4*, *PHT1;8* and *PHT1;9* levels were reduced (0.7-fold), elevated (2.0-fold) and unchanged (1.0-fold), respectively (Figure 5F,H,J) in the root system of PO_4-stressed *drb1* roots. As demonstrated for P⁺ *drb1* shoots, RT-qPCR again revealed that *PHT1;4*, *PHT1;8* and *PHT1;9* expression was mildly altered in P⁺ *drb2* shoots by 0.8-, 1.0- and 3.4-fold, respectively. In non-stressed *drb2* roots however, the expression of all three PO_4 transporters was reduced by 40%, 50% and 60%, respectively, compared to their expression levels in non-stressed Col-0 roots. Furthermore, Figure 5E–J clearly show that the 7-day PO_4 starvation period induced the expression of these three PO_4 transporter encoding genes in both the P⁻ *drb2* shoot and root samples, compared to their expression levels in non-stressed *drb2* shoot and roots. Considering that Pi content of non-stressed and PO_4-stressed *drb4* shoots and roots was determined to be the same as that of the corresponding tissues in P⁺ and P⁻ Col-0 plants, it was unexpected to observe such varied differences in PO_4 transporter expression across both assessed tissues/growth conditions. For example, in P⁺ *drb4* roots, *PHT1;4*, *PHT1;8* and *PHT1;9* levels were each reduced by 60%, compared to P⁺ Col-0 roots (Figure 5F,H,J), yet the Pi content of non-stressed Col-0 and *drb4* roots was identical (Figure 5B).

3. Discussion

A lack of available P in the soil is a key limitation for plant growth globally [3,45] and as a consequence of P limitation, land plants have evolved highly complex regulatory mechanisms to control both the uptake of external P from the soil, primarily in the form of PO_4 (Pi), as well as the remobilization of internal stores of P during periods of low external PO_4 availability [46]. These elaborate P responsive mechanisms allow a plant to attempt to (1) maintain growth and development and (2) regulate cellular P content, regardless of external P concentration [1,2,7]. More contemporary research has focused on the regulatory role played by a suite of PO_4 responsive miRNA sRNAs that either initiate or maintain PO_4 signaling pathways across the plant kingdom [4,20]. Central to this PO_4 responsive miRNA cohort, is miR399, with the miR399 sRNA required to regulate the abundance of the *PHO2* transcript, to in turn regulate the level of the PHO2 protein, an E2 ubiquitin conjugase that mediates the ubiquitin-directed turnover of a group of PO_4 transporter proteins [7,14,47]. The DRB family members, DRB1, DRB2 and DRB4, have each been ascribed a specific functional role in the *Arabidopsis* miRNA pathway [35–40,48,49]. Therefore, we sought to document the involvement of these three DRBs in the production of the PO_4 responsive miRNA, miR399, and to determine the mode of action directed by the miR399 sRNA during PO_4 starvation to regulate *PHO2* abundance in the *drb1*, *drb2* and *drb4* mutant backgrounds. Specifically, we attempted to determine what effect an altered miR399/*PHO2* expression module profile would have on the response of *drb1*, *drb2* or *drb4* mutant plants to the imposed stress in order to establish the contribution of either DRB1, DRB2 and/or DRB4 to the maintenance of P homeostasis in *Arabidopsis*.

3.1. DRB1 is Required to Maintain Phosphorous Homeostasis in Arabidopsis

Here, it was discovered that the maintenance of P homeostasis is impaired in the *drb1* loss-of-function mutant. The most compelling evidence for this was the documented alteration of the shoot to root Pi content ratio in both non-stressed (Figure 5C) and PO_4-stressed *drb1* plants (Figure 5D), relative to wild-type *Arabidopsis* (P^+ or P^- Col-0 plants). Specifically, the shoot Pi content was reduced to a much greater degree in PO_4-stressed *drb1* plants than the observed reduction to Pi content in P^- Col-0 shoots. Furthermore, Pi was demonstrated to over-accumulate in the roots of both P^+ and P^- *drb1* plants (Figure 5A,B), compared to the Pi content of the corresponding tissue, and growth regime, of Col-0 plants. The maintenance of appropriate P content in plant tissues is essential for the production of macromolecules, energy trafficking and for numerous signaling pathways [1,2,46]. Therefore, alterations to the P content of the shoot and root tissues of *drb1* plants indicated that in the absence of functional DRB1, P partitioning is impaired. Assessment of the expression of PO_4 transporters, *PHT1;4*, *PHT1;8* and *PHT1;9*, revealed that the abundance of each transporter was highly elevated by 24.5- 359.2- and 242.5-fold respectively, in the shoot tissue of P^- *drb1* plants. Phosphate transporter expression was also demonstrated to be altered in both P^+ (*PHT1;4* reduced by 1.7-fold and *PHT1;9* reduced by 1.5-fold) and P^- (*PHT1;4* reduced by 1.5-fold and *PHT1;8* elevated by 2.0-fold) *drb1* roots, expression alterations that when taken together indicated that incorrect Pi partitioning in *drb1* plants potentially results from defective PO_4 transport from the root system to the aerial tissue in this mutant background.

Defective root to shoot PO_4 transport in the *drb1* mutant was further evidenced by the unique phenotypic response displayed by the *drb1* shoot to PO_4 stress. Specifically, the fresh weight of the shoot of 15-day old P^- *drb1* plants was only reduced by 21.6% compared to its non-stressed counterpart (Figure 1C). The rosette area of P^- *drb1* plants was also demonstrated to only be reduced by 29.3% post the 7-day PO_4 stress treatment (Figure 1D). Both responses were comparatively mild compared to the 36.6% and 60.1% reductions to fresh weight and rosette area respectively, documented for Col-0 shoots post the application of PO_4 stress. In addition, anthocyanin failed to change in abundance in the shoot tissues of P^- *drb1* plants compared to the shoots of non-stressed P^+ *drb1* plants (Figure 1E). Anthocyanin production is a general response to a range of abiotic stresses, including PO_4 starvation [19,50]. The impaired ability of *drb1* shoots to produce anthocyanin in response to PO_4 stress may implicate DRB1, and the functional partnership DRB1 forms with DCL1, in the induction of PO_4 responsive gene expression pathways. Considering these mild responses displayed by *drb1* shoots, it was therefore surprising to observe that chlorophyll *a* and *b* overaccumulation was promoted to the greatest extent in the aerial tissues of *drb1* plants starved of PO_4. Altered chlorophyll content in P^+ *drb1* shoots indicated that (1) *drb1* shoots are indeed negatively impacted by the imposed PO_4 stress, and (2) that DRB1 may potentially mediate a PO_4-directed role in regulating photosynthesis in *Arabidopsis* chloroplasts.

Considering the well-established role of the DRB1/DCL1 functional partnership in the production of the majority of miRNAs that accumulate in *Arabidopsis* tissues, it was unsurprising to observe that the miR399 precursors, *PRE-MIR399A*, *PRE-MIR399C*, *PRE-MIR399D*, *PRE-MIR399E* and *PRE-MIR399F*, over-accumulated to the greatest extent in P$^+$ *drb1* shoots (Figure 2A–E). In addition, precursors *PRE-MIR399A*, *PRE-MIR399C*, *PRE-MIR399D* and *PRE-MIR399F* were further demonstrated to be most highly abundant in the shoot tissues of PO$_4$-stressed *drb1* plants. The enhanced abundance of miRNA precursor transcripts in the *drb1* mutant background is most likely the result of inefficient precursor transcript processing by DCL1 in the absence of DRB1 functional assistance, with DRB1 accurately positioning DCL1 on each miRNA precursor to direct accurate processing [48,49]. In spite of the readily observable evidence of inefficient miR399 precursor transcript processing in P$^+$ *drb1* shoots, miR399 levels were only reduced by 10% (Figure 5G). Similarly, although miR399 precursor transcript abundance was elevated to a much greater degree in P$^-$ *drb1* shoots due to a combination of (1) *MIR399* gene expression induction in response to PO$_4$ starvation, and (2) inefficient precursor transcript processing in the absence of DRB1 activity, miR399 abundance was again demonstrated to be only mildly elevated by 2.3-fold in the shoots of PO$_4$-stressed *drb1* plants (Figure 5G). Further, the abundance of the miR399 target transcript, *PHO2*, was only mildly elevated by 1.2-fold in response to the 10% reduction in miR399 levels in P$^+$ *drb1* shoots (Figure 2I). Surprisingly, *PHO2* transcript abundance was elevated by 1.5-fold in response to the 2.3-fold elevation in miR399 accumulation in P$^-$ *drb1* shoots, and not reduced as expected. However, in P$^+$ Col-0 shoots, and as expected, the 2.9-fold enhancement to miR399 abundance led to a 50% reduction in *PHO2* expression (Figure 5G,I). Therefore, elevated *PHO2* abundance in response to enhanced miR399 levels in P$^-$ *drb1* shoots, readily demonstrates that miR399-directed *PHO2* transcript cleavage, to regulate *PHO2* expression, is defective in the absence of DRB1 activity.

Altered PO$_4$ transporter expression in *drb1* roots indicated that the response of the root system of the *drb1* mutant to PO$_4$ stress would differ to that of the root system of wild-type *Arabidopsis*. Accordingly, the fresh weight of PO$_4$-stressed *drb1* roots was reduced by 25.0% compared to the mild 6.5% reduction to the fresh weight of P$^-$ Col-0 roots, a 3.8-fold enhancement to the severity of this phenotypic response (Figure 3C). It was therefore curious to observe a similar degree of reduction to primary root length in P$^-$ *drb1* (46.7%) and P$^-$ Col-0 (51.2%) plants (Figure 3D). A greater degree of reduction to the fresh weight of P$^-$ *drb1* roots, compared to P$^-$ Col-0 roots, could be partially explained by the observation that the induction of lateral root formation by PO$_4$ stress was completely defective in P$^-$ *drb1* roots, compared to a 44.0% increase in lateral root number in P$^-$ Col-0 roots (Figure 3D). Considering that the measurement of fresh weight is largely assessing the moisture content of a plant, the observed reduction to fresh weight of P$^-$ *drb1* roots could potentially be indicating that under PO$_4$ stress conditions, DRB1 is somehow involved in regulating the moisture content of the root system of *Arabidopsis*. However, this was not assessed in this study with the mechanism driving the enhancement of fresh weight reductions requiring further investigation in the future.

Similar to its establishment as the primary DRB protein required to regulate miR399 production from the *PRE-MIR399A*, *PRE-MIR399C*, *PRE-MIR399D*, *PRE-MIR399E* and *PRE-MIR399F* precursors in the aerial tissues of non-stressed *Arabidopsis* plants, DRB1 was again demonstrated to be the primary DRB protein required to regulate miR399 production from the *PRE-MIR399A* and *PRE-MIR399C* precursor transcripts in the *Arabidopsis* root system with both precursors demonstrated to accumulate to the greatest degree in P$^+$ and P$^-$ *drb1* roots (Figure 4B,C). Reduced *PRE-MIR399A* and *PRE-MIR399C* processing efficiency in the absence of DRB1 activity, reduced miR399 abundance by 30% in P$^+$ *drb1* roots (Figure 4F), and in turn, this moderate reduction to miR399 levels led to a mild elevation (1.2-fold) in the expression of the miR399 target gene, *PHO2* (Figure 4G). As documented in P$^-$ *drb1* shoots, the 1.8-fold elevation to miR399 levels in P$^-$ *drb1* roots, resulted in a moderate elevation to *PHO2* transcript abundance (1.6-fold), and not a reduction in target gene expression as would be expected for a miRNA that regulates the expression of its targeted genes solely via a mRNA cleavage mode of RNA silencing. However, considering that a similar miRNA/target gene expression profile of elevated miR399 abundance (2.5-fold), together with enhanced *PHO2* expression (1.9-fold) was also observed in PO_4-stressed Col-0 roots, this curious finding indicates that miR399-directed *PHO2* transcript cleavage may not be the predominant mechanism of target gene expression regulation directed by the miR399 sRNA in the *Arabidopsis* root system. Alternatively, elevated *PHO2* expression in P^+ Col-0 and P^+ *drb1* roots when miR399 abundance is also elevated may result from the enhanced expression of the eTM of miR399 activity, *IPS1*. In P^- Col-0 shoots for example, where elevated miR399 abundance was demonstrated to direct enhanced expression repression of the *PHO2* transcript (Figure 2G,I), *IPS1* abundance was elevated by 75.7-fold, compared to its abundance in P^+ Col-0 shoots (Figure 2H). In PO_4-stressed roots, however, *IPS1* expression was elevated to a much greater degree, by 1076-fold (Figure 4F). This 14.2-fold greater promotion to *IPS1* expression in P^- Col-0 roots, than that observed in P^- Col-0 shoots, would be expected to sequester a higher amount of miR399, which in turn, could have led to the observed elevation in *PHO2* expression in P^- Col-0 roots in the presence of 2.5-fold greater abundance of the *PHO2* targeting miRNA, miR399.

3.2. DRB2 is Required to Maintain Phosphate Homeostasis in Arabidopsis

As documented for the *drb1* mutant, P homeostasis was determined to be defective in the *drb2* mutant. Specific to *drb2* plants however, was the 27.8% reduction to the Pi content of non-stressed *drb2* shoots (Figure 5A). Of the four *Arabidopsis* plant lines assessed in this study, *drb2* was the only line determined to have a reduced aerial tissue Pi content when cultivated under standard growth conditions. Furthermore, in P^+ *drb2* shoots, *PHT1;4* (Figure 5E) and *PHT1;8* (Figure 5G) expression was determined to be reduced and elevated by 1.2- and 3.4-fold respectively, compared to the expression of these two PO_4 transporters in P^+ Col-0 shoots. In addition, Pi was determined to over-accumulate by 36.1% in P^+ *drb2* roots. In P^+ *drb2* roots, *PHT1;4*, *PHT1;8* and *PHT1;9* expression was reduced by 1.7-, 2.0- and 2.4-fold respectively, compared to their expression levels in P^+ Col-0 roots. Together, (1) reduced Pi content of P^+ *drb2* shoots, (2) elevated Pi content in P^+ *drb2* roots, and (3) reduced PO_4 transporter gene expression in P^+ *drb2* roots, indicated that PO_4 root to shoot transport is defective in non-stressed *drb2* plants. Based on this finding, it was curious to observe a similar Pi content in P^- *drb2* shoots and P^- Col-0 shoots (Figure 5A), especially considering the document enhancement to *PHT1;4* and *PHT1;9* expression in P^- *drb2* shoots, with the expression of these two PO_4 transporters elevated by 2.8- and 7.0-fold respectively, compared to the degree of expression induction observed in P^- Col-0 roots (Figure 5E,I). However, and as demonstrated for P^+ *drb2* shoots and roots, the Pi content of the root system of PO_4-stressed *drb2* plants was altered, reduced by 48% compared to the Pi content of P^- Col-0 roots. Interestingly, RT-qPCR revealed similar degrees of elevated *PHT1;8* (Figure 5H) and *PHT1;9* (Figure 5J) expression in PO_4-stressed Col-0 and *drb2* roots with only the *PHT1;4* transcript returning a slight difference in its expression in P^- Col-0 roots (elevated by 1.2-fold compared to P^+ Col-0 roots) and P^- *drb2* roots (reduced by 1.1-fold compared to P^+ Col-0 roots). The PO_4 transporters, PHT1;1 and PHT1;4, have been demonstrated to be responsible for the import of more than half of the Pi that is taken up from the soil [51]. It therefore seems unlikely that the mild 10% reduction to *PHT1;4* transcript abundance documented in PO_4-stressed *drb2* roots, is the sole cause of the considerable reduction to the Pi content of the root system in the *drb2* mutant background.

Considering that the Pi content of PO_4-stressed Col-0 and *drb2* shoots was determined to be similar, it was unsurprising to document a similar degree of reduction to fresh weight of the shoot tissues of P^- Col-0 (36.6%) and P^- *drb2* (39.1%) plants (Figure 1C). Rosette area was also decreased by a similar degree in P^- Col-0 (60.1%) and P^- *drb2* (48.0%) plants (Figure 1D). However, compared to PO_4-stressed Col-0 shoots, anthocyanin accumulated to considerably higher levels in the aerial tissues of *drb2* plants when exposed to PO_4 stress (Figure 1E). The induction of anthocyanin production is a well-characterized response to PO_4 starvation [19,50]. Therefore, the considerable enhancement of anthocyanin accumulation in P^- *drb2* shoots, compared to the shoot tissues of PO_4-stressed Col-0 plants, suggests that this P-responsive pathway is hyperactivated in the absence of DRB2 activity, as well as potentially implicating DRB2 in mediating a regulatory role in a range of other P-responsive pathways in *Arabidopsis* aerial tissues that were not assessed in this study.

We have previously demonstrated a role for DRB2 in the production stage of the *Arabidopsis* miRNA pathway with the abundance of specific miRNA cohorts altered in the *drb2* mutant background [37]. More specifically, DRB2 can either be antagonistic or synergistic to DRB1 function in the DRB1/DCL1 partnership for the production of specific miRNAs, resulting in miRNA abundance either being enhanced (antagonistic) or reduced (synergistic) in *drb2* plants [37,38]. Reduced precursor transcript abundance in non-stressed *drb2* shoots, indicated that DRB2 plays a secondary role in regulating miR399 production from the *PRE-MIR399A*, *PRE-MIR399E* and *PRE-MIR399F* precursors, potentially via antagonism of DRB1 function (Figure 2B,E,F). The antagonism of DRB2 on the DRB1/DCL1 partnership becomes more readily apparent via the profiling of miR399 precursor transcript expression in P^- *drb2* shoots, with lower degrees of expression induction observed for the *PRE-MIR399C*, *PRE-MIR399D*, *PRE-MIR399E* and *PRE-MIR399F* precursors (Figure 2C–F). Reduced precursor transcript abundance in P^- *drb2* shoots, compared to the respective abundance of each precursor in either P^- Col-0 or P^- *drb1* shoots, indicates that in the absence of DRB2 activity, precursor transcript processing efficiency is enhanced due to more precursor transcript being freely available to enter the canonical DRB1/DCL1 production pathway.

As demonstrated in P^+ *drb1* shoots, significantly altered precursor transcript abundance in P^+ *drb2* shoots, failed to have a strong influence on the accumulation of miR399, with miR399 levels only mildly elevated by 10% in P^+ *drb2* shoots, compared to P^+ Col-0 shoots (Figure 5G). However, DRB2 antagonism was still evidenced by this mild increase to miR399 abundance compared to the 10% reduction in miR399 levels observed in P^+ *drb1* shoots. The antagonism of DRB2 on miR399 production was further evidenced by the enhanced expression repression of *PHO2* in P^- *drb2* shoots (Figure 2I). The abundance of miR399 was elevated by 2.7-fold in P^- *drb2* shoots, and therefore, a further degree of reduced *PHO2* expression in P^- *drb2* shots, compared to P^- *drb1* shoots where miR399 levels were elevated by 2.3-fold and *PHO2* expression was enhanced by 1.5-fold, clearly demonstrated enhanced DRB1-mediated, miR399-directed, *PHO2* transcript cleavage in the absence of DRB2 antagonism. Similarly, it is important to note here that *IPS1* transcript abundance was enhanced to a much lower degree in P^- *drb2* shoots (27.1-fold) compared to *IPS1* abundance induction in either PO_4-stressed Col-0 (75.7-fold) or *drb1* (85.4-fold) shoots. This unexpected observation again indicated that in the absence of DRB2 activity, miR399-directed target transcript cleavage was enhanced. Although *IPS1* has been identified as a non-cleavable eTM of miR399 activity, the *IPS1* expression trends presented in Figure 5H suggest that miR399 may well be capable of directing miRISC-catalyzed cleavage of the *IPS1* transcript in addition to solely being sequestered by *IPS1*.

Compared to the mild 6.5% reduction to the fresh weight of P^- Col-0 roots, the negative response of the root system of the *drb2* mutant to PO_4 stress was considerably more pronounced at 25.8% (Figure 3C). Considering that the correct regulation of Pi content is dysfunctional in both control and PO_4-stressed *drb2* roots, differing responses to PO_4 stress in *drb2* roots, compared to P^- Col-0 roots, was not surprising. Similarly, inhibition of the primary root length of P^- *drb2* plants at 20.3% was comparatively mild compared to the severe 51.2% inhibition to the primary root length observed for P^- Col-0 plants (Figure 5D). The degree of lateral root induction also differed between PO_4-stressed Col-0 and *drb2* roots (Figure 5E), specifically; lateral root formation was enhanced by ~44% in P^- Col-0 plants, and in PO_4-stressed *drb2* plants, lateral root formation was further promoted by 17% with P^- *drb2* plants developing ~61% more lateral roots than their non-stressed counterparts. When these phenotypic responses of the root system of PO_4-stressed *drb2* plants are considered together, including a lower degree of primary root length inhibition (2.5-fold less than P^- Col-0 plants), and a more pronounced enhancement to lateral root formation (1.4-fold more than P^- Col-0 plants), it was highly surprising that the fresh weight of P^- *drb2* roots was reduced by a 4.0-fold greater degree than documented for P^- Col-0 roots.

Similar levels of expression of *PRE-MIR399A* in both non-stressed and PO_4-stressed Col-0 and *drb2* roots revealed that DRB2 does not play a role in regulating miR399 processing from this precursor transcript (Figure 4B). Reduced expression of *PRE-MIR399C* in P^+ *drb2* roots (compared to P^- Col-0 and P^- *drb1* roots) and a lower level of precursor over-accumulation in P^- *drb2* roots (compared to P^- *drb1* roots), identified DRB2 as playing a secondary role in regulating miR399 production from this precursor transcript in the *Arabidopsis* root system (Figure 4C) via antagonism of DRB1 function. The expression trend of *PRE-MIR399D* in P^- *drb2* roots additionally identified a secondary role for DRB2 in regulating miR399 production from the third miR399 precursor transcript detected in the root system of the four *Arabidopsis* plants lines assessed in this study. However, for the *PRE-MIR399D* precursor, DRB2 appears to be antagonistic to the DRB4/DCL4 partnership, and not to the canonical DRB1/DCL1 partnership demonstrated to be required for the production of the majority of *Arabidopsis* miRNAs. DRB2 has been demonstrated previously to be antagonistic to DRB4 function in the DRB4/DCL4 partnership for the production of a small subset of newly evolved *Arabidopsis* miRNAs processed from precursor transcripts that fold to form highly complementary stem-loop structures [39,40]. Considering that in P^+ *drb2* roots, *PRE-MIR399A* and *PRE-MIR399D* remained at their approximate wild-type levels, and that the *PRE-MIR399C* precursor was reduced in its abundance by 1.7-fold, a finding that initially indicated that this precursor is more efficiently processed by DRB1/DCL1 in the absence of DRB2 activity, the 2.0-fold reduction to miR399 abundance alternatively indicated that *MIR399C* gene expression may in fact be reduced in PO_4-stressed *drb2* roots. It was therefore curious to observe *PHO2* expression to be reduced by 1.3-fold in P^+ *drb2* roots, and not elevated in response to reduced miR399 abundance as expected. However, this observation is potentially demonstrating that in spite of being reduced in abundance, this lower level of miR399 directs more efficient cleavage of the *PHO2* transcript in the absence of DRB2 activity. In P^- *drb2* roots, miR399 abundance was determined to be elevated by 2.6-fold compared to its abundance in P^- Col-0 roots (Figure 4E). As observed in P^+ *drb2* roots, *PHO2* expression scaled in accordance with elevated miR399 abundance, with *PHO2* expression increased by 4.5-fold in PO_4-stressed *drb2* roots. It is interesting to note here that *PHO2* expression scaled with miR399 abundance in six out the eight root tissue samples molecularly assessed by RT-qPCR in this study. We have previously demonstrated that DRB2-dependent miRNAs direct a translational repression mode of miRNA-directed target gene expression repression [52], and scaling of miRNA target transcripts together with their targeting miRNA, has been previously reported for miRNA sRNAs that direct a translational repression mode of target gene expression regulation [52–54].

3.3. DRB4 is Required For miR399 Production in Arabidopsis Roots

Profiling of PO_4 transporter expression in the shoots and roots of P^+ and P^- *drb4* plants revealed considerable alteration to *PHT1;4*, *PHT1;8* and *PHT1;9* transcript abundance across both assessed tissues and growth regimes (Figure 5E–J). However, in spite of these documented differences in PO_4 transporter gene expression in *drb4* shoots and roots, the Pi content of non-stressed and PO_4-stressed *drb4* tissues remained at levels comparable to P^+ and P^- Col-0 shoots and roots (Figure 5A,B). Considering this finding, it was unsurprising that the developmental progression of Col-0 and *drb4* plants was impeded to the same extent when cultivated in the absence of PO_4 for a 7-day period. Specifically, the fresh weight of both P^- Col-0 and P^- *drb4* shoots was reduced by ~36% compared to their non-stressed counterparts of the same age (Figure 1C). In addition, anthocyanin, chlorophyll *a* and chlorophyll *b* were all elevated to the same degree in PO_4-stressed Col-0 and *drb4* shoots, compared to their respective non-stressed counterparts. It was therefore surprising that the rosette area of P^- *drb4* plants was only reduced by 38.7% compared to the more severe 60.1% reduction observed for P^- Col-0 plants. Although an unexpected finding, this result clearly indicated that some of the responses of the *drb4* mutant to PO_4 starvation differ to those of wild-type *Arabidopsis*.

Considering the well-established role of the DRB4/DCL4 partnership in *trans*-acting siRNA (tasiRNA) [55,56] and p4-siRNA [40] production, and for the processing of a small number of newly evolved miRNAs from their highly complementary precursor transcripts [39], it was highly surprising to additionally establish the widespread involvement of DRB4 in regulating the production of the highly conserved miRNA, miR399, in *Arabidopsis* shoots (Figure 2). Specifically, DRB4 was determined to play a secondary role to DRB1 in regulating the efficiency of miR399 production from all five precursors detectable by RT-qPCR in non-stressed *Arabidopsis* shoots. As demonstrated for DRB2, the involvement of DRB4 in miR399 production in *Arabidopsis* shoots is most likely via antagonism of the canonical DRB1/DCL1 partnership. Antagonism of the DRB1/DCL1 partnership by DRB4 was again demonstrated by the accumulation profiles of precursors, *PRE-MIR399C*, *PRE-MIR399D*, *PRE-MIR399E* and *PRE-MIR399F*, in the shoot tissues of PO_4-stressed *drb4* plants (Figure 2C–F). Although precursor transcript abundance was highly variable in *drb4* shoots, miR399 levels were only mildly elevated by 1.2- and 2.4-fold in P^+ *drb4* and P^- *drb4* shoots, respectively (Figure 2G). Surprisingly, in spite of the 20% elevation to miR399 levels in P^+ *drb4* shoots, *PHO2* expression was elevated to a similar degree (30% increase), and not reduced as expected (Figure 5I). In P^- *drb4* shoots, however, the 2.4-fold elevated abundance of the miR399 sRNA was determined, as expected, to reduce the expression of *PHO2* by 2.5-fold. This result clearly indicated that in the absence of DRB4 activity in *Arabidopsis* shoots, the efficiency of DRB1-mediated, miR399-directed cleavage of the *PHO2* transcript is enhanced.

The fresh weight of P^- *drb4* roots was reduced by 18.6% compared to the fresh weight of P^+ *drb4* roots, a 2.9-fold further enhancement of this phenotypic response to PO_4 stress, compared to the mild response of P^- Col-0 roots (6.5% fresh weight reduction compared to P^+ Col-0 roots). The response of the primary root of the *drb4* mutant to PO_4 stress also differed to that of wild-type roots. Namely, the length of P^- *drb4* primary root was only reduced by 10.3% compared to the significant 51.2% reduction to the length of the primary root of P^- Col-0 plants (Figure 3D). Although lateral root development was induced to the same degree (44%) in the root system of PO_4-stressed Col-0 and *drb4* plants, the considerable differences observed for the fresh weight of the *drb4* root system, and the lack of inhibition to primary root length in P^- *drb4* plants, clearly revealed that the *drb4* mutant background is defective in some of its responses to PO_4 starvation, compared to the responses of the Col-0 root system to this stress.

At the molecular level, the wild-type-like expression of the *PRE-MIR399A* precursor in the roots of non-stressed and PO_4-stressed *drb4* plants indicated that DRB4 does not play a role in regulating miR399 production from this precursor in *Arabidopsis* roots. Expression analysis of *PRE-MIR399C* did however identify a secondary role for DRB4 in regulating miR399 production from this precursor, potentially via antagonism of DRB1 function (Figure 4C). Of particular interest stemming from miR399 precursor transcript profiling in non-stressed and PO_4-stressed *Arabidopsis* roots is the unexpected finding that DRB4 appears to be the primary DRB required to regulate miR399 production from the *PRE-MIR399D* precursor (Figure 4D), with the abundance of the *PRE-MIR399D* precursor over-accumulating to its highest levels in both P^+ and P^- *drb4* roots. Curiously, assessment of the stem-loop folding structures of the six precursors from which the miR399 sRNA is liberated does not readily distinguish the *PRE-MIR399D* structure from the folding structures of the other five miR399 precursor transcripts. Therefore, the establishment of a role for DRB4 in regulating miR399 processing efficiency from its precursor transcripts was a highly unexpected finding, a finding that requires additional experimentation in the future to identify the precursor transcript-based sequence and/or structural features that recruits the involvement of DRB4 to the miR399/*PHO2* expression module.

The elevated abundance of the *PRE-MIR399C* and *PRE-MIR399D* precursors in P^+ *drb4* roots indicated reduced precursor transcript processing efficiency in the absence of DRB4. Accordingly, a 30% reduction to miR399 accumulation was observed in P^+ *drb4* roots (Figure 4E). Surprisingly, this 1.4-fold reduction to miR399 levels in P^+ *drb4* roots led to a 2.0-fold reduction to *PHO2* expression (Figure 4G). This result suggested that although miR399 levels were reduced in non-stressed *drb4* roots, the reduced amount of the miR399 sRNA was actually directing enhanced *PHO2* expression repression via unimpeded DRB1-mediated, miR399-directed, *PHO2* cleavage. However, enhanced miR399-directed *PHO2* cleavage appeared to be lost in PO_4-stressed *drb4* roots with both miR399 and *PHO2* levels elevated by 2.0- and 5.1-fold, respectively (Figure 4E,G). Therefore, when taken together, although miR399-directed *PHO2* cleavage appeared to be enhanced in P$^+$ *drb4* roots, the scaling of *PHO2* expression together with miR399 abundance in PO$_4$-stressed *drb4* roots, potentially suggests that in a cell type with altered physiology, and where DRB4 function is defective, the miR399 sRNA changed from directing an mRNA cleavage mode of RNA silencing, to directing a translational repression mode of RNA silencing.

3.4. DRB1, DRB2 and DRB4 Are Required to Regulate the miR399/PHO2 Expression Module in Arabidopsis Shoots and Roots

Here we demonstrate that the phenotypic and molecular response to PO$_4$ starvation were unique to each *drb* mutant background assessed due to the hierarchical contribution of DRB1, DRB2 and DRB4 to the regulation of the miR399/*PHO2* expression module. Specifically, the molecular profiling of miR399 precursor transcript expression identified DRB1 as the primary DRB required for efficient miR399 production from each precursor in non-stressed and PO$_4$-stressed shoots and roots. Deregulated miR399 precursor transcript processing efficiency in the absence of DRB1 activity was demonstrated to result in defective P homeostasis maintenance, altering the shoot to root ratio of Pi content in the *drb1* mutant background. The maintenance of P homeostasis was also defective in *drb2* plants, with the Pi content shoot to root ratio altered in this mutant background, both under standard growth conditions and in conditions of PO$_4$ starvation. An altered Pi content in *drb2* tissues appeared to result from defective PO$_4$ transport between the root system and aerial tissues in the absence of DRB2 function. Further, DRB2 was determined to play a secondary role to DRB1 in regulating miR399 production from the profiled *PRE-MIR399* precursor transcripts. The secondary role of DRB2 in regulating miR399 production from the assessed *PRE-MIR399* precursor transcripts was revealed to most likely be via antagonism of DRB1 function. DRB4 was also determined to play a secondary role in regulating the miR399/*PHO2* expression module in *Arabidopsis* shoots and roots, and as demonstrated for the secondary role of DRB2 in providing additional regulatory complexity to this expression module, DRB4 also appeared to be antagonistic to the primary functional role of DRB1 in regulating miR399 precursor transcript processing efficiency. Furthermore, DRB4 also appeared to be the primary DRB required for miR399 production from the *PRE-MIR399D* precursor in non-stressed and PO$_4$-stressed *Arabidopsis* roots. When taken together, the hierarchical contribution of DRB1, DRB2 and DRB4 to the regulation of the miR399/*PHO2* expression module documented here, readily demonstrates the crucial importance of maintaining P homeostasis in *Arabidopsis* tissues to ensure the maintenance of a wide range of cellular processes to which P is essential.

4. Materials and Methods

4.1. Plant Material and Phosphate Stress Treatment

The T-DNA insertion knockout mutant lines used in this study, including the *drb1* (*drb1-1*; SALK_064863), *drb2* (*drb2-1*; GABI_348A09) and *drb4* (*drb4-1*; SALK_000736) mutants, have been described previously [42]. The seeds of these three *drb* mutant lines, and of wild-type *Arabidopsis* (ecotype Columbia-0 (Col-0)) plants, were sterilized using chlorine gas and post-sterilization, seeds were plated out onto standard *Arabidopsis* plant growth media (half-strength Murashige and Skoog (MS) salts), and stratified for 48 h at 4 °C in the dark. Post-stratification, the sealed plates were transferred to a temperature-controlled growth cabinet (A1000 Growth Chamber, Conviron® Australia) and cultivated for an 8-day period under a standard growth regime of 16 h light / 8 h dark, and a day/night temperature of 22 °C / 18 °C. Post this initial 8-day cultivation period, equal numbers of Col-0, *drb1*, *drb2* and *drb4* seedlings were transferred under sterile conditions to either fresh standard *Arabidopsis* plant growth media that contained 1.0 mM of PO$_4$ (P$^+$ plants; non-stressed controls) or to *Arabidopsis* plant growth media where the PO$_4$ had been replaced with an equivalent molar amount (1.0 mM) of potassium chloride (KCl) (P$^-$ plants; PO$_4$ stress treatment). Post seedling transfer, the P$^+$ and P$^-$ plates for each plant line were returned to the temperature-controlled growth cabinet for an additional 7-day cultivation period. For the tissue-specific analyses performed here, namely the root tissue assessments, additional Col-0, *drb1*, *drb2* and *drb4*, 8-day old seedlings were treated exactly as outlined above, except for the 7-day treatment period, where P$^+$ and P$^-$ plates were orientated for vertical growth. Unless stated otherwise, all the phenotypic and molecular analyses reported here were conducted on 15-day old plants.

4.2. Phenotypic and Physiological Assessments

The fresh weight of 8-day old Col-0, *drb1*, *drb2* and *drb4* whole plants germinated and cultivated on standard *Arabidopsis* plant growth media was initially determined to establish the effect of loss of DRB1, DRB2 or DRB4 activity on *Arabidopsis* development. The fresh weight of 15-day old Col-0, *drb1*, *drb2* and *drb4* plants was also determined to establish the effect of the 7-day PO$_4$ stress treatment on the development of each plant line. The area of the rosette and the length of the primary root of 15-day old Col-0, *drb1*, *drb2* and *drb4* plants was determined via the assessment of photographic images using the ImageJ software. The same photographic images were also used to establish the number of lateral roots formed by P$^+$ and P$^-$ Col-0, *drb1*, *drb2* and *drb4* plants post the 7-day stress treatment period.

A standard methanol:HCl (99:1 v/v) extraction method was applied to extract anthocyanin from P$^+$ and P$^-$ plants, and post extraction, anthocyanin content was determined using a spectrophotometer (Thermo Scientific, Australia) at an absorbance wavelength of 535 nanometers (A$_{535}$). The 99:1 (v/v) methanol:HCl extraction buffer was used as the blanking solution and the A$_{535}$ of each sample was next divided by the fresh weight of the sample to calculate the relative anthocyanin content per gram of fresh weight (A$_{535}$/g FW).

For chlorophyll *a* and *b* content quantification, rosette leaves of 15-day old P$^+$ and P$^-$ Col-0, *drb1*, *drb2* and *drb4* plants were sampled and incubated in 80% acetone for 24 h in the dark. Post incubation, rosette leaf tissue was clarified via centrifugation at 15,000 × *g* for 7 min at room temperature. The resulting supernatants were immediately transferred to a spectrophotometer and the absorbance of these solutions assessed at wavelengths 646 nm (A$_{646}$) and 663 nm (A$_{663}$) using 80% acetone as the blanking solution. The chlorophyll *a* and *b* content of each sample was then determined using the Lichtenthaler's equations exactly as outlined in [57], and these initially determined values were subsequently converted to micrograms per gram of fresh weight (µg/g FW).

The shoot and root tissue of 15-day old P$^+$ and P$^-$ Col-0, *drb1*, *drb2* and *drb4* plants were carefully separated from each other and then ground into a fine powder under liquid nitrogen (LN$_2$). One milliliter (1.0 mL) of 10% acetic acid (v/v in H$_2$O) was added to the ground powder and the powder thoroughly resuspended via vigorous vortexing. The resulting resuspension was then centrifuged at 15,000 × *g* for 5 min at room temperature, and post centrifugation, 700 µL of the resulting supernatant was mixed with an equivalent volume of Ames Assay Buffer (6 parts 0.5% ammonium molybdite (v/v in H$_2$O) to 1 part of 2.5% sulphuric acid (v/v in 10% acetic acid)) and incubated at room temperature for 1 h in the dark. The absorbance of each solution was determined using a spectrophotometer at wavelength 820 nm (A$_{820}$) and the Pi content (µmol/gFW) of each sample subsequently determined via the construction of a Pi standard curve.

4.3. Total RNA Extraction for Quantitative Molecular Assessments

For each molecular assessment reported here, total RNA was extracted from four biological replicates (each biological replicate contained tissue sampled from eight individual plants) of 15-day old P$^+$ and P$^-$ Col-0, *drb1*, *drb2* and *drb4* plants using TRIzol™ Reagent according to the manufacturer's (Invitrogen™) instructions. The quality of the extracted total RNA was visually assessed via a standard electrophoresis approach on a 1.2% (w/v) ethidium bromide stained agarose gel and the quantity of total RNA extracted determined using a NanoDrop spectrophotometer (NanoDrop® ND-1000, Thermo Scientific, Australia).

For the synthesis of a miR399-specific complementary DNA (cDNA), 200 nanograms (ng) of total RNA was treated with 0.2 units (U) of DNase I (New England Biolabs, Australia) according to the manufacturer's instructions. The DNase I-treated total RNA was next used as template for cDNA synthesis with 1.0 U of ProtoScript® II Reverse Transcriptase (New England Biolabs, Australia) and the cycling conditions of 1 cycle of 16 °C for 30 min; 60 cycles of 30 °C for 30 s, 42 °C for 30 s, and 50 °C for 2 s, and; 1 cycle of 85 °C for 5 min.

A global, high molecular weight cDNA library for gene expression quantification was constructed via the initial treatment of 5.0 µg of total RNA with 5.0 U of DNase I according to the manufacturer's protocol (New England Biolabs, Australia). The DNase I-treated total RNA was next purified using an RNeasy Mini Kit (Qiagen, Australia) and 1.0 µg of this preparation used as template for cDNA synthesis along with 1.0 U of ProtoScript® II Reverse Transcriptase (New England Biolabs, Australia) and 2.5 mM of oligo dT$_{(18)}$, according to the manufacturer's instructions.

All generated, single-stranded cDNAs were next diluted to a working concentration of 50 ng/µL in RNase-free H$_2$O prior to their use as a template for the quantification of the abundance of either the miR399 sRNA or of gene transcripts. In addition, all RT-qPCRs used the same cycling conditions of 1 cycle of 95 °C for 10 min, followed by 45 cycles of 95 °C for 10 s and 60 °C for 15 s, and the GoTaq® qPCR Master Mix (Promega, Australia) was used as the fluorescent reagent for all performed RT-qPCR experiments. miR399 abundance and gene transcript expression was quantified using the $2^{-\Delta\Delta CT}$ method with the small nucleolar RNA, *snoR101*, and *UBIQUITIN10* (*UBI10*; *AT4G05320*) used as the respective internal controls to normalize the relative abundance of each assessed transcript. The sequence of each DNA oligonucleotide used in this study either for the synthesis of a miR399-specific cDNA, or to quantify transcript abundance via RT-qPCR is provided in Supplemental Table S1.

Supplementary Materials: The following are available online at http://www.mdpi.com/2223-7747/8/5/124/s1, Table S1: Sequences of the DNA oligonucleotides used in this study for the synthesis of miRNA-specific cDNAs and the RT-qPCR based quantification of miRNA abundance, miRNA target gene expression, or the assessment of standard gene expression.

Author Contributions: J.L.P., C.P.G. and A.L.E. conceived and designed the research. J.L.P. and J.M.O. performed the experiments and analyzed the data. J.L.P., C.P.G., J.M.O. and A.L.E. authored the manuscript and, J.L.P., C.P.G., J.M.O. and A.L.E. have read and approved the final version of the manuscript.

Funding: This research received no external funding.

Acknowledgments: The authors would like to thank fellow members of the Centre for Plant Science at the University of Newcastle for their guidance with plant growth care and RT-qPCR experiments.

Conflicts of Interest: The authors declare no conflict of interest.

References

1. Abel, S.; Ticconi, C.A.; Delatorre, C.A. Phosphate sensing in higher plants. *Planta* **2002**, *115*, 1–8. [CrossRef]
2. Hammond, J.P.; Bennett, M.J.; Bowen, H.C.; Broadley, M.R.; Eastwood, D.C.; May, S.T.; Rahn, C.; Swarup, R.; Woolaway, K.E.; White, P.J. Changes in Gene Expression in *Arabidopsis* Shoot during Phosphate Starvation and the Potential for Developing Smart Plants. *Society* **2003**, *132*, 578–596. [CrossRef] [PubMed]
3. Raghothama, K.G. Phosphate transport and signaling. *Curr. Opin. Plant Biol.* **2000**, *3*, 182–187. [CrossRef]
4. Hackenberg, M.; Shi, B.J.; Gustafson, P.; Langridge, P. Characterization of phosphorus-regulated miR399 and miR827 and their isomirs in barley under phosphorus- sufficient and phosphorus-deficient conditions. *BMC Plant Biol.* **2013**, *13*, 214. [CrossRef]
5. Chiou, T.J.; Lin, S.I. Signaling network in sensing phosphate availability in plants. *Annu. Rev. Plant Biol.* **2011**, *62*, 185–206. [CrossRef] [PubMed]
6. Wu, P.; Ma, L.; Hou, X.; Wang, M.; Wu, Y.; Liu, F.; Deng, X.W. Phosphate Starvation Triggers Distinct Alterations of Genome Expression in *Arabidopsis* Roots and Leaves. *Plant Physiol.* **2003**, *132*, 1260–1271. [CrossRef] [PubMed]
7. Lin, S.I.; Chiang, S.F.; Lin, W.Y.; Chen, J.W.; Tseng, C.Y.; Wu, P.C.; Chiou, T.J. Regulatory network of microRNA399 and PHO2 by systemic signaling. *Plant Physiol.* **2008**, *147*, 732–746. [CrossRef] [PubMed]
8. Jones, D.L. Organic acids in the rhizosphere—A critical review. *Plant Soil* **1998**, *205*, 25–44. [CrossRef]
9. Poirier, Y.; Thoma, S.; Somerville, C.; Schiefelbein, J. Mutant of *Arabidopsis* deficient in xylem loading of phosphate. *Plant Physiol.* **1991**, *97*, 1087–1093. [CrossRef]

10. Stefanovic, A.; Arpat, A.B.; Bligny, R.; Gout, E.; Vidoudez, C.; Bensimon, M.; Poirier, Y. Over-expression of PHO1 in *Arabidopsis* leaves reveals its role in mediating phosphate efflux. *Plant J.* **2011**, *66*, 689–699. [CrossRef]
11. Wang, Y.; Ribot, C.; Rezzonico, E.; Poirier, Y. Structure and Expression Profile of the *Arabidopsis* PHO1 Gene Family Indicates a Broad Role in Inorganic Phosphate Homeostasis 1. *Plant Physiol.* **2004**, *135*, 400–411. [CrossRef]
12. Delhaize, E.; Randall, P.J. Characterization of a Phosphate-Accumulator Mutant of *Arabidopsis thaliana*. *Plant Physiol.* **1995**, *107*, 207–213. [CrossRef]
13. Dong, B.; Rengel, Z.; Delhaize, E. Uptake and translocation of phosphate by *pho2* mutant and wild-type seedlings of *Arabidopsis thaliana*. *Planta* **1998**, *205*, 251–256. [CrossRef]
14. Park, B.S.; Seo, J.S.; Chua, N.H. Nitrogen Limitation Adaptation recruits PHOSPHATE2 to target the phosphate transporter PT2 for degradation during the regulation of *Arabidopsis* phosphate homeostasis. *Plant Cell* **2014**, *26*, 454–464. [CrossRef]
15. Aung, K. *pho2*, a Phosphate Overaccumulator, Is Caused by a Nonsense Mutation in a MicroRNA399 Target Gene. *Plant Physiol.* **2006**, *141*, 1000–1011. [CrossRef]
16. Nilsson, L.; Müller, R.; Nielsen, T.H. (2007). Increased expression of the MYB-related transcription factor, PHR1, leads to enhanced phosphate uptake in *Arabidopsis thaliana*. *Plant Cell Environ.* **2007**, *30*, 1499–1512. [CrossRef]
17. Rubio, V.; Linhares, F.; Solano, R.; Mart'in, A.C.; Iglesias, J.; Leyva, A.; Paz-Ares, J. A conserved MYB transcription factor involved in phosphate starvation signaling both in vascular plants and in unicellular algae. *Genes Dev.* **2001**, *15*, 2122–2133. [CrossRef]
18. Fujii, H.; Chiou, T.J.; Lin, S.I.; Aung, K.; Zhu, J.K. A miRNA involved in phosphate-starvation response in *Arabidopsis*. *Curr. Biol.* **2005**, *15*, 2038–2043. [CrossRef]
19. Hsieh, L.C.; Lin, S.I.; Shih, A.C.C.; Chen, J.W.; Lin, W.Y.; Tseng, C.Y.; Li, W.H.; Chiou, T.J. Uncovering small RNA-mediated responses to phosphate deficiency in *Arabidopsis* by deep sequencing. *Plant Physiol.* **2009**, *151*, 2120–2132. [CrossRef]
20. Buhtz, A.; Springer, F.; Chappell, L.; Baulcombe, D.C.; Kehr, J. Identification and characterization of small RNAs from the phloem of *Brassica napus*. *Plant J.* **2008**, *53*, 739–749. [CrossRef]
21. Pant, B.D.; Buhtz, A.; Kehr, J.; Scheible, W.R. MicroRNA399 is a long distance signal for the regulation of plant phosphate homeostasis. *Plant J.* **2008**, *53*, 731–738. [CrossRef]
22. Wang, Y.-H.; Garvin, D.F.; Kochian, L.V. Rapid induction of regulatory and transporter genes in response to phosphorus, potassium, and iron deficiencies in tomato roots. Evidence for cross talk and root/rhizosphere-mediated signals. *Plant Physiol.* **2002**, *130*, 1361–1370. [CrossRef]
23. Shin, R.; Berg, R.H.; Schachtman, D.P. Reactive oxygen species and root hairs in *Arabidopsis* root response to nitrogen, phosphorus and potassium deficiency. *Plant Cell Physiol.* **2005**, *46*, 1350–1357. [CrossRef]
24. Kant, S.; Peng, M.; Rothstein, S.J. Genetic regulation by NLA and microRNA827 for maintaining nitrate-dependent phosphate homeostasis in *Arabidopsis*. *PLoS Genet.* **2011**, *7*, e1002021. [CrossRef] [PubMed]
25. Liang, G.; He, H.; Yu, D. Identification of Nitrogen Starvation-Responsive MicroRNAs in *Arabidopsis thaliana*. *PLoS ONE* **2012**, *7*, e48951. [CrossRef] [PubMed]
26. Doerner, P. Phosphate starvation signaling: A threesome controls systemic Pi homeostasis. *Curr. Opin. Plant Biol.* **2008**, *11*, 536–540. [CrossRef]
27. Bari, R.; Pant, B.D.; Stitt, M.; Golm, S.P. PHO2, MicroRNA399, and PHR1 Define a Phosphate-Signaling Pathway in Plants. *Plant Physiol.* **2006**, *141*, 988–999. [CrossRef] [PubMed]
28. Berkowitz, O.; Jost, R.; Kollehn, D.O.; Fenske, R.; Finnegan, P.M.; O'Brien, P.A.; Hardy, G.E.S.J.; Lambers, H. Acclimation responses of *Arabidopsis thaliana* to sustained phosphite treatments. *J. Exp. Bot.* **2013**, *64*, 1731–1743. [CrossRef] [PubMed]
29. Chiou, T.J.; Aung, K.; Lin, S.I.; Wu, C.C.; Chiang, S.F.; Su, C.L. Regulation of phosphate homeostasis by MicroRNA in *Arabidopsis*. *Plant Cell* **2006**, *18*, 412–421. [CrossRef] [PubMed]
30. Huang, T.K.; Han, C.L.; Lin, S.I.; Chen, Y.J.; Tsai, Y.C.; Chen, Y.R.; Chen, J.W.; Lin, W.Y.; Chen, P.M.; Liu, T.Y.; et al. Identification of downstream components of ubiquitin-conjugating enzyme PHOSPHATE2 by quantitative membrane proteomics in *Arabidopsis* roots. *Plant Cell* **2013**, *25*, 4044–4060. [CrossRef]

31. Rouached, H.; Arpat, A.B.; Poirier, Y. Regulation of phosphate starvation responses in plants: Signaling players and cross-talks. *Mol. Plant* **2010**, *3*, 288–299. [CrossRef] [PubMed]
32. Martin, A.C.; Del Pozo, J.C.; Iglesias, J.; Rubio, V.; Solano, R.; De La Peña, A.; Leyva, A.; Paz-Ares, J. Influence of cytokinins on the expression of phosphate starvation responsive genes in *Arabidopsis*. *Plant J.* **2000**, *24*, 559–567. [CrossRef] [PubMed]
33. Franco-Zorrilla, J.M.; Valli, A.; Todesco, M.; Mateos, I.; Puga, M.I.; Rubio-Somoza, I.; Leyva, A.; Weigel, D.; García, J.A.; Paz-Ares, J. Target mimicry provides a new mechanism for regulation of microRNA activity. *Nat. Genet.* **2007**, *39*, 1033–1037. [CrossRef] [PubMed]
34. Mallory, A.C.; Bouché, N. MicroRNA-directed regulation: To cleave or not to cleave. *Trends Plant Sci.* **2008**, *13*, 359–367. [CrossRef] [PubMed]
35. Han, M.H.; Goud, S.; Song, L.; Fedoroff, N. (2004). The *Arabidopsis* double-stranded RNA-binding protein HYL1 plays a role in microRNA-mediated gene regulation. *Proc. Natl. Acad. Sci. USA* **2004**, *101*, 1093–1098. [CrossRef]
36. Eamens, A.L.; Smith, N.A.; Curtin, S.J.; Wang, M.B.; Waterhouse, P.M. The *Arabidopsis thaliana* double-stranded RNA binding protein DRB1 directs guide strand selection from microRNA duplexes. *RNA* **2009**, *15*, 2219–2235. [CrossRef]
37. Eamens, A.L.; Kim, K.W.; Curtin, S.J.; Waterhouse, P.M. DRB2 is required for microRNA biogenesis in *Arabidopsis thaliana*. *PLoS ONE* **2012**, *7*, e35933. [CrossRef]
38. Eamens, A.L.; Kim, K.W.; Waterhouse, P.M. DRB2, DRB3 and DRB5 function in a non-canonical microRNA pathway in *Arabidopsis thaliana*. *Plant Signal. Behav.* **2012**, *7*, 1224–1229. [CrossRef]
39. Rajagopalan, R.; Vaucheret, H.; Trejo, J.; Bartel, D.P. A diverse and evolutionarily fluid set of microRNAs in *Arabidopsis thaliana*. *Genes Dev.* **2006**, *20*, 3407–3425. [CrossRef]
40. Pélissier, T.; Clavel, M.; Chaparro, C.; Pouch-Pélissier, M.N.; Vaucheret, H.; Deragon, J.M. Double-stranded RNA binding proteins DRB2 and DRB4 have an antagonistic impact on polymerase IV-dependent siRNA levels in *Arabidopsis*. *RNA* **2011**, *17*, 1502–1510. [CrossRef]
41. Lu, C.; Fedoroff, N. A mutation in the *Arabidopsis* HYL1 gene encoding a dsRNA binding protein affects responses to abscisic acid, auxin, and cytokinin. *Plant Cell* **2000**, *12*, 2351–2366. [CrossRef]
42. Curtin, S.J.; Watson, J.M.; Smith, N.A.; Eamens, A.L.; Blanchard, C.L.; Waterhouse, P.M. The roles of plant dsRNA-binding proteins in RNAi-like pathways. *FEBS Lett.* **2008**, *582*, 2753–2760. [CrossRef]
43. Luo, Q.J.; Mittal, A.; Jia, F.; Rock, C.D. An autoregulatory feedback loop involving PAP1 and TAS4 in response to sugars in *Arabidopsis*. *Plant Mol. Biol.* **2012**, *80*, 117–129. [CrossRef]
44. Williamson, L.C.; Ribrioux, S.; Fitter, A.H.; Leyser, H.M.O. Phosphate availability regulates root system architecture in *Arabidopsis*. *Plant Physiol.* **2001**, *126*, 875–882. [CrossRef]
45. Schachtman, D.P.; Reid, R.J.; Ayling, S.M. Phosphorus Uptake by Plants: From Soil to Cell. *Plant Physiol.* **1998**, *116*, 447–453. [CrossRef]
46. Yang, X.J.; Finnegan, P.M. Regulation of phosphate starvation responses in higher plants. *Ann. Bot.* **2010**, *105*, 513–526. [CrossRef]
47. Remy, E.; Cabrito, T.R.; Batista, R.A.; Teixeira, M.C.; Sá-Correia, I.; Duque, P. The Pht1;9 and Pht1;8 transporters mediate inorganic phosphate acquisition by the *Arabidopsis thaliana* root during phosphorus starvation. *New Phytol.* **2012**, *195*, 356–371. [CrossRef]
48. Vazquez, F.; Vaucheret, H.; Rajagopalan, R.; Lepers, C.; Gasciolli, V.; Mallory, A.C.; Hilbert, J.L.; Bartel, D.P.; Crété, P. Endogenous *trans*-acting siRNAs regulate the accumulation of *Arabidopsis* mRNAs. *Mol. Cell* **2004**, *16*, 69–79. [CrossRef]
49. Kurihara, Y.; Takashi, Y.; Watanabe, Y. The interaction between DCL1 and HYL1 is important for efficient and precise processing of pri-miRNA in plant microRNA biogenesis. *RNA* **2006**, *12*, 206–212. [CrossRef]
50. Jiang, C.; Gao, X.; Liao, L.; Harberd, N.P.; Fu, X. Phosphate starvation root architecture and anthocyanin accumulation responses are modulated by the gibberellin-DELLA signaling pathway in *Arabidopsis*. *Plant Physiol.* **2007**, *145*, 1460–1470. [CrossRef]
51. Shin, H.; Shin, H.S.; Dewbre, G.R.; Harrison, M.J. Phosphate transport in *Arabidopsis*: Pht1;1 and Pht1;4 play a major role in phosphate acquisition from both low- and high-phosphate environments. *Plant J.* **2004**, *39*, 629–642. [CrossRef] [PubMed]
52. Reis, R.S.; Hart-Smith, G.; Eamens, A.L.; Wilkins, M.R.; Waterhouse, P.M. Gene regulation by translational inhibition is determined by Dicer partnering proteins. *Nat. Plants* **2015**, *1*, 14027. [CrossRef]

53. Yang, L.; Wu, G.; Poethig, R.S. Mutations in the GW-repeat protein SUO reveal a developmental function for microRNA-mediated translational repression in *Arabidopsis*. *Proc. Natl. Acad. Sci. USA* **2012**, *109*, 315–320. [CrossRef] [PubMed]
54. Xu, M.; Hu, T.; Zhao, J.; Park, M.Y.; Earley, K.W.; Wu, G.; Yang, L.; Poethig, R.S. Developmental Functions of miR156-Regulated *SQUAMOSA PROMOTER BINDING PROTEIN-LIKE (SPL)* Genes in *Arabidopsis thaliana*. *PLoS Genet.* **2016**, *12*, e1006263. [CrossRef]
55. Adenot, X.; Elmayan, T.; Laursssergues, D.; Boutet, S.; Bouché, N.; Gasciolli, V.; Vaucheret, H. DRB4-Dependent *TAS3 trans*-Acting siRNAs Control Leaf Morphology through AGO7. *Curr. Biol.* **2006**, *16*, 927–932. [CrossRef] [PubMed]
56. Nakazawa, Y.; Hiraguri, A.; Moriyama, H.; Fukuhara, T. The dsRNA-binding protein DRB4 interacts with the Dicer-like protein DCL4 in vivo and functions in the *trans*-acting siRNA pathway. *Plant Mol. Biol.* **2007**, *63*, 777–785. [CrossRef] [PubMed]
57. Lichtenthaler, H.K.; Wellburn, A.R. Determination of total carotenoids and chlorophylls *a* and *b* of leaf extracts in different solvents. *Biochem. Soc. Trans.* **1983**, *11*, 591–592. [CrossRef]

 © 2019 by the authors. Licensee MDPI, Basel, Switzerland. This article is an open access article distributed under the terms and conditions of the Creative Commons Attribution (CC BY) license (http://creativecommons.org/licenses/by/4.0/).

Article

Profiling the Abiotic Stress Responsive microRNA Landscape of *Arabidopsis thaliana*

Joseph L. Pegler, Jackson M. J. Oultram, Christopher P. L. Grof and Andrew L. Eamens *

Centre for Plant Science, School of Environmental and Life Sciences, Faculty of Science, University of Newcastle, Callaghan 2308, Australia; Joseph.Pegler@uon.edu.au (J.L.P.); Jackson.Oultram@uon.edu.au (J.M.J.O.); chris.grof@newcastle.edu.au (C.P.L.G.)
* Correspondence: andy.eamens@newcastle.edu.au; Tel.: +61-249-217-784

Received: 5 February 2019; Accepted: 6 March 2019; Published: 10 March 2019

Abstract: It is well established among interdisciplinary researchers that there is an urgent need to address the negative impacts that accompany climate change. One such negative impact is the increased prevalence of unfavorable environmental conditions that significantly contribute to reduced agricultural yield. Plant microRNAs (miRNAs) are key gene expression regulators that control development, defense against invading pathogens and adaptation to abiotic stress. *Arabidopsis thaliana* (*Arabidopsis*) can be readily molecularly manipulated, therefore offering an excellent experimental system to alter the profile of abiotic stress responsive miRNA/target gene expression modules to determine whether such modification enables *Arabidopsis* to express an altered abiotic stress response phenotype. Towards this goal, high throughput sequencing was used to profile the miRNA landscape of *Arabidopsis* whole seedlings exposed to heat, drought and salt stress, and identified 121, 123 and 118 miRNAs with a greater than 2-fold altered abundance, respectively. Quantitative reverse transcriptase polymerase chain reaction (RT-qPCR) was next employed to experimentally validate miRNA abundance fold changes, and to document reciprocal expression trends for the target genes of miRNAs determined abiotic stress responsive. RT-qPCR also demonstrated that each miRNA/target gene expression module determined to be abiotic stress responsive in *Arabidopsis* whole seedlings was reflective of altered miRNA/target gene abundance in *Arabidopsis* root and shoot tissues post salt stress exposure. Taken together, the data presented here offers an excellent starting platform to identify the miRNA/target gene expression modules for future molecular manipulation to generate plant lines that display an altered response phenotype to abiotic stress.

Keywords: *Arabidopsis thaliana*; abiotic stress; heat stress; drought stress; salt stress; microRNAs (miRNAs); miRNA target gene expression; RT-qPCR

1. Introduction

Anthropogenically driven climate change is a rapidly growing concern globally, forcing interdisciplinary research collaborations to provide solutions that address and/or negate the numerous negative consequences of a changing climate, with the provision of sustainable food security the overarching goal of contemporary agricultural research [1–3]. Throughout the last half-century, agriculture has attempted to continue to achieve the food demands of an ever-growing global population via the unsustainable practice of clearing biodiverse terrestrial ecosystems for additional cultivation of traditional cropping species, an alarming practice that further contributes to the global carbon footprint and climate change [4,5]. Considering the capability limitation of the current maximum annual global crop yield to land area ratio, it is obvious that alternate strategies are now required if cropping agriculture is to continue to ensure food security, whilst terminating unsustainable farming practices, and while achieving these goals in an ever increasingly unfavorable environment [4].

Lacking the mobility of metazoa, the sessile nature of a plant requires intricate and interrelated gene expression networks to mediate the plant's ability to physiologically and phenotypically respond to its surrounding environment. Such multilayered molecular networks are especially important to a plant's adaptive and/or defensive response when either abiotic or biotic stress is encountered [6,7]. Elucidating the gene expression cascades that underpin the ability of a plant to adapt to, or mitigate, the negative impact of abiotic stress is the first key step in the development of new plant lines harboring molecular modifications which allow the plant to display an altered response phenotype when exposed to abiotic stress. The genetic model plant, *Arabidopsis thaliana* (*Arabidopsis*), is readily amenable to molecular modification, thereby offering plant biology researchers an excellent experimental system to validate which introduced molecular modifications mediate the expression of abiotic stress tolerance phenotypes.

Since their initial identification in *Arabidopsis* in 2002 [8], plant microRNAs (miRNAs), small non-protein-coding regulatory RNAs, have been repeatedly demonstrated to be key regulators of gene expression across all phases of plant development [9,10], in mediating a defense response against invading viral, bacterial or fungal pathogens [11,12], or to direct a plant's adaptive response to exposure to abiotic stress, including the stresses of heat, drought and salt stress [13,14]. Each *Arabidopsis* miRNA is processed from a stem-loop structured precursor transcript, a non-protein-coding RNA that has folded back upon itself to form this structure, post RNA polymerase II (Pol II)-catalyzed transcription from a unique *MICRORNA* (*MIR*) gene [15–17]. Like protein coding loci, the promoter regions of many *MIR* genes harbor *cis*-elements that contribute to the control of *MIR* gene expression in response to numerous signals external to the cell, including the signals that stem from abiotic stress [18–20]. Altered *MIR* gene expression, and therefore altered mature miRNA abundance, in turn leads to changes in miRNA target gene expression, with each miRNA loaded by the miRNA-induced silencing complex (miRISC) to be used as a sequence specificity guide to modulate target gene expression via either a messenger RNA (mRNA) cleavage or translational repression mechanism of RNA silencing [9,21–23]. To date, in *Arabidopsis*, numerous miRNA/target gene expression modules have been demonstrated to be responsive to abiotic stress, with alteration to the molecular profile of some expression modules further shown to assist the plant to adapt to abiotic stress due to molecular-driven changes to key pathways, such as the photosynthesis, sugar signaling, stomatal control and hormone signaling pathways [6,7]. Of particular interest is the demonstration that considerable numbers of *MIR* gene families identified as abiotic stress responsive in *Arabidopsis*, play a conserved functional role across phylogenetically diverse dicotyledonous and monocotyledonous species, including many of the major monocot grasses (such as *Zea mays*, *Oryza sativa* and *Triticum aestivum*) cultivated to provide much of the daily calorific intake of the world's population [24–28]. This demonstration identifies the use of *Arabidopsis* as an ideal experimental system to molecularly modify the profile of such conserved abiotic stress responsive miRNA expression modules to determine if the introduced modifications enable *Arabidopsis* to display an altered response phenotype to abiotic stress.

High throughput sequencing was therefore employed here to profile the miRNA landscape of wild-type *Arabidopsis* plants exposed to heat, drought and salt stress. Sequencing identified large miRNA cohorts responsive to each applied stress, with 121, 123 and 118 miRNA sRNAs determined to have a greater than 2.0-fold abundance change post heat, drought and salt stress treatment of *Arabidopsis* whole seedlings, respectively. For each assessed stress, a quantitative reverse transcriptase polymerase chain reaction (RT-qPCR)-based approach was used to experimentally confirm the abundance of five miRNA sRNAs. For each miRNA experimentally validated to have altered abundance post stress exposure, RT-qPCR was additionally used to document reciprocal target gene expression profiles to further confirm each miRNA/target gene expression module as abiotic stress responsive. Sequencing, and the initial RT-qPCR profiling of miRNA abundance and miRNA target gene expression was performed on whole seedling samples, therefore RT-qPCR was next employed to confirm the documented whole seedling expression trends in *Arabidopsis* root and shoot tissues post salt stress exposure. This analysis showed that the initial abiotic stress responsive miRNA expression profiles identified in whole plant samples were an accurate representation of the tissue-specific profile of each assessed expression module. Taken together, the data presented here identifies numerous miRNA/target gene expression modules that could be targeted for future molecular modification to determine if such modification allows *Arabidopsis* to display an altered response to abiotic stress. Furthermore, the information gathered in *Arabidopsis* using such a molecular approach could be potentially translated to an agronomically important cropping species for the future generation of plant lines that display an adaptive phenotypic response to abiotic stress.

2. Results

2.1. Response of Wild-type Arabidopsis Seedlings to Heat, Drought and Salt Stress

Post germination and cultivation on standard growth media, 8 day old wild-type *Arabidopsis* seedlings (ecotype Columbia-0 (Col-0)) were exposed to a 7-day period of either heat, drought or salt stress. Figure 1A displays the phenotypes expressed by heat, drought and salt stressed Col-0 plants, compared to that of 15 day old, non-stressed wild-type *Arabidopsis*. The growth of drought (mannitol supplemented media) and salt stressed plants was significantly repressed, to differing degrees, after 7 days of exposure to both stress treatments, as readily demonstrated by a reduction to rosette tissue fresh weight (Figure 1B), and rosette diameter (Figure 1C). Furthermore, salt stress treatment induced the accumulation of anthocyanin (Figure 1D), an antioxidant produced by plants to combat the cellular stress caused by reactive oxygen species [29,30], specifically in the shoot apex of salt stressed *Arabidopsis* whole seedlings (Figure 1A). Unlike the reductions observed to rosette tissue fresh weight and the diameter of the rosette of drought and salt stressed Col-0 plants, the 7-day heat stress treatment resulted in the promotion of both of these phenotypic parameters. Specifically, compared to non-stressed control plants, the rosette leaf fresh weight (Figure 1B), and rosette diameter (Figure 1C), were increased by 75% and 127% respectively, in heat stressed Col-0 plants. Promotion of specific aerial tissue growth parameters, namely hypocotyl and rosette leaf petiole elongation, has been reported previously for *Arabidopsis* plants cultivated under conditions of elevated temperature [31,32]. Although promotion of aerial tissue growth suggested that the 7-day heat stress treatment had a positive influence on *Arabidopsis* development, the 177% increase in anthocyanin accumulation observed in parallel (Figure 1D), alternatively suggested that this treatment actually induced high levels of stress in *Arabidopsis* cells, specifically the cells of the shoot apex and rosette leaf petioles (Figure 1A). Therefore, to additionally demonstrate that each applied stress was influencing *Arabidopsis* at the molecular level, the expression of the well characterized stress-induced gene, *Δ1-PYRROLINE-5-CARBOXYLATE SYNTHETASE1* (*P5CS1*; *AT2G39800*) [33–35] was quantified by RT-qPCR. This analysis revealed that *P5CS1* transcript abundance was upregulated 3.0-, 1.7- and 45-fold in heat, drought and salt stressed *Arabidopsis* whole seedlings, respectively (Figure 1E).

2.2. Profiling of the microRNA Landscape of Heat, Drought and Salt Stressed Arabidopsis Whole Seedlings

Demonstrated induction of *P5CS1* expression (Figure 1E), a well characterized [33–35] stress-induced gene in *Arabidopsis*, post exposure to heat, drought and salt stress, suggested that all three applied stresses were inducing molecular responses in *Arabidopsis*. Therefore, total RNA was extracted from non-stressed plants, and from heat, drought and salt stressed *Arabidopsis* whole seedlings, and the sRNA fraction of each analyzed via high throughput sequencing to profile the respective miRNA landscapes (Figure 2A). In total, 333 miRNA sRNAs were identified by sequencing across the control and stress treatments (see Supplemental Table S1). Sequencing further revealed a greater than 2-fold abundance change for 121, 123 and 118 mature miRNA sRNAs for heat, drought and salt stressed *Arabidopsis* whole seedlings, respectively (Figure 2B). More specifically, heat stress promoted the accumulation of 90 miRNAs (miR395a abundance was increased to the greatest degree at 89.4-fold) and repressed the abundance of 17 miRNAs (miR3932b levels showed the greatest degree of reduction at -19.6-fold). Exposure of wild-type *Arabidopsis* whole seedlings to drought stress enhanced the abundance of 111 miRNAs and reduced the levels of 2 miRNAs, with the accumulation of miRNAs, miR851 (31.1-fold) and miR397b (-7.8-fold), determined to be influenced to the greatest degree by drought stress treatment. Post salt stress exposure, 86 miRNAs were determined to have a greater than 2-fold elevated abundance (miR778 was upregulated to the greatest degree at 34.0-fold), and further, 22 miRNAs were determined to be reduced in their abundance (miR169g showed the greatest degree of reduction at -8.7-fold). It was also of interest to document reciprocal abundance trends for an additional 5, 5 and 1 miRNA sRNAs upon comparison of each of the applied stresses, namely the heat/drought, heat/salt and drought/salt stress comparisons, respectively. In addition, a further 4 miRNA sRNAs, including miR169f, miR169h, miR397b and miR857, were determined to have an opposing change in abundance upon exposure to each of the three abiotic stresses assessed (Figure 2B). These findings indicate that the promoter regions of the encoding loci of these miRNAs harbor multiple *cis*-elements that direct the changes in *MIR* gene expression which would be required to result in the observed changes in the abundance of this miRNA cohort post exposure to different abiotic stresses [18–20].

A modified RT-qPCR approach [36] was next employed to experimentally validate the sequencing determined abundance of five miRNAs for each stress treatment. For heat stressed *Arabidopsis* whole seedlings, sequencing determined that the abundance of miRNAs, miR169, miR395 and miR396, was altered -7.4, 37.8 and 2.9-fold, respectively, compared to their abundance in non-stressed plants. The altered abundance trend of all three miRNAs was confirmed by RT-qPCR with quantified fold changes of -3.2-, 2.2- and 2.9-fold for the miR169, miR395 and miR396 sRNAs, respectively (Figure 3A–C). Although the abundance changes determined by RT-qPCR were not as dramatic as those determined via sequencing for miRNAs miR169 and miR395 (especially for miR395), the obtainment of a matching abundance trend for each quantified miRNA post heat stress exposure was highly encouraging. Therefore, RT-qPCR was again applied to confirm the sequencing identified abundance fold changes of -2.7-, 4.0- and 2.7-fold for miRNAs, miR857, miR156 and miR399, respectively (Figure 3D–F). Fold changes of -4.3-, 3.0- and 3.2-fold were determined for the miR857, miR156 and miR399 sRNAs respectively, post the 7-day drought stress treatment of *Arabidopsis* whole seedlings by RT-qPCR. RT-qPCR also confirmed the sequencing identified miRNA abundance trends for salt stressed *Arabidopsis* whole seedlings. Fold changes of -2.5-, 2.9- and 3.9-fold were determined by RT-qPCR for miRNAs, miR169, miR399 and miR778 respectively (Figure 3G–I), compared to the abundance fold changes of -4.8-, 4.0- and 34.0-fold determined via sequencing for these three miRNAs in response to salt stress treatment. In addition, miR839 and miR855 abundance in heat, drought and salt stressed plants was also quantified via RT-qPCR due to sequencing indicating that the level of both of these miRNAs did not vary significantly post application of each stress (Figure 3J,K). RT-qPCR confirmed that the levels of these two miRNAs varied less than 0.5-fold post stress exposure, a finding that suggests that neither miRNA is abiotic stress responsive. Taken together, the data presented in Figure 3 demonstrated that the high throughput sequencing employed here (Figure 2) was a reliable tool for profiling miRNA abundance changes in abiotic stressed *Arabidopsis* whole seedlings, and that once identified, RT-qPCR quantification provides a more biologically accurate reflection of the changes in sRNA abundance post exposure to each stress treatment.

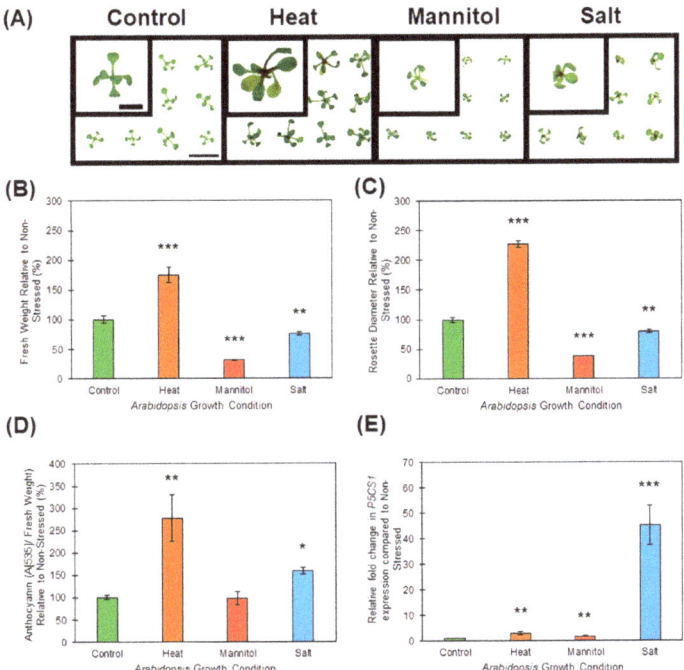

Figure 1. Phenotypic and physiological consequence of heat, drought and salt stress treatment of 15-day-old wild-type *Arabidopsis* whole seedlings. (**A**) Phenotypes displayed by 15-day old wild-type *Arabidopsis* whole seedlings post a 7-day treatment regime with heat, drought or salt stress, compared to non-stressed seedlings of the same age (left panel). Scale bar = 1.0 centimeter (cm) on larger sized panels and 0.5 cm on the superimposed images of a single representative seedling. (**B**) Whole seedling fresh weight of heat, drought (mannitol) and salt stressed *Arabidopsis* compared to their non-stressed counterparts of the same age. (**C**) Rosette diameter of 15-day-old *Arabidopsis* whole seedlings post 7-day exposure to heat, drought (mannitol) and salt stress compared to the non-stressed control. (**D**) Anthocyanin accumulation in heat, drought (mannitol) and salt stressed *Arabidopsis* whole seedlings compared to non-stressed whole seedlings of the same age (15 days). (**E**) RT-qPCR assessment of the expression of the stress induced gene, *Δ1-PYRROLINE-5-CARBOXYLATE SYNTHETASE1* (*P5CS1*; *AT2G39800*) expression in 15-day-old *Arabidopsis* whole seedlings post a 7-day heat, drought (mannitol) and salt stress treatment regime compared to the abundance of the *P5CS1* transcript in non-stressed *Arabidopsis* whole seedlings of the same age. (**B**–**E**) Error bars represent the standard deviation of four biological replicates and each biological replicate consisted of a pool of six individual plants. The presence of an asterisk above a column represents a statistically significantly difference between the stress treated sample and the non-stressed control sample (p-value: * < 0.05; ** < 0.005; *** < 0.001).

2.3. Assessment of microRNA Target Gene Expression in Heat, Drought and Salt Stressed Arabidopsis Whole Seedlings

It has been extensively documented in *Arabidopsis* that miRNA sRNAs direct expression regulation of their targeted gene(s) via either a mRNA cleavage or translational repression mode of miRNA-directed RNA silencing [9,21–23]. Therefore, to identify the mode of target gene expression regulation directed by the miRNAs experimentally validated here to be responsive to heat, drought or salt stress, RT-qPCR was next employed to reveal the changes in miRNA target gene transcript abundance post exposure of *Arabidopsis* whole seedlings to these three abiotic stresses. For miRNAs, miR169, miR395 and miR396, the three miRNAs determined to be responsive to heat stress treatment via their RT-qPCR-determined, -3.2-, 2.2- and 2.9-fold change in abundance, the transcript level of a single target gene for each miRNA was demonstrated to have a reciprocally altered trend in abundance to that of their targeting miRNA (Figure 4A–C). Specifically, the miR169 target, *NUCLEAR FACTOR Y, SUBUNIT A5* (*NFYA5*; *AT1G54160*) was determined to have a 32.9-fold elevation in expression in heat stressed wild-type *Arabidopsis* compared to its levels in non-stressed whole seedlings (Figure 4A). In addition, *ATP SULFURYLASE1* (*ATPS1*; *AT3G22890*) and *GROWTH REGULATING FACTOR7* (*GRF7*; *AT5G53660*), the target genes of the heat-induced miRNAs, miR395 and miR396, respectively, were determined to have 2.4- (Figure 4B) and 2.2-fold (Figure 4C) reduced expression. RT-qPCR identified similar trends in altered expression for *LACCASE7* (*LAC7*; *AT3G09220*), *SQUAMOSA PROMOTER BINDING PROTEIN-LIKE9* (*SPL9*; *AT2G42200*) and *PHOSPHATE2* (*PHO2*; *AT2G33770*), the target genes of drought responsive miRNAs, miR857, miR156 and miR399, respectively. That is, *LAC7* expression was elevated 2.5-fold (Figure 4D) in response to the 2.7-fold reduction in miR857 levels (Figure 3D), and *SPL9* (Figure 4E) and *PHO2* (Figure 4F) expression was repressed by 2.8- and 4.5-fold respectively, in accordance with the documented 4.0- and 2.7-fold elevated abundance of the targeting miRNAs, miR156 (Figure 3E) and miR399 (Figure 3F), post drought stress treatment of *Arabidopsis* whole seedlings.

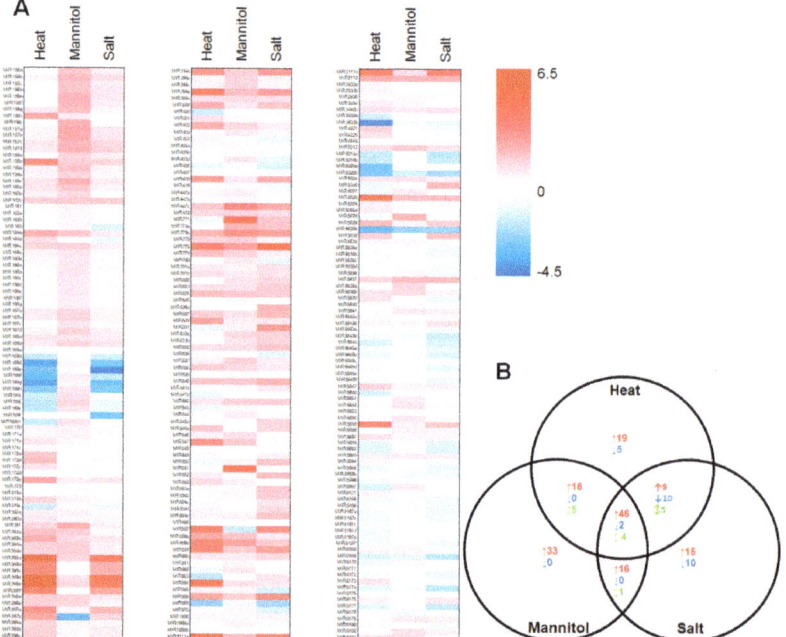

Figure 2. Profiling of the miRNA landscape of heat, drought and salt stressed 15-day-old wild-type *Arabidopsis* whole seedlings. (**A**) Red (up) and blue (down) shaded tiles represent a Log2 fold change in abundance of the *Arabidopsis* miRNA sRNAs detected via high throughput sequencing (see Supplemental Table S1 for the normalized read numbers used to determine fold change values). (**B**) The number of miRNA sRNAs determined to have a greater than 2-fold change in abundance in heat, drought and salt stressed 15-day-old wild-type *Arabidopsis* whole seedlings compared to the abundance of each detected miRNA sRNA in non-stressed control plants of the same age. Red colored up arrows indicate the number of miRNAs with elevated abundance under each assessed stress, blue colored down arrows represent the number of miRNAs with reduced abundance post stress treatment and green colored up/down arrows state the number of miRNA sRNAs with a differing abundance trend between the individual stress treatments.

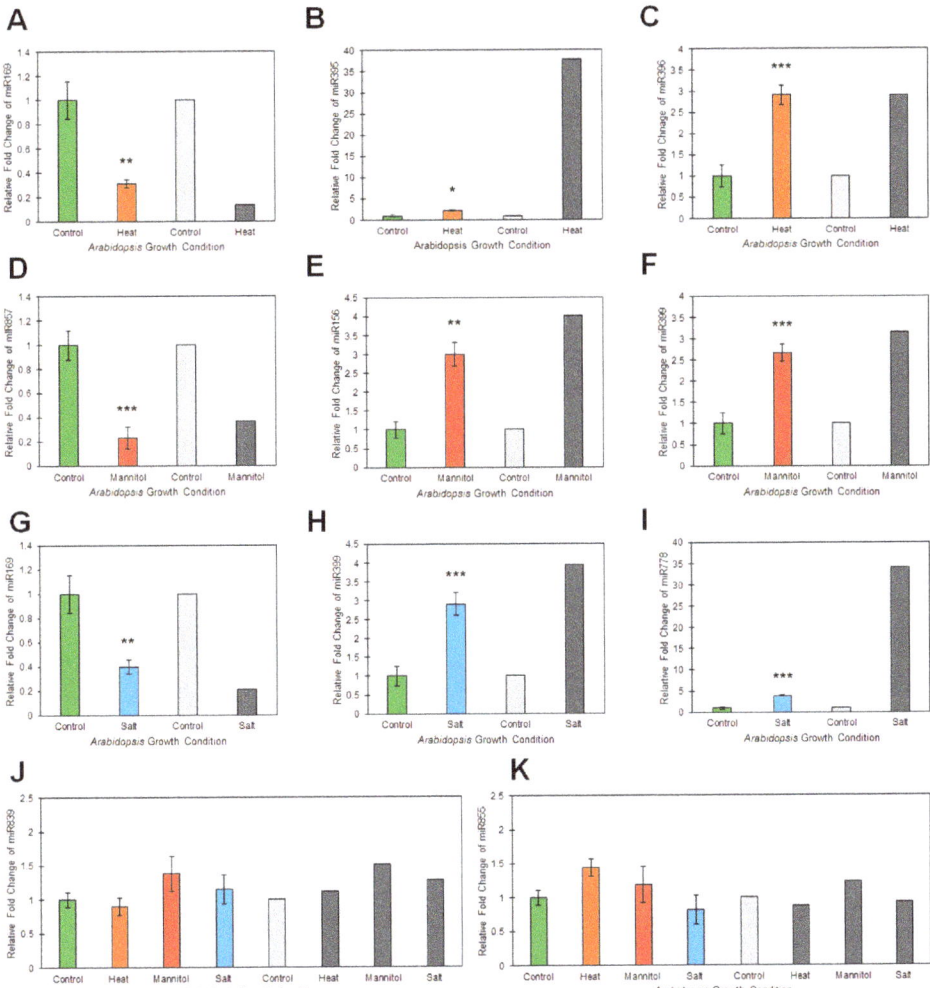

Figure 3. Quantification of miRNA abundance via RT-qPCR analysis of 15-day-old wild-type *Arabidopsis* whole seedlings post exposure to heat, drought and salt stress treatment. (**A–C**) RT-qPCR assessment of miR169 (**A**), miR395 (**B**) and miR396 (**C**) abundance in heat stressed *Arabidopsis* whole seedlings. (**D–F**) RT-qPCR assessment of miR857 (**D**), miR156 (**E**) and miR399 (**F**) abundance in drought (mannitol) stressed *Arabidopsis* whole seedlings. (**G–I**) RT-qPCR assessment of miR169 (**G**), miR399 (**H**) and miR778 (**I**) abundance in salt stressed *Arabidopsis* whole seedlings. (**J,K**) RT-qPCR assessment of miR839 (**J**) and miR855 (**K**) abundance across heat, drought (mannitol) and salt stressed *Arabidopsis* whole seedlings. (**A–K**) Colored columns (green = non-stressed control; orange = heat stress; red = drought stress, and; blue = salt stress) represent RT-qPCR determined abundance of each quantified miRNA sRNA and the light (control) and dark grey (stress) shaded columns present the fold changes in miRNA abundance as determined via high throughput sequencing. Error bars represent the standard deviation of four biological replicates and each biological replicate consisted of a pool of six individual plants. The presence of an asterisk above a column represents a statistically significantly difference between the stress treated sample and the non-stressed control sample (*p*-value: * < 0.05; ** < 0.005; *** < 0.001).

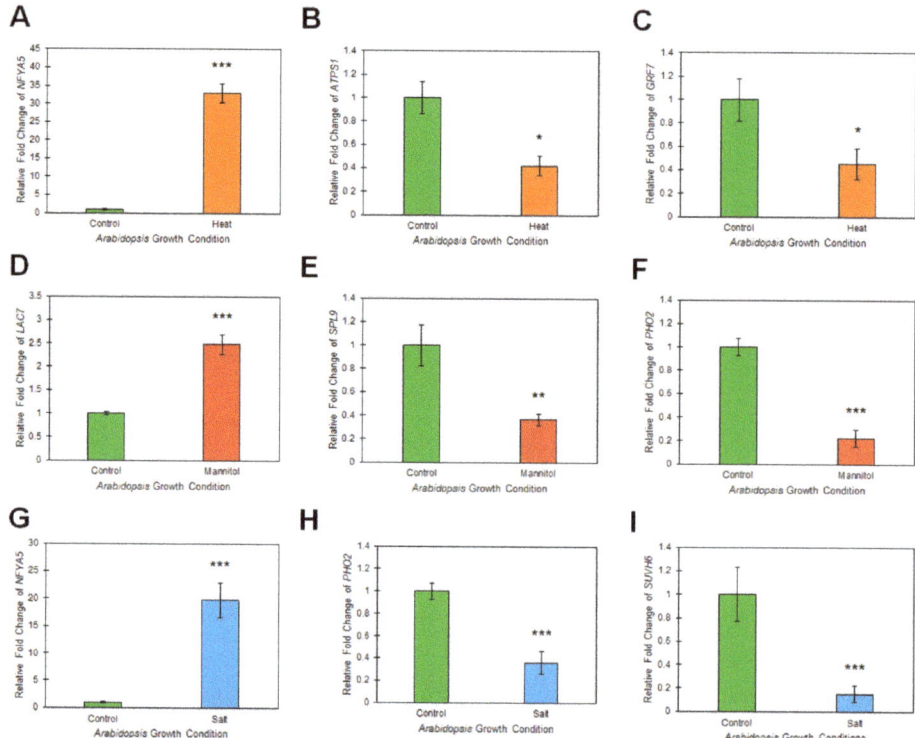

Figure 4. Determination of miRNA target gene expression via RT-qPCR analysis of 15-day-old wild-type *Arabidopsis* whole seedlings post exposure to heat, drought and salt stress treatment. (**A–C**) RT-qPCR assessment of *NFYA5* (**A**), *ATPS1* (**B**) and *GRF7* (**C**) miRNA target gene expression in heat stressed *Arabidopsis* whole seedlings. (**D–F**) RT-qPCR assessment of *LAC7* (**D**), *SPL9* (**E**) and *PHO2* (**F**) miRNA target gene expression in drought (mannitol) stressed *Arabidopsis* whole seedlings. (**G–I**) RT-qPCR assessment of *NFYA5* (**G**), *PHO2* (**H**) and *SUVH6* (**I**) miRNA target gene expression in salt stressed *Arabidopsis* whole seedlings. (**A–I**) Colored columns (green = non-stressed control; orange = heat stress; red = drought stress, and; blue = salt stress) represent RT-qPCR quantified expression of a single target gene for each miRNA assessed via RT-qPCR analysis in Figure 3. Error bars represent the standard deviation of four biological replicates and each biological replicate consisted of a pool of six individual plants. The presence of an asterisk above a column represents a statistically significantly difference between the stress treated sample and the non-stressed control sample (p-value: * < 0.05; ** < 0.005; *** < 0.001).

Reciprocal expression trends were again observed post RT-qPCR assessment of target gene expression in salt stressed samples. Namely, *NFYA5* transcript abundance was significantly elevated 19.7-fold (Figure 4G) in response to the RT-qPCR documented 2.5-fold reduction in miR169 levels (Figure 3G). Further, the transcript abundance of *PHO2* and *SU(VAR)3-9 HOMOLOG6* (*SUVH6*; *AT2G22740*) was reduced by -2.8- (Figure 4H) and -6.6-fold (Figure 4I), respectively. Reduced *PHO2* and *SUVH6* expression in salt stressed *Arabidopsis* whole seedlings was not a surprising observation considering that the abundance of their targeting miRNAs, miR399 and miR778, was determined to be elevated by 2.9- and 3.9-fold (Figure 3H,I), respectively. Taken together, the target gene expression data presented in Figure 4 indeed suggested that altered miRNA abundance in response to each assessed stress was in turn leading to changes in miRNA target gene transcript abundance. In addition, demonstration of reciprocal trends in abundance for the miRNAs determined to be abiotic stress responsive in Figure 3 compared to the miRNA target gene expression profiles presented in Figure 4, strongly suggested that each abiotic stress responsive miRNA was regulating the expression of its assessed target gene via a mRNA cleavage mode of miRNA-directed RNA silencing.

2.4. Profiling of Salt Responsive microRNA Expression Modules in Arabidopsis Root and Shoot Tissues

To determine whether the documented alterations to abiotic stress responsive miRNA expression modules identified in *Arabidopsis* whole seedlings was an accurate indication of the changes occurring in specific and developmentally distinct *Arabidopsis* tissues, the three miRNA expression modules determined to be salt responsive in *Arabidopsis* whole seedlings (the miR169/*NFYA5*, miR399/*PHO2* and miR778/*SUVH6* expression modules), were profiled in *Arabidopsis* root and shoot tissue by RT-qPCR post exposure to salt stress. Prior to performing this molecular analysis however, the root architecture of wild-type *Arabidopsis* plants cultivated on vertically orientated control and salt stress growth media was assessed. It has been demonstrated previously that the major phenotypic response of the *Arabidopsis* root system to exposure to salt stress is reduced expansion of the primary root [37]. Figure 5A clearly shows that compared to non-stressed wild-type *Arabidopsis* seedlings, the primary phenotypic response of the root system of Col-0 plants exposed to the 7-day salt stress regime was inhibition of primary root elongation, with primary root length reduced by ~60% in salt stressed plants compared to the primary root length of non-stressed control plants (Figure 5A,B).

Inhibition of primary root elongation, coupled with the vertically cultivated salt stressed plants again displaying reductions to the overall size of aerial tissue (i.e., rosette size; Figure 5A), as demonstrated in Figure 1A for *Arabidopsis* plants cultivated on horizontally orientated growth media, led us to next assess *P5SC1* expression in the vertically cultivated salt stressed *Arabidopsis* root and shoot tissue (Figure 5C). Compared to its levels in non-stressed roots and shoots, RT-qPCR revealed *P5CS1* expression to be significantly induced with transcript abundance elevated by 9.1- and 44.0-fold respectively, in salt stressed *Arabidopsis* roots and shoots (Figure 5C). This finding strongly suggested that both tissues types were indeed '*stressed*' by the 7 days of vertical cultivation on plant growth media supplemented with 150 mM sodium chloride. RT-qPCR was therefore next employed to profile the miR169/*NFYA5*, miR399/*PHO2* and miR778/*SUVH6* expression modules in salt stressed root and shoot samples and revealed an opposing trend in abundance for each profiled expression module across both assessed tissues. For example, miR169 abundance was determined to be reduced by 1.6- and 2.2-fold in salt stressed roots and shoots respectively (Figure 5D), while *NFYA5* expression was elevated 4.6- and 16.5-fold in these two tissues (Figure 5E). Similar altered trends in abundance for the miR399 sRNA and its targeted transcript, the *PHO2* mRNA, were also revealed by RT-qPCR, namely; miR399 abundance was elevated by 3.7- and 3.0-fold in salt stressed roots and shoots (Figure 5F), and *PHO2* target gene expression was repressed accordingly in the corresponding tissues by 3.6- and 2.3-fold (Figure 5G), respectively. In addition, RT-qPCR revealed that the abundance of the *SUVH6*-targeting miRNA, miR778, was elevated in both salt stressed root and shoot tissue (Figure 5H). The 3.9- and 2.1-fold elevated abundance of the miR778 sRNA in *Arabidopsis* roots and shoots following the salt stress treatment was determined to result in repressed target gene expression, with the abundance of the *SUVH6* transcript reduced by ~2.0-fold in both assessed tissues (Figure 5I). Taken together, the data presented in Figure 5 confirmed that for the three miRNAs determined to be responsive to salt stress, via their profiling in *Arabidopsis* whole seedlings using a high throughput sequencing approach, provided an accurate reflection of the altered abundance of these three miRNAs in developmentally distinct tissues.

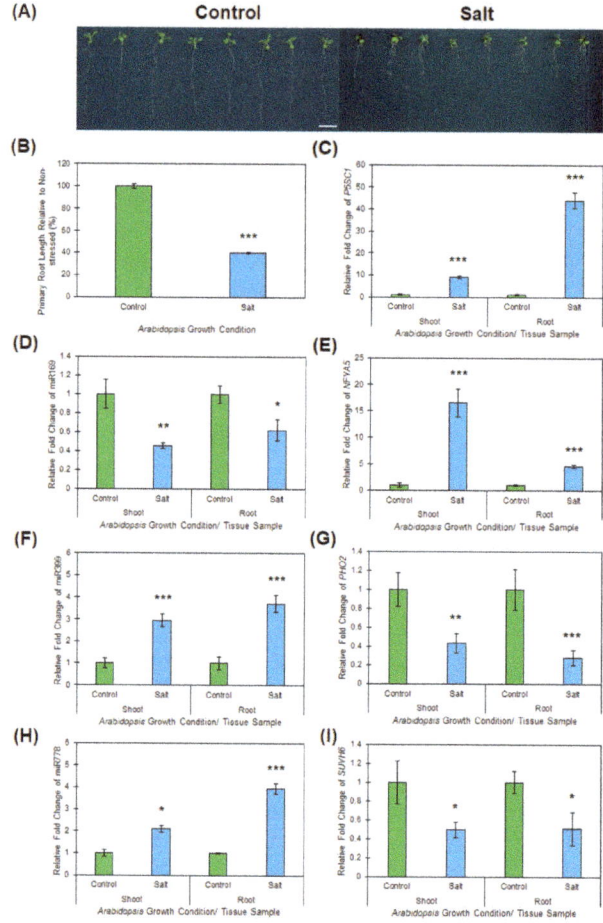

Figure 5. Phenotypic and molecular assessment of the root and shoot tissues of 15-day-old wild-type *Arabidopsis* plants post the 7-day salt stress treatment regime. (**A**) Root and shoot architecture of 15-day-old wild-type *Arabidopsis* seedlings post a 7-day salt stress treatment (right panel) during which the growth media plates were orientated for vertical growth. Scale bar = 1.0 cm. (**B**) Primary root length of 15-day-old *Arabidopsis* whole seedlings cultivated on vertically oriented media growth plates that contained either standard *Arabidopsis* growth media (non-stressed control) or growth media that had been supplemented with 150 mM sodium chloride (stress treatment). (**C**) RT-qPCR assessment of the expression of the stress induced gene, *P5CS1*, expression in 15-day-old *Arabidopsis* root and shoot material post 7-day salt stress treatment compared to the abundance of the *P5CS1* transcript in non-stress control plants of the same age. (**D,E**) RT-qPCR quantification of miR169 abundance (**D**) and *NFYA5* target gene expression (**E**) in salt stressed *Arabidopsis* root and shoot tissues. (**F,G**) RT-qPCR quantification of miR399 abundance (**F**) and *PHO2* target gene expression (**G**) in salt stressed *Arabidopsis* root and shoot tissues. (**H,I**) RT-qPCR quantification of miR778 abundance (**H**) and *SUVH6* target gene expression (**I**) in salt stressed *Arabidopsis* root and shoot tissues. (**B–I**) Colored columns represent the values obtained for non-stressed control plants (green colored columns) and the salt stressed samples (blue colored columns). Error bars represent the standard deviation of four biological replicates and each biological replicate consisted of a pool of six individual plants. The presence of an asterisk above a column represents a statistically significantly difference between the salt stress sample and the non-stressed controls (*p*-value: * < 0.05; ** < 0.005; *** < 0.001).

3. Discussion

In an attempt to provide sustainable food security into the future, it is essential that the complex, fundamental molecular networks that underpin the ability of a plant to maintain yield, particularly during extended periods of abiotic stress, are elucidated. This would provide the foundation for plant biology researchers to use a molecular approach to develop new plant lines that are readily able to adapt to, or mitigate the negative impacts that result from exposure to abiotic stress. With the miRNA class of sRNA demonstrated to be a key regulator of all aspects of plant development, as well as playing a central role in the ability of a plant to mount a defensive response against invading pathogens, or to mediate an adaptive response to abiotic stress, there currently remains a significant lack of resource datasets available for *Arabidopsis* to allow researchers to identify candidate miRNA expression modules for molecular modification as part of the future development of new plant lines that display adaptive phenotypes to abiotic stress. Towards this goal, here the genetic model plant species, *Arabidopsis thaliana*, was used to profile the miRNA landscape that potentially underpins, in part, the physiological and phenotypic responses of *Arabidopsis* to exposure to the abiotic stresses, heat, drought and salt stress. Most notably, sRNA sequencing revealed that the abundance of 121, 123 and 118 mature miRNA sRNAs was significantly (>2.0-fold) up- or down-regulated in response to heat, drought and salt stress treatment of *Arabidopsis* whole seedlings, respectively. The subsequent experimental validation of the miRNA abundance changes identified via high throughput sequencing by RT-qPCR, in combination with the additional use of RT-qPCR to document reciprocal trends in transcript abundance for each assessed miRNA target gene, was essential to confidently identify the miRNA expression modules responsive to each assessed stress.

In response to heat stress, a significant reduction to miR169 abundance (-3.2-fold) was observed in *Arabidopsis* whole seedlings (Figure 3A). Reduced miR169 levels have been reported previously for *Arabidopsis* post exposure to either drought stress or nitrogen starvation [38,39]. Further, [39] went on to demonstrate that reduced miR169 abundance in drought stressed *Arabidopsis* resulted in deregulated *NFYA5* expression, a target gene expression trend also observed here for heat stressed *Arabidopsis* (Figure 4A). The authors also revealed that overexpression of *NFYA5* in *Arabidopsis* resulted in these molecularly modified plant lines displaying reduced leaf water loss, due to reduced stomatal aperture, and drought stress tolerance [39]. Documentation of similar alterations to the miR169/*NFYA5* expression module in this study post heat stress treatment of *Arabidopsis* whole seedlings (Figure 3A, Figure 4A), to those previously reported for drought stressed *Arabidopsis* [39], suggests that these molecular changes are potentially in part driving the physical adaptation of *Arabidopsis* to both stresses, namely, alteration of stomatal aperture to promote water retention during exposure to such stress. It is also important to note here that the expression of several of the *MIR169* gene loci from which the miR169 precursor transcripts are transcribed were demonstrated to be induced in *Arabidopsis* by heat stress [40]. Induction of *MIR169* gene expression would be expected to result in elevated mature miR169 sRNA accumulation, and not reduced miR169 abundance, as observed here (Figure 3A). Curiously however, [40] did not report on whether *MIR169* gene expression induction actually resulted in elevated miR169 abundance in heat stressed *Arabidopsis* plants. The noted reduction to miR169 abundance reported here for heat stressed *Arabidopsis* plants, suggests that in *Arabidopsis*, heat stress represses *MIR169* gene expression, rather than promote transcription from these loci as reported by [40]. The opposing effect of heat stress exposure on miR169 accumulation in *Arabidopsis* reported here in Figure 3A, to that previously reported [40], could potentially be the result of differences in the application of the stress. Specifically, in [40], two-week-old *Arabidopsis* plants were transferred to moistened filter paper and exposed to the 40 °C heat stress treatment for a duration of either 3 or 6 h, whereas here, 8-day-old *Arabidopsis* seedlings were exposed to a prolonged 7-day heat stress treatment of elevated day/night (16/8 h) temperatures of 32 °C/28 °C. Nonetheless, the detection of elevated target gene (*NFYA5*) expression (Figure 4A), in accordance with the documented reduction in the abundance of the miR169 sRNA (Figure 3A), indicates that the alterations to the miR169/*NFYA5* expression module observed here in heat stressed *Arabidopsis*, are biologically relevant.

Elevated abundance has previously been reported for miRNAs, miR395 and miR396, post exposure of *Arabidopsis* to heat stress [41,42]. Similar abundance changes for the miR395 (Figure 3B) and miR396 (Figure 3C) sRNAs were observed here for heat stressed wild-type *Arabidopsis* whole seedlings. In addition, elevated miRNA abundance was further demonstrated to direct enhanced miRNA-directed target gene expression repression, with both the miR395 and miR396 target genes, *ATPS1* (Figure 4B) and *GRF7* (Figure 4C) respectively, determined to have reduced transcript abundance in heat stressed *Arabidopsis*. Taken together, comparison of the findings reported here, to those reported previously for miR395 and miR396 [41,42], strongly suggest that these two miRNAs are indeed heat stress responsive miRNAs, and further, that enhanced miR395- and miR396-directed expression repression of *ATPS1* and *GRF7*, respectively, potentially forms part of the adaptive response of *Arabidopsis* to elevated temperature.

Here, mannitol was used as an osmoticum to stimulate osmotic stress in *Arabidopsis* whole seedlings in an attempt to replicate drought stress conditions in a tightly controlled growth environment (i.e., sealed plant tissue culture plates). Post stress treatment, miR857 abundance was revealed to be reduced in *Arabidopsis* whole seedlings via both high throughput sequencing and RT-qPCR (Figure 2A, Figure 3D). The miR857 sRNA has previously been demonstrated to post-transcriptionally regulate the expression of *LAC7*, a laccase enzyme involved in mediating lignin deposition in the secondary xylem [43]. In addition to revealing reduced miR857 abundance, RT-qPCR showed that *LAC7* expression was elevated in drought stressed *Arabidopsis* whole seedlings (Figure 4D). Considering its documented role in secondary xylem development, the observed alterations to the miR857/*LAC7* expression module may potentially mediate an adaptive response to osmotic stress in *Arabidopsis*, potentially directing a change to tissue architecture in response to drought stress. Unlike miR857, miR156 and miR399 abundance was elevated by the mannitol-induced drought stress treatment (Figure 3E,F). In accordance, RT-qPCR showed that the transcript abundance of *SPL9* (Figure 4E) and *PHO2* (Figure 4F), the target genes of miR156 and miR399, respectively, was reduced in response to the elevated abundance of their targeting miRNA sRNAs. Interestingly, the miR156/*SPL9* expression module, together with the downstream gene, *DIHYDROFLAVONOL-4-REDUCTASE* (*DFR*; AT5G42800) have been demonstrated previously to play a role in anthocyanin metabolism [44], and Figure 1D shows that anthocyanin accumulation remained at its non-stressed levels in *Arabidopsis* plants cultivated on standard growth media supplemented with 200 mM mannitol, in spite of these plants displaying reductions to their fresh weight and rosette diameter, in addition to elevated expression of the stress induced gene, *P5CS1* (Figure 1E). Furthermore, in *Arabidopsis*, both of these miRNAs (miR156 and miR399) have been previously demonstrated to be responsive to mannitol-induced drought stress [41,44], findings that when taken together with those presented here in Figures 3 and 4, strongly suggest that the miR156/*SPL9* and miR399/*PHO2* expression modules are indeed responsive to mannitol-induced drought stress.

High throughput sRNA sequencing and RT-qPCR revealed miR169 abundance to be reduced by -4.8 and -2.5-fold respectively, post exposure to salt stress. This abundance change opposes that reported previously for the miR169 sRNA in rice and cotton [45,46], where miR169 accumulation was demonstrated to be induced by salt stress. However, the observed differences in miR169 abundance post salt stress exposure in rice [45], cotton [46] and *Arabidopsis* (Figure 3G), is most likely the result of unique *cis*-element landscapes of the promoter regions of *MIR169* loci across these three species [19]. In *Arabidopsis*, miR169 abundance has been previously demonstrated to be reduced by drought stress [40], conditions of limited phosphate [47], and nitrogen starvation [38]. The findings of these reports [38,40,47], together with those presented here, namely deregulated *NFYA5* target gene expression in *Arabidopsis* whole seedlings (Figure 4G), roots and shoots (Figure 5E), due to loss of miR169-directed *NFYA5* expression repression in these tissues, indicates that the miR169/*NFYA5* expression module potentially plays a central role in mediating the response of *Arabidopsis* to a range of abiotic stresses, potentially even forming a '*crosstalk junction*' to link the highly complicated molecular networks that are required to directed the physiological and phenotypic responses of *Arabidopsis* to abiotic stress.

Salt stress treatment was shown to enhance miR399 sRNA abundance in *Arabidopsis* whole seedlings (Figure 3H) as well as in root and shoot tissue (Figure 5F), a previously reported finding [41]. Furthermore, and using a molecular approach in *Arabidopsis*, [48] revealed *MIR399F* gene expression to be induced by salt stress and that *Arabidopsis* plants modified to constitutively overexpress the *MIR399F* gene were more tolerant to salt stress than unmodified wild-type plants. Given *PHO2* targeting by miR399, and the previously documented role for phosphate in modulating root system architecture alterations under salt stress conditions in *Arabidopsis*, the elevated abundance of miR399 shown here, in conjunction with the demonstrated reductions to the level of the *PHO2* target transcript (Figure 4H, Figure 5G), are consistent with the proposed role of the miR399/*PHO2* expression module in the complex phosphate-salt regulatory network in *Arabidopsis* tissues [49,50]. Like miR399, miR778 has previously been classed as a phosphate responsive miRNA in *Arabidopsis* [47,51,52]. Here we demonstrate that miR778 abundance is also elevated in response to salt stress in *Arabidopsis* whole seedlings (Figure 3I), roots and shoots (Figure 5H). Accordingly, via a RT-qPCR approach, we further revealed that elevated miR778 abundance resulted in enhanced expression repression of the miR778 target gene, *SUVH6*, in salt stressed *Arabidopsis* tissues (Figure 4I, Figure 5I). Interestingly, the miR778 target, SUVH6, is involved in directing methylation of the lysine 9 residue of histone H3 (H3K9 methylation), and further, *SUVH6* expression repression via the constitutive overexpression of the *MIR778* precursor transcript resulted in the modified *Arabidopsis* plants displaying moderately enhanced primary and lateral root growth, and elevated levels of free phosphate and anthocyanin accumulation in the aerial tissues of these plants when cultivated in a phosphate deficient growth environment [52]. These findings, together with the alterations to both the miR399/*PHO2* and miR778/*SUVH6* expression module reported here for salt stressed *Arabidopsis* (Figures 3 and 5), add further weight to the importance of phosphate-mediated responses in *Arabidopsis* tissues as part of the adaptive response of *Arabidopsis* to salt stress.

Altered miRNA abundance, and miRNA target gene expression, have been identified as key molecular responses to an array of abiotic stresses across an evolutionary diverse range of plant species [6,14,24,41,48,49]. Here we have specifically assessed alterations to the miRNA landscapes of heat, drought and salt stressed wild-type *Arabidopsis* whole seedlings and identified large miRNA cohorts responsive to each stress. Alteration to a select number of miRNA/target gene expression modules for the heat, drought and salt stress treatments were experimentally validated via an RT-qPCR approach. Considering that many abiotic responsive miRNAs have been demonstrated to play a conserved functional role across phylogenetically diverse plant species [14,24,25], it is envisaged that the dataset generated in this study forms a valuable resource for the wider plant biology research community; a resource that can be used as the starting point to identify the specific miRNA expression modules to be molecularly manipulated in plant species amenable to genetic modification as part of the future development of plant lines with an altered miRNA and/or miRNA target gene abundance that display a tolerance phenotype to either heat, drought or salt stress. Alternatively, for plant species that are not readily amenable to genetic modification, this dataset can additionally be used to identify the specific miRNA expression modules to be targeted for rapid high throughput screening (via RT-qPCR) across diverse germplasm of a specific species to select those genotypes that harbor natural alterations to the molecular profile of the miRNA expression module of interest.

4. Materials and Methods

4.1. Plant Material

The seeds of wild-type *Arabidopsis thaliana* (*Arabidopsis*), ecotype Columbia-0 (Col-0), were surface sterilized using chlorine gas and post sterilization, seeds were plated out onto standard *Arabidopsis* plant growth media (half strength Murashige and Skoog (MS) salts) and stratified in the dark at 4°C for 48 h. Post stratification, sealed plates containing the surface sterilized seeds were transferred to a temperature-controlled growth cabinet (A1000 Growth Chamber, Conviron® Australia) and cultivated for 8 days under a standard growth regime of 16 h light/8 h dark and a 22 °C/18 °C day/night temperature. Following this 8-day cultivation period, equal numbers of Col-0 seedlings were transferred under sterile conditions to either fresh standard *Arabidopsis* plant growth media (control treatment), or to plant growth media that had been supplemented with 200 millimolar (mM) mannitol (drought stress treatment) or 150 mM of sodium chloride (salt stress treatment). Post seedling transfer, the non-stressed control and the drought and salt stress treatment plates were returned to the growth cabinet and cultivated for an additional 7-day period under standard growth conditions. For the heat stress treatment, 8-day-old seedlings were also transferred under sterile conditions to standard growth media, however the 16/8 h day/night temperature was elevated to 32 °C/28 °C for the duration of the 7-day stress treatment period. At the end of the 7-day treatment period, all of the phenotypic and molecular assessments reported here were conducted on 15 day old *Arabidopsis* whole seedlings. For the tissue specific analyses reported in Figure 5, plants were treated exactly as outlined above, except for the 7-day treatment period, when 8 day old seedlings were transferred and cultivated on control and salt stress media plates that were orientated for vertical growth.

4.2. Phenotypic and Physiological Assessments

All phenotypic assessments reported here were conducted on 15 day old *Arabidopsis* seedlings. The performance of wild-type *Arabidopsis* plants exposed to each assessed stress is therefore presented relative to non-stressed control plants. More specifically, each phenotypic measurement collected for *Arabidopsis* seedlings exposed to each assessed abiotic stress regime is presented as a percentage of the corresponding measurement determined for non-stressed control seedlings cultivated under standard growth conditions for the duration of the 7-day stress treatment period. Rosette diameter and primary root length analysis was determined via assessment of photographic images using the ImageJ software. A standard 99:1 (v/v) methanol:HCl extraction protocol was used to extract anthocyanin from control and stress treated Col-0 plants. Post extraction, anthocyanin content was determined using a spectrophotometer (Thermo Scientific, Australia) at an absorbance wavelength of 535 nanometers (A_{535}) and using the 99:1 (v/v) methanol:HCl solution as the blank.

4.3. Total RNA Extraction and High Throughput Sequencing of the small RNA Fraction

Total RNA was isolated from four biological replicates (each biological replicate contained 6 individual plants) of 15 day old Col-0 whole seedlings cultivated under normal growth conditions for the duration of the experimental period, or post 7-days of heat, drought or salt stress treatment, using TRIzol™ Reagent (Invitrogen™) according to the manufacturer's instructions. The quality and quantity of the isolated total RNA was assessed using a Nanodrop spectrophotometer (NanoDrop® ND-1000, Thermo Scientific, Australia) and via standard electrophoresis on a 1.2% (w/v) ethidium bromide-stained agarose gel to allow for RNA visualization. Next, 5.0 micrograms (µg) of each of the four biological replicates for each treatment, were pooled together and diluted in RNase-free water to obtain a final preparation of 25 microliters (µL) of total RNA at a concentration of 800 nanograms (ng) per µL. Samples were shipped to the Australian Genome Research Facility (AGRF; Melbourne node, Australia) with the AGRF performing all subsequent preparatory steps prior to sequencing the small RNA fraction of each sample on an Illumina HiSeq 2500 platform.

4.4. Bioinformatic Assessment of the microRNA Landscape of Arabidopsis Whole Seedlings

Using the Qiagen CLC Genomics Workbench (11) software, next-generation sequencing adapter sequences were removed prior to performing sequence quality trimming to remove any sRNA reads that were either shorter than 15 nucleotides (nts), or longer than 35 nts in length. Additionally, parameters within the CLC Genomic Workbench were applied to remove any ambiguous nucleotides at either the 5′ or 3′ terminus of each sequencing read (i.e., the removal of any 'N' nucleotides on sequence ends), or to 'trim' low quality sequences using a modified '*Mott trimming*' algorithm. The remaining sequences that aligned perfectly (i.e., zero mismatches) to known *Arabidopsis* miRNAs listed in miRBase 22 were then annotated. The values determined for the; (1) raw read count of each detected miRNA sRNA across the four treatments (control, heat, drought and salt); (2) Log2 fold change in abundance for each miRNA sRNA per stress treatment, compared to the non-stressed control values; (3) total number of high quality raw reads per library; (4) total number of miRNA sRNA raw reads per library, and; (5) percentage of the total library size that the miRNA class of sRNA represents, is presented in Supplemental Table S1.

4.5. Quantitative Reverse Transcriptase Polymerase Chain Reaction Analyses

Quantitative reverse transcriptase polymerase chain reaction (RT-qPCR) assessment of miRNA sRNA and miRNA target gene transcript abundance was conducted on 4 biological replicates: the same four biological replicates that were pooled together to perform the high throughput sequencing analysis of the sRNA fraction of each sample. The synthesis of miRNA-specific complementary DNA (cDNA) was conducted using 200 ng of DNase I treated (New England BioLabs, Australia) total RNA as template and 1.0 unit (U) of ProtoScript® II Reverse Transcriptase (New England BioLabs, Australia) according to manufacturer's instructions. The cycling conditions for miRNA-specific cDNA synthesis were: 1 cycle at 16°C for 30 min; 60 cycles of 30°C for 30 s, 42°C for 30 s 50°C for 2 s, and; 1 cycle of 85°C for 5 min. To generate a global high molecular weight cDNA library for the quantification of miRNA target gene expression, 5.0 µg of total RNA was treated with 5.0 U of DNase I (New England BioLabs, Australia) according to manufacturer's instructions. Post DNase I treatment, the total RNA was purified using an RNeasy Mini Kit according to the manufacturer's instructions (Qiagen, Australia), and then 1.0 µg of this preparation was used as template for cDNA synthesis with 1.0 U of ProtoScript® II Reverse Transcriptase according to the manufacturer's instructions (New England Biolabs, Australia) along with 2.5 µM of oligo $dT_{(18)}$. All single stranded cDNA preparations were next diluted to 50 ng/µL in RNase-free water prior to RT-qPCR quantification of miRNA sRNA abundance or miRNA target gene expression. The GoTaq® qPCR Master Mix (Promega, Australia) was used as the fluorescent reagent for all performed RT-qPCRs, and all RT-qPCRs had the same cycling conditions of: 1 cycle of 95°C for 10 min, followed by 45 cycles of 95°C for 10 s and 60°C for 15 s. The abundance of each assessed miRNA sRNA and the expression of each examined miRNA target gene was determined using the $2^{-\Delta\Delta CT}$ method with the small nucleolar RNA, *snoR101*, and *UBIQUITIN10* (*UBI10*; *AT4G05320*) used as the respective internal controls to normalize the relative abundance of each assessed transcript. All DNA oligonucleotides used for either miRNA-specific cDNA synthesis or the quantification of miRNA target gene expression are provided in Supplemental Table S2. For the synthesis of miRNA-specific cDNA of a miRNA sRNA that belongs to a multimember family, and where multiple family members were detected via the high throughput sequencing approach, a miRNA family consensus sequence was determined and the primer designed to hybridize will all detected family members.

Supplementary Materials: The following are available online at http://www.mdpi.com/2223-7747/8/3/58/s1, Table S1: Raw miRNA reads and the Log2 fold change in abundance of each *Arabidopsis thaliana* miRNA sRNA detected via high throughput sRNA sequencing, Table S2: Sequences of the DNA oligonucleotides used in this study for the synthesis of miRNA-specific cDNAs and the RT-qPCR based quantification of miRNA abundance or miRNA target gene expression.

Author Contributions: C.P.L.G. and A.L.E. conceived and designed the research. J.L.P. and J.M.J.O. performed the experiments and analyzed the data. J.L.P., C.P.L.G., J.M.J.O. and A.L.E. authored the manuscript and, J.L.P., C.P.L.G., J.M.J.O. and A.L.E. have read and approved the final version of the manuscript.

Funding: This research received no external funding.

Acknowledgments: The authors would like to thank fellow members of the Centre for Plant Science for their guidance with plant growth care and RT-qPCR experiments.

Conflicts of Interest: The authors declare no conflict of interest.

References

1. Erb, K.H.; Haberl, H.; Plutzar, C. Dependency of global primary bioenergy crop potentials in 2050 on food systems, yields, biodiversity conservation and political stability. *Energy Policy* **2012**, *47*, 260–269. [CrossRef] [PubMed]
2. Godfray, H.C.J.; Beddington, J.R.; Crute, I.R.; Haddad, L.; Lawrence, D.; Muir, J.F.; Pretty, J.; Robinson, S.; Thomas, S.M.; Toulmin, C. Food security: The challenge of feeding 9 billion people. *Science* **2010**, *327*, 812–818. [CrossRef] [PubMed]
3. Tilman, D.; Balzar, C.; Hill, J.; Befort, B.L. Global food demand and the sustainable intensification of agriculture. *Proc. Natl. Acad. Sci. USA* **2011**, *108*, 20260–20264. [CrossRef] [PubMed]
4. Foley, J.A.; Ramankutty, N.; Brauman, K.A.; Cassidy, E.S.; Gerber, J.S.; Johnston, M.; Mueller, N.D.; O'Connell, C.; Ray, D.K.; West, P.C.; et al. Solutions for a cultivated planet. *Nature* **2011**, *478*, 337–342. [CrossRef] [PubMed]
5. Ramankutty, N.; Evan, A.T.; Monfreda, C.; Foley, J.A. Farming the plant: 1. Geographic distribution of global agricultural lands in the year 2000. *Glob. Biogeochem. Cycles* **2008**, *22*. [CrossRef]
6. Khraiwesh, B.; Zhu, J.K.; Zhu, J. Role of miRNAs and siRNAs in biotic and abiotic stress responses of plants. *Biochim. Biophys. Acta* **2012**, *1819*, 137–148. [CrossRef] [PubMed]
7. Sunkar, R.; Chinnusamy, V.; Zhu, J.; Zhu, J.K. Small RNAs as big players in plant abiotic stress responses and nutrient deprivation. *Trends Plant Sci.* **2007**, *12*, 301–309. [CrossRef] [PubMed]
8. Reinhart, B.J.; Weinstein, E.G.; Rhoades, M.W.; Bartel, B.; Bartel, D.P. MicroRNAs in plants. *Genes Dev.* **2002**, *16*, 1616–1626. [CrossRef] [PubMed]
9. Chen, X. A microRNA as a translational repressor of *APETALA2* in *Arabidopsis* flower development. *Science* **2004**, *303*, 2022–2025. [CrossRef] [PubMed]
10. Palatnik, J.F.; Allen, E.; Wu, X.; Schommer, C.; Schwab, R.; Carrington, J.C.; Weigel, D. Control of leaf morphogenesis by microRNAs. *Nature* **2003**, *425*, 257–263. [CrossRef] [PubMed]
11. Kasschau, K.D.; Xie, Z.; Allen, E.; Llave, C.; Chapman, E.J.; Krizan, K.A.; Carrington, J.C. P1/HC-Pro, a viral suppressor of RNA silencing, interferes with *Arabidopsis* development and miRNA function. *Dev. Cell* **2003**, *4*, 205–217. [CrossRef]
12. Li, Y.; Zhang, Q.; Zhang, J.; Wu, L.; Qi, Y.; Zhou, J.M. Identification of microRNAs involved in pathogen-associated molecular pattern-triggered plant innate immunity. *Plant Physiol.* **2010**, *152*, 2222–2231. [CrossRef] [PubMed]
13. Liu, H.H.; Tian, X.; Li, Y.J.; Wu, C.A.; Zheng, C.C. Microarray-based analysis of stress-regulated microRNAs in *Arabidopsis thaliana*. *RNA* **2008**, *14*, 836–843. [CrossRef] [PubMed]
14. Sunkar, R.; Zhu, J.K. Novel and stress-regulated microRNAs and other small RNAs from *Arabidopsis*. *Plant Cell* **2004**, *16*, 2001–2019. [CrossRef] [PubMed]
15. Bartel, D.P. MicroRNAs: Genomics, biogenesis, mechanism, and function. *Cell* **2004**, *116*, 281–297. [CrossRef]
16. Kurihara, Y.; Takashi, Y.; Watanabe, Y. The interaction between DCL1 and HYL1 is important for efficient and precise processing of pri-miRNA in plant microRNA biogenesis. *RNA* **2006**, *12*, 206–212. [CrossRef] [PubMed]
17. Vazquez, F.; Gasciolli, V.; Crété, P.; Vaucheret, H. The nuclear dsRNA binding protein HYL1 is required for microRNA accumulation and plant development, but not posttranscriptional transgene silencing. *Curr. Biol.* **2004**, *14*, 346–351. [CrossRef] [PubMed]
18. Megraw, M.; Baev, V.; Rusinov, V.; Jensen, S.T.; Kalantidis, K.; Hatzigeorgiou, A.G. MicroRNA promoter element discovery in *Arabidopsis*. *RNA* **2006**, *21*, 1612–1619. [CrossRef] [PubMed]
19. Pegler, J.L.; Grof, C.P.L.; Eamens, A.L. Profiling of the differential abundance of drought and salt stress-responsive microRNAs across grass crop and genetic model plant species. *Agronomy* **2018**, *8*, 118. [CrossRef]
20. Zhao, X.; Li, L. Comparative analysis of microRNA promoters in *Arabidopsis* and rice. *Genom. Proteom. Bioinform.* **2013**, *11*, 56–60. [CrossRef] [PubMed]
21. Baulcombe, D.C. RNA silencing in plants. *Nature* **2004**, *431*, 356–363. [CrossRef] [PubMed]
22. Baumberger, N.; Baulcombe, D.C. *Arabidopsis* ARGONAUTE1 is an RNA Slicer that selectively recruits microRNAs and short interfering RNAs. *Proc. Natl. Acad. Sci. USA* **2005**, *102*, 11928–11933. [CrossRef] [PubMed]

23. Reis, R.S.; Hart-Smith, G.; Eamens, A.L.; Wilkins, M.R.; Waterhouse, P.M. Gene regulation by translational inhibition is determined by Dicer partnering proteins. *Nat. Plants* **2015**, *1*, 14027. [CrossRef] [PubMed]
24. Noman, A.; Fahad, S.; Aqeel, M.; Ali, U.; Amanullah; Anwar, S.; Baloch, S.K.; Zainab, M. miRNAs: Major modulators for crop growth and development under abiotic stresses. *Biotechnol. Lett.* **2017**, *39*, 685–700. [CrossRef] [PubMed]
25. Shriram, V.; Kumar, V.; Devarumath, R.M.; Khare, T.S.; Wani, S.H. MicroRNAs as potential targets for abiotic stress tolerance in plants. *Front. Plant Sci.* **2016**, *14*, 817. [CrossRef] [PubMed]
26. Zhang, B. MicroRNA: A new target for improving plant tolerance to abiotic stress. *J. Exp. Bot.* **2015**, *66*, 1749–1761. [CrossRef] [PubMed]
27. Zhang, B.; Unver, T. A critical and speculative review on microRNA technology in crop improvement: Current challenges and future directions. *Plant Sci.* **2018**, *274*, 193–200. [CrossRef] [PubMed]
28. Liu, H.; Able, A.J.; Able, J.A. SMARTER De-Stressed Cereal Breeding. *Trends Plant Sci.* **2016**, *21*, 909–925. [CrossRef] [PubMed]
29. Chalker-Scott, L. Environmental significance of anthocyanins in plant stress responses. *Photochem. Photobiol.* **1999**, *70*, 1–9. [CrossRef]
30. Lotkowska, M.E.; Tohge, T.; Fernie, A.R.; Xue, G.P.; Balazadeh, S.; Mueller-Roeber, B. The *Arabidopsis* transcription factor MYB112 promotes anthocyanin formation during salinity and under high light stress. *Plant Physiol.* **2015**, *169*, 1862–1880. [CrossRef] [PubMed]
31. Gray, W.M.; Östin, A.; Sandberg, G.; Romano, C.P.; Estelle, M. High temperature promotes auxin-mediated hypocotyl elongation in *Arabidopsis*. *Proc. Natl. Acad. Sci. USA* **1998**, *95*, 7197–7202. [CrossRef] [PubMed]
32. Koini, M.A.; Alvey, L.; Allen, T.; Tilley, C.A.; Harberd, N.P.; Whitelam, G.C.; Franklin, K.A. High temperature-mediated adaptations in plant architecture require the bHLH transcription factor PIF4. *Curr. Biol.* **2009**, *19*, 408–413. [CrossRef] [PubMed]
33. Lv, W.T.; Lin, B.; Zhang, M.; Hua, X.J. Proline accumulation is inhibitory to *Arabidopsis* seedlings during heat stress. *Plant Physiol.* **2011**, *156*, 1921–1933. [CrossRef] [PubMed]
34. Urano, K.; Maruyama, K.; Ogata, Y.; Morishita, Y.; Takeda, M.; Sakurai, N.; Suzuki, H.; Saito, K.; Shibata, D.; Kobayashi, M.; et al. Characterization of the ABA-regulated global responses to dehydration in *Arabidopsis* by metabolomics. *Plant J.* **2009**, *56*, 106–1078. [CrossRef] [PubMed]
35. Strizhov, N.; Abrahám, E.; Okrész, L.; Blickling, S.; Zilberstein, A.; Schell, J.; Koncz, C.; Szabados, L. Differential expression of two P5CS genes controlling proline accumulation during salt-stress requires ABA and is regulated by ABA1, ABI1 and AXR2 in *Arabidopsis*. *Plant J.* **1997**, *12*, 557–569. [CrossRef] [PubMed]
36. Mestdagh, P.; Feys, T.; Bernard, N.; Guenther, S.; Chen, C.; Speleman, F.; Vandesompele, J. High-throughput stem-loop RT-qPCR miRNA expression profiling using minute amounts of input RNA. *Nucleic Acids Res.* **2008**, *36*, e143. [CrossRef] [PubMed]
37. Potters, G.; Pasternak, T.P.; Guisez, Y.; Palme, K.J.; Jansen, M.A.K. Stress-induced morphogenic responses: Growing out of trouble? *Trends Plant Sci.* **2007**, *12*, 98–105. [CrossRef] [PubMed]
38. Zhao, M.; Ding, H.; Zhu, J.K.; Zhang, F.; Li, W.X. Involvement of miR169 in the nitrogen-starvation resposnses in *Arabidopsis*. *New Phytol.* **2011**, *190*, 906–915. [CrossRef] [PubMed]
39. Li, W.X.; Oono, Y.; Zhu, J.; He, X.J.; Wu, J.M.; Iida, K.; Lu, X.Y.; Cui, X.; Jin, H.; Zhu, J.K. The *Arabidopsis* NFYA5 transcription factor is regulated transcriptionally and posttranscriptionally to promote drought resistance. *Plant Cell* **2008**, *20*, 2238–2251. [CrossRef] [PubMed]
40. Li, Y.; Fu, Y.; Ji, L.; Wu, C.; Zheng, C. Characterization and expression analysis of the *Arabidopsis* miR169 family. *Plant Sci.* **2010**, *178*, 271–280. [CrossRef]
41. Barciszewska-Pacak, M.; Milanowska, K.; Knop, K.; Bielewicz, D.; Nuc, P.; Plewka, P.; Pacak, A.M.; Vazquez, F.; Karlowski, W.; Jarmolowski, A.; et al. *Arabidopsis* microRNA expression regulation in a wide range of abiotic stress responses. *Front. Plant Sci.* **2015**, *6*, 410. [CrossRef] [PubMed]
42. Zhong, S.H.; Liu, J.Z.; Jin, H.; Lin, L.; Li, Q.; Chen, Y.; Yuan, Y.X.; Wang, Z.Y.; Huang, H.; Qi, Y.J.; et al. Warm temperatures induce transgenerational epigenetic release of RNA silencing by inhibiting siRNA biogenesis in *Arabidopsis*. *Proc. Natl. Acad. Sci. USA* **2013**, *110*, 9171–9176. [CrossRef] [PubMed]
43. Zhao, Y.; Lin, S.; Qiu, Z.; Cao, D.; Wen, J.; Deng, X.; Wang, X.; Lin, J.; Li, X. MicroRNA857 is involved in the regulation of secondary growth of vascular tissues in *Arabidopsis*. *Plant Physiol.* **2015**, *169*, 2539–2552. [CrossRef] [PubMed]

44. Cui, L.G.; Shan, J.X.; Shi, M.; Gao, J.P.; Lin, H.X. The *miR156-SPL9-DFR* pathway coordinates the relationship between development and abiotic stress tolerance in plants. *Plant J.* **2014**, *80*, 1108–1117. [CrossRef] [PubMed]
45. Zhao, B.; Ge, L.; Liang, R.; Li, W.; Ruan, K.; Lin, H.; Jin, Y. Members of the miR-169 family are induced by high salinity and transiently inhibit the NF-YA transcription factor. *BMC Mol. Biol.* **2009**, *10*, 29. [CrossRef] [PubMed]
46. Yin, Z.; Li, Y.; Yu, J.; Liu, Y.; Li, C.; Han, X.; Shen, F. Difference in miRNA expression profiles between two cotton cultivars with distinct salt sensitivity. *Mol. Biol. Rep.* **2012**, *39*, 4961–4970. [CrossRef] [PubMed]
47. Hsieh, L.C.; Lin, S.I.; Shih, A.C.; Chen, J.W.; Lin, W.Y.; Tseng, C.Y.; Li, W.H.; Chiou, T.J. Uncovering small RNA-mediated responses to phosphate deficiency in *Arabidopsis* by deep sequencing. *Plant Physiol.* **2009**, *151*, 2120–2132. [CrossRef] [PubMed]
48. Baek, D.; Chun, H.J.; Kang, S.; Shin, G.; Park, S.J.; Hong, H.; Kim, C.; Kim, D.H.; Lee, S.Y.; Kim, M.C.; et al. A role for *Arabidopsis miR399f* in salt, drought, and ABA signaling. *Mol. Cell* **2015**, *39*, 111–118.
49. Bari, R.; Pant, B.D.; Stitt, M.; Scheible, W.-R. PHO$_2$, microRNA399, and PHR1 define a phosphate-signaling pathway in plants. *Plant Physiol.* **2006**, *141*, 988–999. [CrossRef] [PubMed]
50. Kawa, D.; Julkowska, M.M.; Sommerfield, H.M.; Ter Horst, A.; Haring, M.A.; Testerink, C. Phosphate-dependent root system architecture responses to salt stress. *Plant Physiol.* **2016**, *172*, 690–706. [CrossRef] [PubMed]
51. Pant, B.D.; Musialak-Lange, M.; Nuc, P.; May, P.; Buhtz, A.; Kehr, J.; Walther, D.; Scheible, W.R. Identification of nutrient-responsive *Arabidopsis* and rapeseed microRNAs by comprehensive real-time polymerase chain reaction profiling and small RNA sequencing. *Plant Physiol.* **2009**, *150*, 1541–1555. [CrossRef] [PubMed]
52. Wang, L.; Zeng, J.H.Q.; Song, J.; Feng, S.J.; Yang, Z.M. miRNA778 and *SUVH6* are involved in phosphate homeostasis in *Arabidopsis*. *Plant Sci.* **2015**, *238*, 273–285. [CrossRef] [PubMed]

© 2019 by the authors. Licensee MDPI, Basel, Switzerland. This article is an open access article distributed under the terms and conditions of the Creative Commons Attribution (CC BY) license (http://creativecommons.org/licenses/by/4.0/).

Article

Functional Characterization of microRNA171 Family in Tomato

Michael Kravchik, Ran Stav, Eduard Belausov and Tzahi Arazi *

Institute of Plant Sciences, Agricultural Research Organization, Volcani Center, P.O. Box 6, Bet Dagan 50250, Israel; michael.kravchik@mail.huji.ac.il (M.K.); ranstav@volcani.agri.gov.il (R.S.); eddy@volcani.agri.gov.il (E.B.)
* Correspondence: tarazi@agri.gov.il; Tel.: +972-3-968-34-98

Received: 9 December 2018; Accepted: 28 December 2018; Published: 4 January 2019

Abstract: Deeply conserved plant microRNAs (miRNAs) function as pivotal regulators of development. Nevertheless, in the model crop *Solanum lycopersicum* (tomato) several conserved miRNAs are still poorly annotated and knowledge about their functions is lacking. Here, the tomato miR171 family was functionally analyzed. We found that the tomato genome contains at least 11 *SlMIR171* genes that are differentially expressed along tomato development. Downregulation of sly-miR171 in tomato was successfully achieved by transgenic expression of a short tandem target mimic construct (STTM171). Consequently, sly-miR171-targeted mRNAs were upregulated in the silenced plants. Target upregulation was associated with irregular compound leaf development and an increase in the number of axillary branches. A prominent phenotype of *STTM171* expressing plants was their male sterility due to a production of a low number of malformed and nonviable pollen. We showed that sly-miR171 was expressed in anthers along microsporogenesis and significantly silenced upon *STTM171* expression. Sly-miR171-silenced anthers showed delayed tapetum ontogenesis and reduced callose deposition around the tetrads, both of which together or separately can impair pollen development. Collectively, our results show that sly-miR171 is involved in the regulation of anther development as well as shoot branching and compound leaf morphogenesis.

Keywords: miR171; pollen; STTM; tapetum; callose; tomato

---

1. Introduction

Plant microRNAs (miRNAs) constitute a major class of endogenous small RNAs and trigger the sequence-specific post-transcriptional repression of one to several target mRNAs with high sequence complementarity. The analysis of miRNAs from various land plants species indicated the presence of at least eight deeply conserved miRNA families in all embryophytes [1]. Studies of these miRNAs suggest that most of them act as master regulators of development by negative regulation of the expression of transcription factors that function in critical developmental processes [2,3].

The miR171 family is deeply conserved and exists in all major land plant groups, including bryophytes, one of the oldest groups of land plants [4]. Known plant genomes contain variable number of *MIR171* genes: from only two in *Citrus sinensis* to staggering 21 in *Glycine max* (miRBase release 22). Members of a miR171 family contain one or more nucleotide changes similar to members from other miRNA families, but unlike other conserved miRNAs they may be offset by three nucleotides relative to each other [5]. This atypical sequence offset may result in different target specificities for certain miR171 members [6]. Hitherto, miR171 members have been demonstrated to guide the cleavage of mRNAs coding for GRAS domain SCARECROW-like transcription factors that belong to the HAIRY MERISTEM (HAM) or NODULATION SIGNALING PATHWAY (NSP) clades [6–8].

In *Nicotiana benthamiana*, spatial characterization of miR171 expression by in-situ hybridization has shown that it is expressed in a wide variety of tissues including the shoot apical meristem

(SAM), leaf primordia, anthers and ovaries, thus hinting on its involvement in their development [9]. In *A. thaliana* (Arabidobsis), ath-miR171b expression was shown to oscillate during the diurnal cycle suggesting a potential role for light in its accumulation [10]. In *Medicago truncatula*, miR171h expression is upregulated in roots during their colonization by arbuscular mycorrhizal fungi or in response to lipochito-oligosaccharides that are released during fungi pre-symbiotic growth, suggesting that miR171h functions to prevent over-colonization of roots by arbuscular mycorrhizal fungi [8]. In rice, it was demonstrated that reduction of osa-miR171b contributes to Rice stripe virus symptoms whereas osa-miR171b overexpression caused opposite effects, suggesting that expression of miR171-targeted mRNAs may facilitate viral infection [11]. In addition, miR171 has been suggested to be involved in the regulation of abiotic stresses based on its upregulation in Arabidopsis seedlings grown under high salinity, cold, and drought conditions [12].

Target mimic is a non-cleavable miRNA complementary sequence embedded within a longer endogenous or artificial RNA. In contrast to overexpression of a miRNA, which will silence its cognate mRNA targets and hence faithfully report on their functions, the target mimic acts as an "miRNA sponge" that titer out complementary miRNAs and hence is suitable for their functional characterization [13]. Furthermore, ectopic expression of miRNA or its cleavage-resistant mRNA target may lead to deceptive identification of miRNA function, due to incorrect spatio-temporal expression. Overexpression of miR171 resulted in various opposite phenotypes such as dwarfed barley with less tillers and taller rice with more tillers, and even silencing of miR171 can lead to different phenotypes between various species [14]. Ath-miR171a downregulation by single target mimic configurations resulted in a range of phenotypes consistent with ath-miR171a involvement in multiple developmental processes. Common phenotypes included closed buds and reduced pollination due to altered sepal development that bent the carpels, and pale green leaves due to reduced chlorophyll accumulation [15,16]. In addition, ath-miR171a-depleted Arabidopsis had larger rosette leaves, a larger root system during the growth in soil and modified leaf angle under limited light conditions [16]. Nevertheless, a single target mimic sequence can only silence complementary miR171 members, but will not effectively bind miR171 members with sequence offset, thus limiting the efficacy of such configuration for functional analysis of miR171. Recently, expression of two target mimic sequences in a single transcript via short tandem target mimic (STTM) configuration was shown to efficiently induce complementary miRNAs degradation in Arabidopsis and tomato [13,17]. This approach, which may be applied to silence two different miRNAs in parallel, is thus highly suitable for the in planta functional characterization of miR171. Indeed, this approach has been successfully applied in rice. Rice STTM171 plants were semidwarf and had semienclosed panicles and drooping flag leaves. These unique phenotypes suggest divergent functions for miR171 in dicots and monocots [14,18].

Tomato is the number one non-starchy vegetable consumed worldwide and also serves as a primary model for fruit development and ripening. Nevertheless, at present, tomato miRNAs are poorly annotated and the functions of most remain elusive. To date, six miR171 members, sly-miR171a-f, were cloned from tomato (miRBase, release 22) [19]. Previously, we have demonstrated that sly-miR171a and sly-miR171b guide the cleavage of *SlHAM* and *SlHAM2* and sly-miR171b also guides the cleavage of the tomato *NSP2* homolog *SlNSP2L*. Sly-miR171a and sly-miR171b overexpression resulted in over-proliferation of meristematic cells in the periphery of meristems and in the organogenic compound leaf rachis, suggesting that sly-miR171-targeted *SlHAMs* function in meristem maintenance and compound leaf morphogenesis [6]. As part of our continuous effort to unravel the identity and roles of tomato miRNAs, in the current study, we re-annotated the tomato miR171 family and utilized the STTM approach to silence abundant members in the family and investigate their functions. Our results revealed the presence of a much more complex miR171 family than previously documented in tomato and its necessity for vegetative, anther, and pollen development.

2. Results and Discussion

2.1. The Tomato miR171 Family of miRNAs

Six sly-miR171 members, sly-miR171a-f were previously cloned from tomato (miRBase, release 22). To identify additional sly-miR171 members, the small RNAs deposited in the Tomato Functional Genomics Database (TFGD; http://ted.bti.cornell.edu/cgi-bin/TFGD/sRNA/sRNA.cgi) and in-house tomato cv. M82 small RNA data [20] were queried with known miR171 sequences. This search detected eleven differentially abundant 21-nucleotide putative sly-miR171 sequences, including the previously cloned sly-miR171a, b, e, f and a 1-nucleotide longer version of sly-miR171d (Figure 1A). Sequence alignment revealed that identified sly-miR171 sequences can be divided into two groups, which are offset by three nucleotides relative to each other, similar to the Arabidopsis miR171 founder sequences ath-miR171a (group A) and ath-miR171c (group B). By mapping the putative sly-miR171 sequences to the tomato genome followed by alignment of the cloned small RNAs to their predicted pre-miRNAs sequences, we identified the miRNA stars (miRNA*) for all, indicating that they are authentic miRNAs (Table S1). In addition, this alignment revealed that one pre-miRNA (*SlMIR171a,b*) encodes for both sly-miR171a and sly-miR171b and their respective miRNA* strands (Figure 1B; Table S1). Moreover, four additional newly identified miR171 members (iso-sly-miR171a.1, iso-sly-miR171a.2, iso-sly-miR171b, iso-sly-miR171d) were found to be encoded by an identical precursor as that of other miR171 members and overlapped them in sequence (Figure 1B and Table S1), suggesting that they represent iso-miRNAs [21]. Querying the TFGD with these sequences confirmed their expression in tissues other than seedlings supporting their functionality as miRNAs (Figure 1C). It is noteworthy that iso-sly-miR171d miRNA* strand was previously annotated as sly-miR171c (miRBase, release 22). This analysis indicates that the tomato miR171 family is much more complex than previously thought, but it is of medium size compared to other miR171 families such as in *Glycine max* that contains up to 21 members (miRBase, release 22). Nevertheless, such complexity may hint on redundancy and specialization among different sly-miR171 members. Sly-miR171a and sly-miR171b, which represent group A and B sly-miR171 members, respectively, guide the cleavage of *SlHAM* and *SlHAM2*. Sly-miR171b, but not sly-miR171a, also guides the cleavage of the tomato *SlNSP2L* [6]. Prediction of mRNA targets and mining the published tomato degradome data [22] did not reveal strong evidence for additional sly-miR171-guided mRNA cleavage (Supplement 1).

To identify sly-miR171 sites of activity in tomato, analysis of public small RNAseq data at TFGD was performed. This analysis revealed that sly-miR171a, sly-miR171b, iso-sly-miR171d, and sly-miR171e are the most abundant members in the family, but the expression of each is prominent in a distinct tissue or developmental stage: sly-miR171a—leaves, floral buds and anthesis flowers, sly-miR171b—immature green fruit, iso-sly-miR171d—anthesis flowers and sly-miR171e—leaves, mature and ripening fruit (Figure 1C). Hence, the sly-miR171 family functions in vegetative as well as reproductive tissues with possible functional diversification between different members.

2.2. Knockdown of Sly-miR171 Activity Using the STTM Approach

Previously it was demonstrated that STTM configuration, which contains two target mimic sequences separated by a spacer, is very effective in counteracting the activity of several miRNAs in Arabidopsis and tomato [13,17]. Therefore, a similar STTM configuration was chosen to knockdown sly-miR171 family activity and uncover its importance for tomato development. Since group A and group B sly-miR171 members are offset by three nucleotides relative to each other, the STTM171 fragment was designed to comprise two different target mimic sequences, each of which have complementarity suitable to bind most of group A or group B sly-miR171 members, especially the most abundant sly-miR171 members (Figures 2A and S1). The STTM171 fragment was cloned downstream of the CaMV *35S* promoter (*35S:STTM171*) and then transformed into tomato cv. M82. Seventeen independent transgenic *35S:STTM171* plants were regenerated and screened by northern blot for reduced sly-miR171a and sly-miR171b levels. This analysis identified three T0 primary transformants

(9, 17, and 19) with significantly reduced sly-miR171 levels compared to control transgenic *35S:GFP* plants (Figure 2B). Plants *35S:STTM171-9* and *35S:STTM171-19*, which accumulated ~27% and ~21% of the total sly-miR171 levels, respectively, produced only few completely seedless fruits. Compared to the control plants, the *35S:STTM171-17* T0 plant, which accumulated ~30% of total sly-miR171 levels, produced smaller fruit in size and number, most of which were seedless and few contained a small number of seeds (Figure S2A,B). The smaller fruit size of *35S:STTM171-17* plants was probably a secondary effect which emerged due to the reduction in seed number [23]. Transformation efforts to produce additional independent fertile transgenic plants with significantly reduced sly-miR171 levels were not successful (data not shown). The sterility of *35S:STTM171-9* and *35S:STTM171-19* plants prevented further analysis of their progeny. Thus, further characterization of STTM171 plants was performed on *35S:STTM171-17* T2 and T3 progeny. Quantitation of sly-miR171-targeted *SlHAM*, *SlHAM2*, and *SlNSP2* transcripts in young leaves of the *35S:STTM171-17* T2 plants revealed significant ~2–2.5 fold upregulation in all (Figure 2C) indicating that both sly-miR171a and sly-miR171b activities were attenuated in these leaves by *35S:STTM171* expression. Indeed, quantitation of sly-miR171a-b in these leaves confirmed their reduced accumulation (Figure 2D).

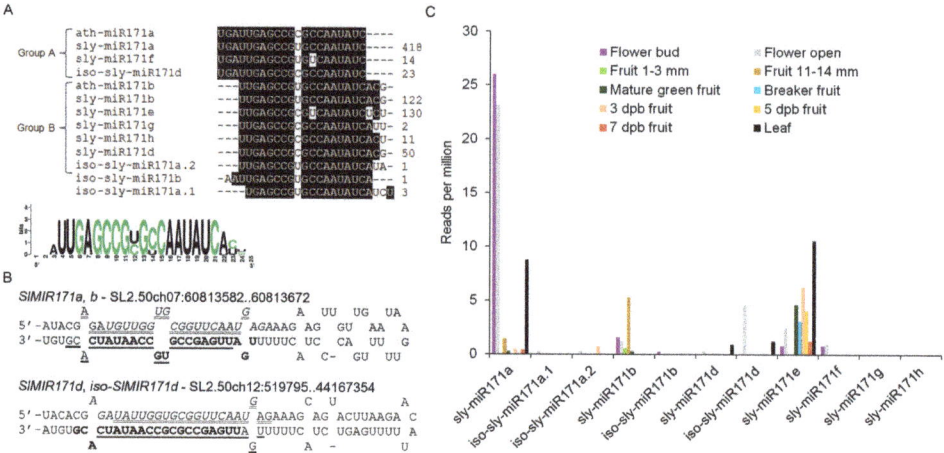

Figure 1. The tomato miR171 family. (**A**) Nucleotide sequence alignment of Arabidopsis (ath-miR171) and tomato (sly-miR171) miR171 members. Relative abundance in seedlings is indicated for each on the right. (**B**) Examples of sly-miR171 precursors that encode two sly-miR171 isoforms. The sequence of *SlMIR171a,b* (SL2.50ch07:60813582..60813672) and *SlMIR171d, iso-SlMIR171d* (SL2.50ch12:519795..519887) stem and loops. The sequences of sly-miR171a/d, sly-miR171a*/d*, sly-miR171b/ sly-miRiso-d and sly-miR171b*/iso-d* are bold-face, italicized, underlined and double-underlined, respectively. (**C**) Accumulation of sly-miR171 members in flower and fruit tissues of tomato cv. Microtome (Flower and fruit) and Heinz (leaf) based on small RNA-seq data deposited in the TFGD database. Dpb—days post breaker.

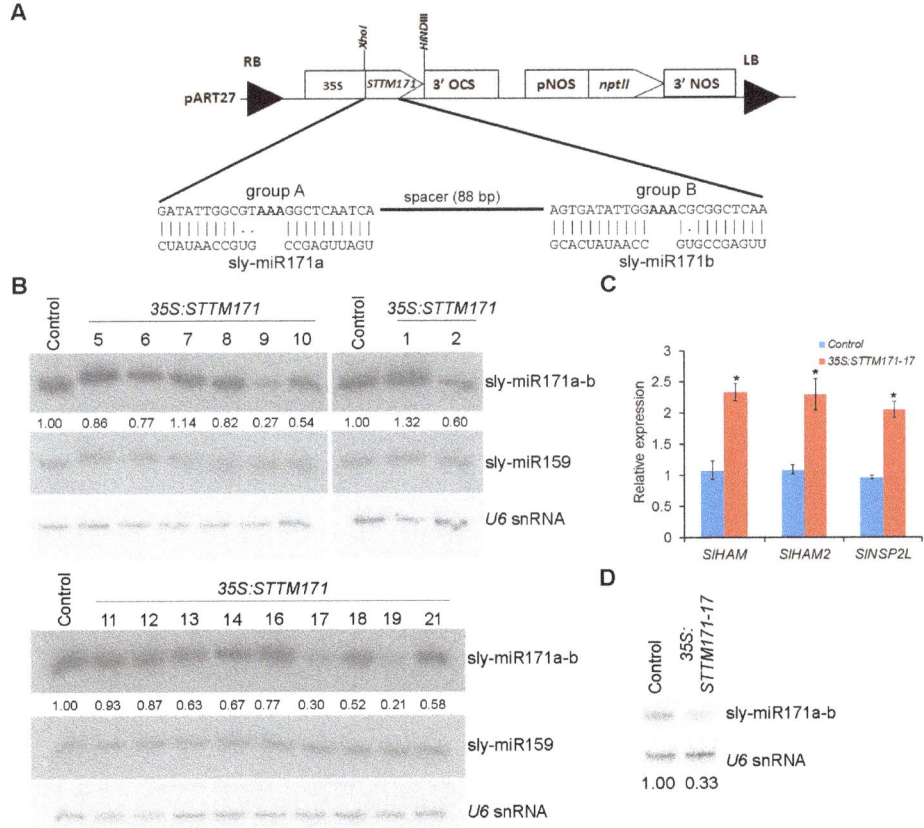

Figure 2. Generation of transgenic STTM171 tomato with reduced sly-miR171 levels. (**A**) A scheme of the Short Tandem Target Mimic construct used for tomato M82 transformation. The Watson–Crick pairings between group A and B target mimic sites and sly-miR171 representative members are shown in the expanded region. (**B**) RNA gel blot analysis of sly-miR171 levels in indicated transgenic T0 plants. Total RNA (5 µg) from leaves was probed by sly-miR171a (sly-miR171a-b), sly-miR159 and *U6* antisense probes. Sly-miR171 expression levels were determined after normalization to sly-miR159 and *U6* snRNA by geometric averaging and are indicated below. (**C**) RT-qPCR analysis of sly-miR171 target transcripts in RNA from young leaves of one-month old T2 35S:STTM171-17 plants. *TIP41* expression values were used for normalization. Error bars indicate ± SD of three biological replicates, each measured in triplicate. Asterisks indicate significant difference relative to *35:GFP* control plants (Tukey–Kramer multiple comparison test; $p < 0.01$). (**D**) RNA gel blot analysis of sly-miR171 in 5 µg total RNA from the samples analyzed in C. The blots were probed with sly-miR171a (sly-miR171a-b) antisense probe. Sly-miR171 expression levels were determined after normalization to *U6* snRNA and are indicated below.

2.3. Sly-miR171 Silencing Affected Compound Leaf Morphogenesis and Increased Branching

During vegetative development the compound leaves of *35S:STTM171-17* plants developed primary leaflets that frequently had a distorted growth angle, were larger and their lobes were deeper than that of the control, implicating sly-miR171 in compound leaf morphogenesis (Figure 3A,B). In addition, compared to control plants, the number of axillary shoots was significantly higher in the *35S:STTM171* plants (Figure 3C). This phenotype is consistent with the increased lateral branch number observed in transgenic tomato plants that ectopically expressed *SlHAM2/SlGRAS24*, which was upregulated in *35S:STTM171* leaves (Figure 2C), and with transgenic Arabidopsis that ectopically expressed the ath-miR171c-ressitant versions of *SCL6-II/HAM1*, *SCL6-III/HAM2*, and *SCL6-IV/HAM4* [24,25]. Both *SlHAM* and *SlHAM2* are abundant in vegetative and reproductive meristems and function in their maintenance [6]. Thus, sly-miR171 may suppress lateral branching by the negative regulation of the expression of *SlHAM2* and apparently also *SlHAM* in axillary meristems.

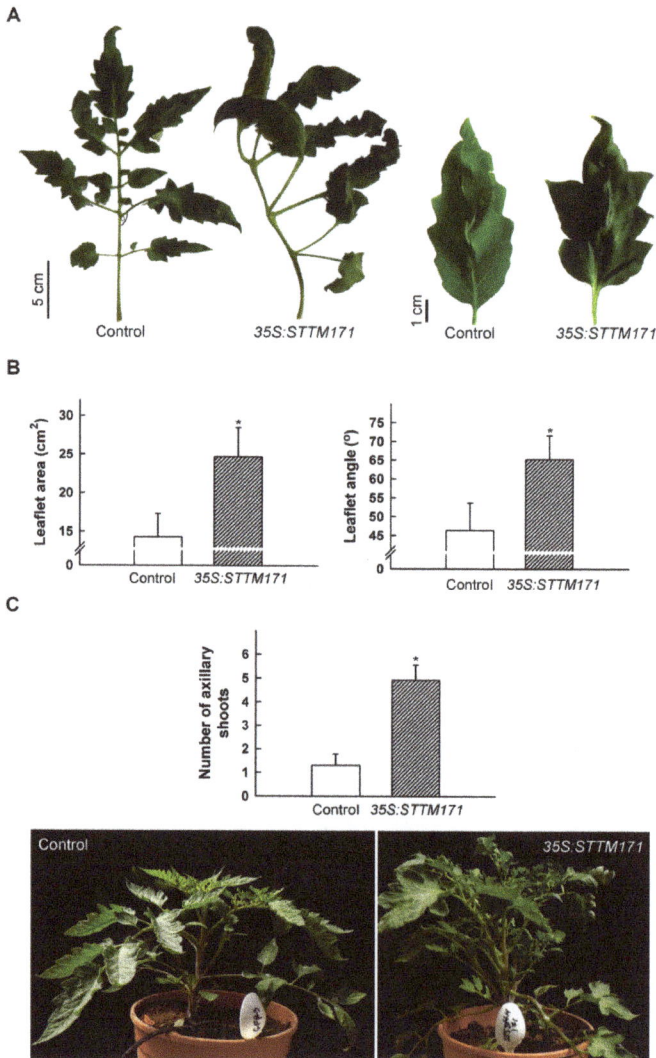

Figure 3. Vegetative phenotypes of *35S:STTM171* plants. (**A**) Photograph of representative fifth leaf and terminal leaflet from 45 DAG plants of indicated genotypes. (**B**) Quantitation of primary leaflet area and petiolule angle (indicated in (A)) in leaves (n = 26) similar to those shown in (A). (**C**) Quantitation of the number of axillary shoots on the main stem ($\geq$0.5 cm) at eight leaf stage plants (n = 13). Error bars indicate ±SD. Asterisks indicate significant difference as determined by Student's *t*-test ($p \leq 0.001$). Representative plant of each genotype is shown below. Pot diameter = 18.8 cm.

2.4. Sly-miR171 Silencing Affected Pollen Morphology and Production

Despite the normal number and morphology of their flowers, *35S:STTM171* plants set only few fruits that were mostly seedless. Whereas manual pollination of *35S:STTM171* flowers with wild-type pollen rarely succeeded, the reciprocal pollination completely failed to induce fruit set, indicating reduced male fertility of the *35S:STTM171* flowers. A similar male sterile phenotype was observed in the transgenic F1 progeny from the cross between wild type and *35S:STTM171* T2 plants suggesting that the *35S:STTM171* transgene caused the phenotype. This is also supported by the finding that overexpression of the sly-miR171 target *SlHAM2/SlGRAS24* reduced seed number due to male sterility [25]. To further understand the basis of the male sterility phenotype, we assessed the productivity and quality of pollen grains in anthesis flowers of *35S:STTM171* by differential Alexander staining [26], which distinguishes between aborted and non-aborted pollen, and by testing pollen germination. This analysis indicated that the average total number of *35S:STTM171* pollen grains was reduced by 42% compared to control flowers. Moreover, the majority of the *35S:STTM171* pollen grains were aborted (60.4%), a fraction that is 4-fold higher than that found in control pollen grains. Consistent with that, the number of germinated pollen grains fell by 5.6-fold (Figure 4A). These data suggest that the *35S:STTM171* plants produce relatively small numbers of poor-quality pollen grains compared to control. This is consistent with the apparent sterility of *35S:STTM171* plants and explains why manual fertilization of wild-type tomato flowers with *35S:STTM171* pollen grains was unsuccessful. The observed almost seedless fruit and reduced pollen viability phenotypes in *35S:STTM171* plants are reminiscent to those described in *SlGRAS24/SlHAM2*-overexpressing plants [25]. This similarity suggested that the negative regulation of *SlHAM2* levels by sly-miR171 may be critical for pollen development.

To determine the cause of pollen abortion in *STTM171* expressing plants we initially analyzed pollen morphology by scanning electron microscope (SEM). Wild-type tomato cv. M82 mature pollen grains are psilate and tricolporate [27]. SEM analysis of *35S:STTM171* mature pollen grains revealed that although they remained psilate, they had deformed shapes. These included a collapsed wall, disordered germinal apertures and instead of being tricolporate many were tetracolpate (Figure 4B,C). In contrast to their morphological abnormalities, DAPI staining of *35S:STTM171* developing pollen nuclei did not reveal any nuclear aberrations during the formation of tetrads (Figure 5A), microspores (Figure 5B,C), and bicellular pollen (Figure 5D), suggesting that morphological and not nuclear aberrations underlie the poor quality of the *35S:STTM171* pollen.

2.5. The 35S:STTM171 Anthers Accumulated Reduced Sly-miR171 Levels Associated with Delayed Tapetum Degeneration and Reduced Callose Deposition

Male sterility is frequently associated with deviations in the development of the anthers that contain the sporogenous tissue and its circumjacent tissues, the tapetum and middle layer, which ultimately gives rise to the pollen grains and support pollen development correspondingly [28]. In situ of miR171 in *N. benthamiana* developing flowers has detected high uniform expression in the pollen sacs and surrounding tissues of young anthers [9], suggesting miR171 involvement in pollen development. Northern analysis with sly-miR171a/b/e validated probes (Supplement 2; Figure S3) showed that sly-miR171a/b and sly-miR171e are abundant in anthers throughout their development until maturity (anthesis flower) (Figure S4A). Anther developmental stages were defined according to the study of flower development of tomato by Brukhin et al. [29] and verified by DAPI staining from the tetrad stage (Figure 5). In agreement with the northern analysis, deep sequencing of small RNAs from developing tomato anthers identified sly-miR171a, its group member iso-sly-miR171d and sly-miR171d, a group member of sly-miR171e [30]. Compared to control anthers, the anthers of *35S:STTM171* accumulated significantly reduced levels of sly-miR171a/b in the meiosis (4 mm, stage 9), tetrad stage (5 mm, stages 10–11), free microspores stage (6 mm, stage 12) and mature pollen stage (12 mm, stages 18–19). Moreover, reduced levels were observed for sly-miR171e for which silencing was even more pronounced than for sly-miR171a/b. Depending on the developmental stage, the *35S:STTM171* anthers accumulated only around 5–20% of the control levels of sly-miR171e (Figure S4B).

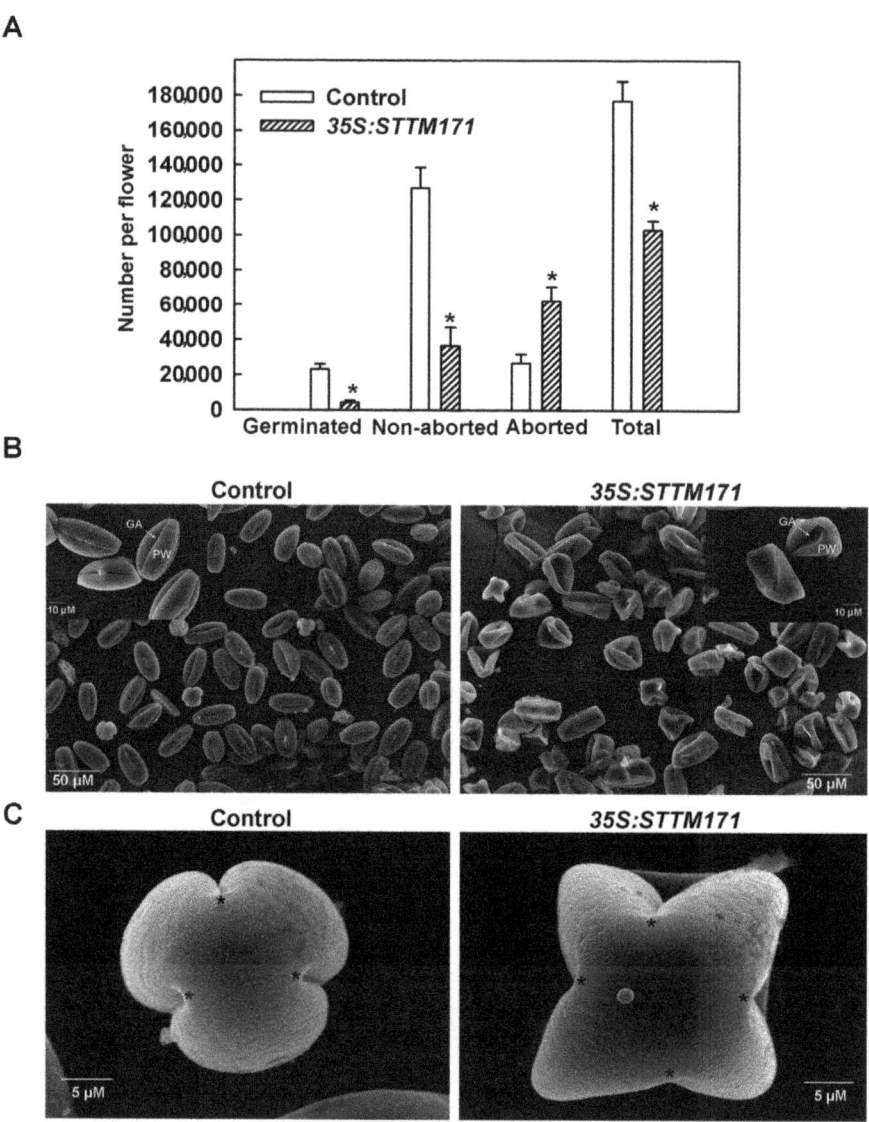

Figure 4. Effect of miR171 family downregulation on pollen quality and quantity. (**A**) Quantitation of total, aborted, non-aborted, and germinated mature pollen grains per anthesis flower of indicated genotype (n = 30). Asterisks indicate significant difference as determined by Student's *t*-test ($p \leq 0.01$). (**B**) Scanning electron micrographs of dehydrated pollen grains from indicated genotypes. Note the high number of collapsed pollen grains in the *35S:STTM171* sample. Inset shows magnified views of few representative pollen grains from each genotype. PW: pollen wall; GA: germinal aperture. (**C**) Scanning electron micrographs of polar view of representative mature pollen grains from indicated genotypes. The locations of the germinal aperture are indicated by asterisks.

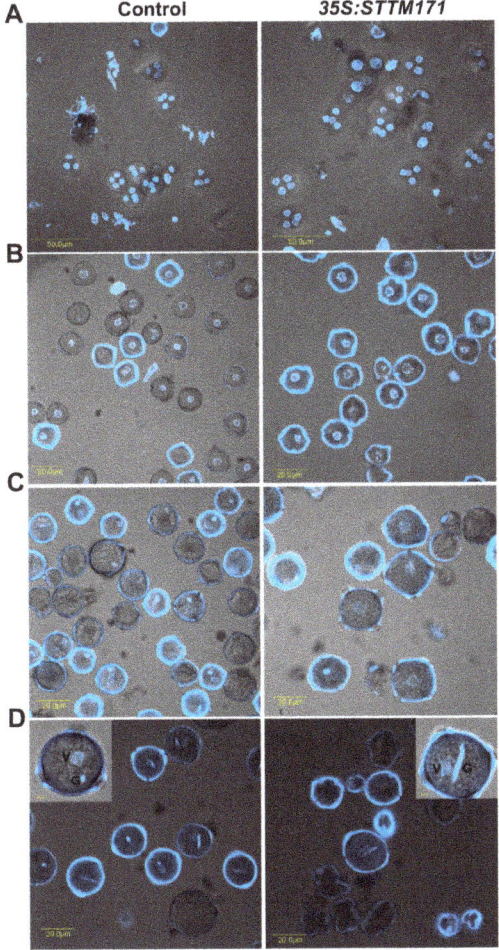

Figure 5. The *35:STTM171* pollen grains contain normal germ unit. DAPI fluorescence micrographs of control and *35:STTM171* tetrads (**A**), microspores (**B**), binucleate microspore (**C**), mature pollen grains (**D**). Inset in (**D**) shows magnified view of a representative pollen grain. The vegetative (V) and generative (G) cells are indicated. Note that the pollen surface fluorescence is due to auto-fluorescence at the same wavelength used for DAPI detection.

Next, we asked whether the reduced accumulation of sly-miR171 in *35S:STTM171* anthers is associated with abnormal development of anther tissues. To answer that, control and *35S:STTM171* anthers at major developmental stages were comparatively examined using transverse section light microscopy. Following examination of anthers under light microscopy showed that at the microsporocyte stage (3 mm buds, stage 8) the control pollen mother cells (PMC) are enclosed by a single layered tapetum (Figure 6A). At this stage, no distinct differences between control and the *35S:STTM171* transgenic anthers were observed (Figure 6B). Morphological differences were initially observed at the meiosis stage (4 mm bud, stage 9, Figure 6C). At that stage the control tapetal layer, which is composed from condensed tapetum cells, as indicated by their shrinkage and deep staining, encloses dividing microsporocytes, whereas in *35S:STTM171* anthers, tapetal cells remain expanded and vacuolated and dividing microsporocytes were not observed (Figure 6D). At the tetrad stage control tapetal cells were completely shrunk (5 mm bud, stage 11, Figure 6E), likely due to the initiation of programmed cell death (PCD) [31], and enclosed separated tetrads. In contrast, in *35S:STTM171* anthers, tapetal cells were not shrunken and instead were enlarged while most tetrads were not separated (Figure 6F). At the microspore stage, degenerated tapetal layer enclosing free microspores was observed in both control (8 mm bud, stages 12–13, Figure 6G) and *35S:STTM171* anthers (Figure 6H), except that in the latter the tapetum was less degenerated. At the bicellular pollen stage the control tapetum was completely degenerated and pollens were mature with characteristic densely stained cytoplasm (10 mm bud, stage 17–18, Figure 6I), whereas in *35S:STTM171* anthers, remnants of the degenerated tapetum were still visible and enclosed aborted pollen (Figure 6J). These observations indicated that the degeneration of the tapetum in the anthers silenced for sly-miR171 was delayed and initiated only after the tetrad stage in comparison to control where tapetum degradation occurred significantly earlier and on time.

Whereas tapetum is still intact at the tetrad stage, its cells export sporopollenin on microspore primexine surface that will provide a basis for future assembling of lipid materials into the pollen coat exine. During this process and earlier meiosis, the PMC and tetrads are surrounded by callose layer that provides protection for developing microspores. Later, the callose layer goes through degradation by tapetum supplied callase and the tapetum goes through PCD aiming to supply additional materials to the microspores [32–34]. The degradation of callose allows microspore release from tetrads and degradation of the tapetum supplies lipidic tapetum-derived materials for pollen exine and nutrients for pollen maturation [33,34]. The timing of tapetal cell death and consistency of the above-described events are critical for pollen development and interference in this process usually results in male sterility [35]. In several studies, delaying tapetum degeneration was shown to affect pollen morphology and results in pollen abortion. The rice mutant *tapetal degeneration retardation* (*Ostdr*) shows delayed tapetal breakdown resulting in a failure of pollen wall deposition and subsequent microspore degeneration [36]. Mutation in rice *OsACOS12* delays PCD-induced tapetum degradation leading to collapsed aborted pollen [37]. Thus, a likely possibility is that the delayed degeneration of *STTM171* expressing tapetum may be responsible at least in part for the deformed morphology of respective pollen grains.

Callose (β-1,3 glucan) protects PMC and later developing microspores from swelling, rupture, impact of diploid tissues and serves as a mold for future exine layer [38,39]. Often, defective tapetum development plan is accompanied with depletion of callose. To test if the delayed degeneration of *35S:STTM171* tapetum cells affected callose dynamics we performed a lacmoid stain of control and of *35S:STTM171* anthers at the meiosis and tetrad stages. We observed strong staining in control anthers at the meiosis stage (Figure 7A) and much weaker staining at the tetrad stage (Figure 7B) probably as a result of initiated callose degradation by tapetum supplied callase. In contrast, significantly weaker staining was observed in the *35S:STTM171* anthers (Figure 7A,B). Moreover, the weak callose staining was associated with enlarged tapetum cells characteristic of those that have not initiated PCD, rather than condensed cells undergoing PCD, as in the wild-type (Figures 6D,F and 7). In transgenic tobacco plants with delayed tapetum development, male sterility was caused because of premature degradation of callose [40]. In *DISFUNCTIONAL TAPETUM1* mutant enlarged tapetum cells and thin callose layer caused pollen collapse which resulted in male sterility [41]. This suggests that the delayed tapetum development in *35S:STTM171* anthers also delayed the callose deposition. Alternatively, callose depletion might be caused independently of tapetum development as in the *CALLOSE SYNTHASE5* mutants, where failure to produce callose resulted in collapsed pollen [33,42] reminiscent of the *35S:STTM171* pollen phenotype. Taken together these results suggest that the silencing of sly-miR171 in the *35S:STTM171* anthers perturbed tapetum development, callose dynamics and as a result pollen ontogenesis. However, additional studies are required to support this suggestion and determine the mechanism by which these miR171 members regulate anther development.

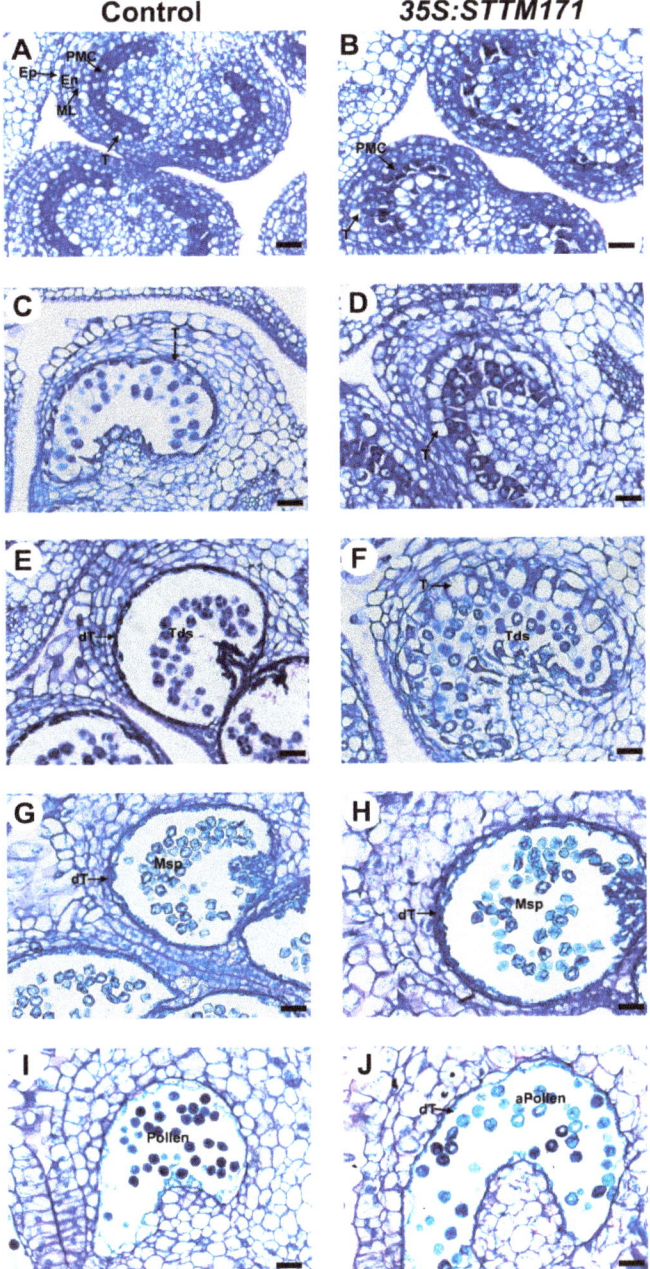

Figure 6. Histological analysis of control and *35S:STTM171* anthers. Pictures of Toluidine-blue stained cross sections of control and *35:STTM171* anthers at subsequent stages of microspore development as follows: (**A,B**) microsporocyte stage, (**C,D**) meiosis stage, (**E,F**) tetrad stage, (**G,H**) microspore stage, (**I,J**) bicellular pollen stage. dT-degenerated tapetum; En-endothecium; Ep-epidermis; ML-middle cell layer; Msp-microspore; MMC-microspore mother cell; T-tapetum; Tds-tetrads; aPollen-aborted pollen. Scale bars = 20 µm.

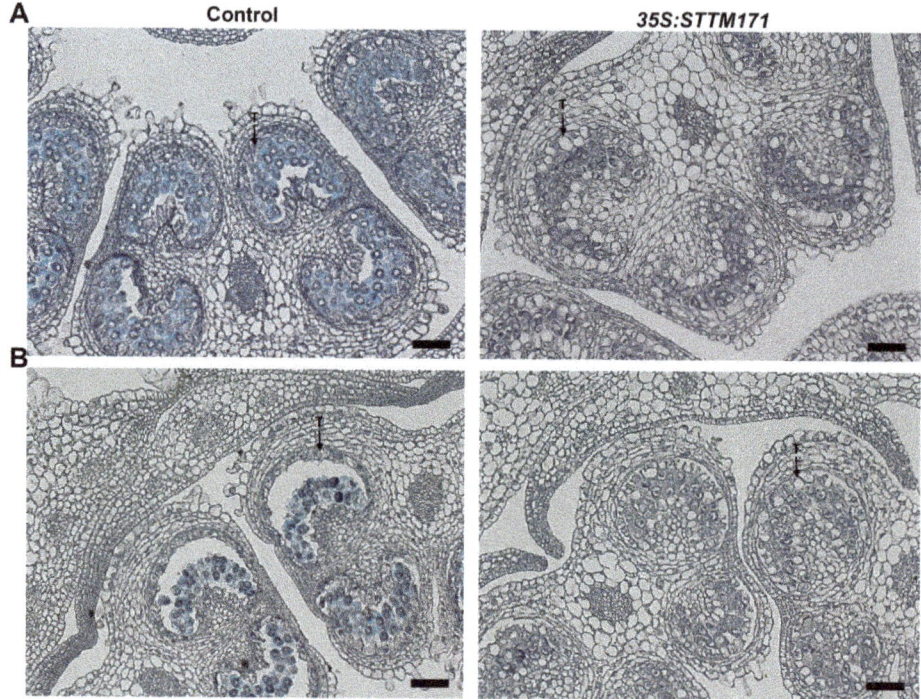

Figure 7. Callose detection in control and *35S:STTM171* anthers. Pictures of Lacmoid stained cross sections of anthers at the meiosis (**A**, 4 mm bud) and tetrad (**B**, 5 mm bud) stages are shown. T-tapetum. Scale bars = 20 μm.

3. Materials and Methods

3.1. Plant Material and Growth Conditions

Tomato cv. M82 plants and seedlings were grown under greenhouse and growth chamber conditions, respectively, as previously described [20].

3.2. Plasmid Construction

The pART27-OP:SlMIR171a and pART27-OP:SlMIR171b responder plasmids were described elsewhere [6]. For the pART27-OP:SlMIR171e responder construct, a 239 bp fragment from *SlMIR171e* including the *pre-miR171e* was amplified with Xho_MIR171e_F and Hind_MIR171e_R which contained *Xho*I and *Hind*III sites at their 5′ ends (for primer sequences, see Supplementary Table S3 online). The amplified fragment was restricted with *Xho*I and *Hind*III and ligated into pART27 binary vector containing the OP array to obtain pART27-OP:SlMIR171e [6]. For the *35S:STTM171* construct, a 136 bp STTM fragment was synthetically synthesized and then PCR amplified with Xho_STTM_171_F and Hind_STTM_171_R primers that contained *Xho*I and *Hind*III sites at their 5′ ends. The amplified fragment was restricted with *Xho*I and *Hind*III and cloned into the appropriate sites of the pART27 binary vector containing the CaMV 35S promoter and *Agrobacterium tumefaciens* octopine synthase terminator (OCS) [17] to obtain pART27-35S:STTM171.

3.3. Transformation of Tomato Plants

The binary vector pART27-35S:STTM171 was transformed into tomato cv. M82 and transgenic plants were selected as described previously [20]. Each kanamycin resistant plant was also subjected to genomic DNA PCR with the primer pair Xho_STTM_171_F and OCS_rev to detect the 35S:STTM171 transgene.

3.4. Total RNA Extraction and RNA Gel-Blot Analysis

Total RNA was extracted from different tomato tissues with Bio-Tri RNA reagent (Bio-Lab, Jerusalem, Israel) according to the manufacturer's protocol. A small-RNA gel-blot analysis of the total RNA was performed as described previously [43] using a complementary radiolabeled oligos as probes (probe sequences are listed in Supplementary Table S3 online).

3.5. cDNA Synthesis and Quantitative RT-PCR Assay

First-strand cDNA was synthesized from 2 µg of total RNA with Maxima first strand cDNA synthesis kit (Thermo Scientific, Waltham, MA, USA) following the manufacturer's instructions. A negative control (-RT) was used to ensure the absence of genomic DNA template in the samples. Three independent biological replicates were used for each sample, and quantification was performed in triplicate. PCR was performed in StepOnePlus Real-Time PCR System (Thermo Fisher Scientific) following the manufacturer's instructions. Primer sequences are listed in Supplementary Table S3 online. Relative expression levels were normalized to *SlTIP41* as a reference gene, and calculated by the standard curve method.

3.6. Analysis of Leaf Morphology and Axillary Shoot Number

Primary leaflet area was calculated by Tomato Analyzer 3.0 software [44]. Leaflet angle, namely the angle between the petiolule and the rachis was measured manually by a protractor. The number of axillary shoots ($\geq$5 mm long) was counted at eight-leaf stage tomato plants.

3.7. Determination of Pollen Quality and Quantity

To determine the pollen quantity and quality, mature pollen was extracted, stained, and counted according to Firon et al. [45]. Briefly, two flowers at the day of anthesis were sampled from control and transgenic plants and three anthers were removed from each flower, sliced in the middle, and immediately placed in a microcentrifuge tube containing germination solution (0.5 mL, 10% sucrose, 2 mM boric acid, 2 mM calcium nitrate, 2 mM magnesium sulfate, and 1 mM potassium nitrate). Then the pollen grains were released by vortex, incubated for 4 h at 25 °C, and stained with Alexander dye, that colors aborted pollen grains in blue-green, and non-aborted pollen grains in magenta-red [26]. The pollen grains were counted under a light microscope in a haemocytometer, eight fields for each sample.

3.8. Scanning Electron Microscopy (SEM) and 4′,6-Diamidino-2-Phenylindole (DAPI) Staining of Pollen Grains

For SEM analysis, pollen grains were collected and placed in FAA (3.7% formaldehyde, 5% acetic acid, 50% EtOH) solution until use. Then the FAA solution was removed and pollen grains were dehydrated in an increasing gradient of ethanol (up to 100%), critical-point-dried, mounted on a copper plate and gold-coated. Samples were viewed in a Jeol 5410 LV microscope (Tokyo, Japan). To stain pollen grains with DAPI, grains at different developmental stages were released by vortex into a DAPI solution (0.1 M sodium phosphate buffer (pH 7), 1 mM EDTA, 0.1% Triton X-100, 0.4 µg/mL DAPI), incubated for 10 min at room temperature and then viewed by Olympus IX81/FV500 laser-scanning confocal microscope (Olympus) at 361 nm maximum absorption, 461 nm maximum emission.

3.9. Histology and Callose Staining

For histological analyses, stamens at different developmental stages were taken and fixed in FAA solution until use, then dehydrated in increasing concentrations of ethanol (70%, 80%, 90%,

95%, and 100%), cleared with histoclear, and embedded in paraffin. Microtome-cut sections (6-µm thick) were spread on microscope slides, and stained with 0.03% Toluidine blue O. For callose staining microtome-cut sections of stamens were stained with 0.2% Lacmoid in 50% ethanol for 48h, then placed in 1% sodium bicarbonate in 50% ethanol for 10 min [46]. Stained slides were examined under bright-field using an Olympus (Olympus, www.olympus-lifescience.com) light microscope equipped with a digital camera.

Supplementary Materials: The following are available online at http://www.mdpi.com/2223-7747/8/1/10/s1, Supplement 1. Prediction of target mRNAs for sly-miR171 and corresponding sly-mi171* strands. Supplement 2. Validation of probe specificity of sly-miR171. Figure S1. Alignment between STTM171 and sly-miR171 sequences. Figure S2. Fruit of *35:STTM171* T2 plants. Figure S3. Determination of the specificity of sly-miR171a, b, e RNA gel blot probes. Figure S4. Developing *35:STTM171* anthers accumulate reduced levels of sly-miR171. Table S1: A list of sly-miR171 precursors and corresponding miRNA/miRNA* pairs. Table S2: psRNATarget analysisa of sly-miR171 members and their star strands. Table S3: Primers and probes used in this study.

Author Contributions: M.K. designed, performed experiments, analyzed data, and wrote the manuscript. R.S. did transgenic plants. E.B assisted in confocal microscopy experiments. T.A. supervised the study and wrote the manuscript.

Funding: This work was supported by the Israel Science Foundation grant 939/12 to TA.

Acknowledgments: We would like to thank Guiliang Tang, Biological Sciences, Michigan Technological University, for his help with the design and cloning of the miR171 STTM construct.

Conflicts of Interest: The authors declare no conflict of interest.

References

1. Axtell, M.J. Classification and comparison of small RNAs from plants. *Annu. Rev. Plant Biol.* **2013**, *64*, 137–159. [CrossRef] [PubMed]
2. Rubio-Somoza, I.; Weigel, D. MicroRNA networks and developmental plasticity in plants. *Trends Plant Sci.* **2011**, *16*, 258–264. [CrossRef] [PubMed]
3. Jones-Rhoades, M.W.; Bartel, D.P. Computational Identification of Plant MicroRNAs and Their Targets, Including a Stress-Induced miRNA. *Mol. Cell* **2004**, *14*, 787–799. [CrossRef] [PubMed]
4. Axtell, M.J.; Bowman, J.L. Evolution of plant microRNAs and their targets. *Trends Plant Sci.* **2008**, *13*, 343–349. [CrossRef] [PubMed]
5. Zhu, X.; Leng, X.; Sun, X.; Mu, Q.; Wang, B.; Li, X.; Wang, C.; Fang, J. Discovery of conservation and diversification of miR171 genes by phylogenetic analysis based on global genomes. *Plant Genome* **2015**, *8*. [CrossRef]
6. Hendelman, A.; Kravchik, M.; Stav, R.; Frank, W.; Arazi, T. Tomato HAIRY MERISTEM genes are involved in meristem maintenance and compound leaf morphogenesis. *J. Exp. Bot.* **2016**, *67*, 6187–6200. [CrossRef]
7. Llave, C.; Kasschau, K.D.; Rector, M.A.; Carrington, J.C. Endogenous and silencing-associated small RNAs in plants. *Plant Cell* **2002**, *14*, 1605–1619. [CrossRef]
8. Lauressergues, D.; Delaux, P.-M.; Formey, D.; Lelandais-Brière, C.; Fort, S.; Cottaz, S.; Bécard, G.; Niebel, A.; Roux, C.; Combier, J.-P. The microRNA miR171h modulates arbuscular mycorrhizal colonization of *Medicago truncatula* by targeting *NSP2*. *Plant J.* **2012**, *72*, 512–522. [CrossRef]
9. Válóczi, A.; Várallyay, É.; Kauppinen, S.; Burgyán, J.; Havelda, Z. Spatio-temporal accumulation of microRNAs is highly coordinated in developing plant tissues. *Plant J.* **2006**, *47*, 140–151. [CrossRef]
10. Siré, C.; Moreno, A.B.; Garcia-Chapa, M.; López-Moya, J.J.; Segundo, B.S. Diurnal oscillation in the accumulation of Arabidopsis microRNAs, miR167, miR168, miR171 and miR398. *FEBS Lett.* **2009**, *583*, 1039–1044. [CrossRef]
11. Tong, A.; Yuan, Q.; Wang, S.; Peng, J.; Lu, Y.; Zheng, H.; Lin, L.; Chen, H.; Gong, Y.; Chen, J.; et al. Altered accumulation of osa-miR171b contributes to rice stripe virus infection by regulating disease symptoms. *J. Exp. Bot.* **2017**, *68*, 4357–4367. [CrossRef]
12. Liu, H.-H.; Tian, X.; Li, Y.-J.; Wu, C.-A.; Zheng, C.-C. Microarray-based analysis of stress-regulated microRNAs in *Arabidopsis thaliana*. *RNA* **2008**, *14*, 836–843. [CrossRef] [PubMed]
13. Yan, J.; Gu, Y.; Jia, X.; Kang, W.; Pan, S.; Tang, X.; Chen, X.; Tang, G. Effective small RNA destruction by the expression of a Short Tandem Target Mimic in Arabidopsis. *Plant Cell* **2012**, *24*, 415–427. [CrossRef]

14. Zhang, H.; Zhang, J.; Yan, J.; Gou, F.; Mao, Y.; Tang, G.; Botella, J.R.; Zhu, J.-K. Short tandem target mimic rice lines uncover functions of miRNAs in regulating important agronomic traits. *Proc. Natl. Acad. Sci. USA* **2017**, *114*, 5277–5282. [CrossRef]
15. Todesco, M.; Rubio-Somoza, I.; Paz-Ares, J.; Weigel, D. A Collection of target mimics for comprehensive analysis of MicroRNA function in *Arabidopsis thaliana*. *PLoS Genet.* **2010**, *6*, e1001031. [CrossRef] [PubMed]
16. Ivashuta, S.; Banks, I.R.; Wiggins, B.E.; Zhang, Y.; Ziegler, T.E.; Roberts, J.K.; Heck, G.R. Regulation of gene expression in plants through miRNA inactivation. *PLoS ONE* **2011**, *6*, e21330. [CrossRef]
17. Damodharan, S.; Zhao, D.; Arazi, T. A common miRNA160-based mechanism regulates ovary patterning, floral organ abscission and lamina outgrowth in tomato. *Plant J.* **2016**, *86*, 458–471. [CrossRef] [PubMed]
18. Peng, T.; Qiao, M.; Liu, H.; Teotia, S.; Zhang, Z.; Zhao, Y.; Wang, B.; Zhao, D.; Shi, L.; Zhang, C.; et al. A resource for inactivation of microRNAs using Short Tandem Target Mimic Technology in model and crop plants. *Mol. Plant* **2018**, *5*, 1400–1417. [CrossRef] [PubMed]
19. Kozomara, A.; Griffiths-Jones, S. miRBase: Annotating high confidence microRNAs using deep sequencing data. *Nucleic Acids Res.* **2014**, *42*, D68–D73. [CrossRef]
20. Kravchik, M.; Sunkar, R.; Damodharan, S.; Stav, R.; Zohar, M.; Isaacson, T.; Arazi, T. Global and local perturbation of the tomato microRNA pathway by a trans-activated DICER-LIKE 1 mutant. *J. Exp. Bot.* **2014**, *65*, 725–739. [CrossRef]
21. Wang, H.; Zhang, X.; Liu, J.; Kiba, T.; Woo, J.; Ojo, T.; Hafner, M.; Tuschl, T.; Chua, N.-H.; Wang, X.-J. Deep sequencing of small RNAs specifically associated with Arabidopsis AGO1 and AGO4 uncovers new AGO functions. *Plant J.* **2011**, *67*, 292–304. [CrossRef]
22. Karlova, R.; van Haarst, J.C.; Maliepaard, C.; van de Geest, H.; Bovy, A.G.; Lammers, M.; Angenent, G.C.; de Maagd, R.A. Identification of microRNA targets in tomato fruit development using high-throughput sequencing and degradome analysis. *J. Exp. Bot.* **2013**, *64*, 1863–1878. [CrossRef]
23. Imanshi, S.; Hiura, I. Relationship between fruit weight and seed content in the tomato. *J. Jpn. Soc. Hortic. Sci.* **1975**, *44*, 33–40. [CrossRef]
24. Wang, L.; Mai, Y.-X.; Zhang, Y.-C.; Luo, Q.; Yang, H.-Q. MicroRNA171c-Targeted SCL6-II, SCL6-III, and SCL6-IV genes regulate shoot branching in Arabidopsis. *Mol. Plant* **2010**, *3*, 794–806. [CrossRef]
25. Huang, W.; Peng, S.; Xian, Z.; Lin, D.; Hu, G.; Yang, L.; Ren, M.; Li, Z. Overexpression of a tomato miR171 target gene SlGRAS24 impacts multiple agronomic traits via regulating gibberellin and auxin homeostasis. *Plant Biotechnol. J.* **2016**, *15*, 472–488. [CrossRef]
26. Alexander, M.P. Differential staining of aborted and nonaborted pollen. *Stain Technol.* **1969**, *44*, 117–122. [CrossRef]
27. Kapp, R. *How to Know Pollen and Spores*; W. C. Brown Co. Publishers: Dubuque, IA, USA, 1969.
28. Gorman, S.W.; McCormick, S.; Rick, D.C. Male sterility in tomato. *Crit. Rev. Plant Sci.* **1997**, *16*, 31–53. [CrossRef]
29. Brukhin, V.; Hernould, M.; Gonzalez, N.; Chevalier, C.; Mouras, A. Flower development schedule in tomato Lycopersicon esculentum cv. sweet cherry. *Sex. Plant Reprod.* **2003**, *15*, 311–320.
30. Omidvar, V.; Mohorianu, I.; Dalmay, T.; Fellner, M. Identification of miRNAs with potential roles in regulation of anther development and male-sterility in 7B-1 male-sterile tomato mutant. *BMC Genom.* **2015**, *16*, 878. [CrossRef]
31. Goldberg, R.B.; Beals, T.P.; Sanders, P.M. Anther development: Basic principles and practical applications. *Plant Cell* **1993**, *5*, 1217–1229. [CrossRef]
32. Owen, H.A.; Makaroff, C.A. Ultrastructure of microsporogenesis and microgametogenesis in Arabidopsis thaliana (L.) Heynh. ecotype Wassilewskija (Brassicaceae). *Protoplasma* **1995**, *185*, 7–21. [CrossRef]
33. Dong, X.; Hong, Z.; Sivaramakrishnan, M.; Mahfouz, M.; Verma, D.P.S. Callose synthase (CalS5) is required for exine formation during microgametogenesis and for pollen viability in Arabidopsis. *Plant J.* **2005**, *42*, 315–328. [CrossRef]
34. Quilichini, T.D.; Douglas, C.J.; Samuels, A.L. New views of tapetum ultrastructure and pollen exine development in *Arabidopsis thaliana*. *Ann. Bot.* **2014**, *114*, 1189–1201. [CrossRef]
35. Kawanabe, T.; Ariizumi, T.; Kawai-Yamada, M.; Uchimiya, H.; Toriyama, K. Abolition of the tapetum suicide program ruins microsporogenesis. *Plant Cell Physiol.* **2006**, *47*, 784–787. [CrossRef]

36. Li, N.; Zhang, D.-S.; Liu, H.-S.; Yin, C.-S.; Li, X.; Liang, W.; Yuan, Z.; Xu, B.; Chu, H.-W.; Wang, J.; et al. The Rice tapetum degeneration retardation gene is required for tapetum degradation and anther development. *Plant Cell* **2006**, *18*, 2999–3014. [CrossRef]
37. Yang, X.; Liang, W.; Chen, M.; Zhang, D.; Zhao, X.; Shi, J. Rice fatty acyl-CoA synthetase OsACOS12 is required for tapetum programmed cell death and male fertility. *Planta* **2017**, *246*, 105–122. [CrossRef]
38. Zhang, C.; Guinel, F.C.; Moffatt, B.A. A comparative ultrastructural study of pollen development in *Arabidopsis thaliana* ecotype Columbia and male-sterile mutant apt1-3. *Protoplasma* **2002**, *219*, 59–71. [CrossRef]
39. Zhu, J.; Chen, H.; Li, H.; Gao, J.-F.; Jiang, H.; Wang, C.; Guan, Y.-F.; Yang, Z.-N. Defective in Tapetal Development and Function 1 is essential for anther development and tapetal function for microspore maturation in Arabidopsis. *Plant J.* **2008**, *55*, 266–277. [CrossRef]
40. Worrall, D.; Hird, D.L.; Hodge, R.; Paul, W.; Draper, J.; Scott, R. Premature dissolution of the microsporocyte callose wall causes male sterility in transgenic tobacco. *Plant Cell Online* **1992**, *4*, 759–771. [CrossRef]
41. Zhang, W.; Sun, Y.; Timofejeva, L.; Chen, C.; Grossniklaus, U.; Ma, H. Regulation of Arabidopsis tapetum development and function by DYSFUNCTIONAL TAPETUM1 (DYT1) encoding a putative bHLH transcription factor. *Development* **2006**, *133*, 3085–3095. [CrossRef]
42. Nishikawa, S.; Zinkl, G.M.; Swanson, R.J.; Maruyama, D.; Preuss, D. Callose (β-1,3 glucan) is essential for Arabidopsis pollen wall patterning, but not tube growth. *BMC Plant Biol.* **2005**, *5*, 22. [CrossRef]
43. Talmor-Neiman, M.; Stav, R.; Klipcan, L.; Buxdorf, K.; Baulcombe, D.C.; Arazi, T. Identification of trans-acting siRNAs in moss and an RNA-dependent RNA polymerase required for their biogenesis. *Plant J.* **2006**, *48*, 511–521. [CrossRef]
44. Rodríguez, G.R.; Moyseenko, J.B.; Robbins, M.D.; Morejón, N.H.; Francis, D.M.; van der Knaap, E. Tomato Analyzer: A useful software application to collect accurate and detailed morphological and colorimetric data from two-dimensional objects. *J. Vis. Exp. JoVE* **2010**. [CrossRef]
45. Firon, N.; Nepi, M.; Pacini, E. Water status and associated processes mark critical stages in pollen development and functioning. *Ann. Bot.* **2012**, *109*, 1201–1214. [CrossRef]
46. Krishnamurthy, K.V. *Methods in Cell Wall Cytochemistry*; CRC Press: Boca Raton, FL, USA, 1999; pp. 69–70.

© 2019 by the authors. Licensee MDPI, Basel, Switzerland. This article is an open access article distributed under the terms and conditions of the Creative Commons Attribution (CC BY) license (http://creativecommons.org/licenses/by/4.0/).

Article

Genome-Wide Investigation of the Role of MicroRNAs in Desiccation Tolerance in the Resurrection Grass *Tripogon loliiformis*

Isaac Njaci [1], Brett Williams [1], Claudia Castillo-González [2], Martin B. Dickman [3], Xiuren Zhang [2] and Sagadevan Mundree [1,*]

1. Centre for Tropical Crops and Biocommodities, Queensland University of Technology, Brisbane, QLD 4000, Australia; injaci@gmail.com (I.N.); b.williams@qut.edu.au (B.W.)
2. Department of Biochemistry and Biophysics, Institute for Plant Genomics and Biotechnology, Texas A&M University, College Station, TX 77843, USA; castillo.cm@tamu.edu (C.C.-G.); xiuren.zhang@tamu.edu (X.Z.)
3. Department of Plant Pathology and Microbiology, Institute for Plant Genomics and Biotechnology, Texas A&M University, College Station, TX 77843, USA; mbdickman@tamu.edu
* Correspondence: sagadevan.mundree@qut.edu.au; Tel.: + 61-7-3138-2000

Received: 31 July 2018; Accepted: 29 August 2018; Published: 31 August 2018

Abstract: Drought causes approximately two-thirds of crop and yield loss worldwide. To sustain future generations, there is a need to develop robust crops with enhanced water use efficiency. Resurrection plants are naturally resilient and tolerate up to 95% water loss with the ability to revive upon watering. Stress is genetically encoded and resilient species may garner tolerance by tightly regulating the expression of stress-related genes. MicroRNAs (miRNAs) post-transcriptionally regulate development and other stress response processes in eukaryotes. However, their role in resurrection plant desiccation tolerance is poorly understood. In this study, small RNA sequencing and miRNA expression profiling was conducted using *Tripogon loliiformis* plants subjected to extreme water deficit conditions. Differentially expressed miRNA profiles, target mRNAs, and their regulatory processes were elucidated. Gene ontology enrichment analysis revealed that development, stress response, and regulation of programmed cell death biological processes; Oxidoreductase and hydrolyase molecular activities; and SPL, MYB, and WRKY transcription factors were targeted by miRNAs during dehydration stress, indicating the indispensable regulatory role of miRNAs in desiccation tolerance. This study provides insights into the molecular mechanisms of desiccation tolerance in the resurrection plant *T. loliiformis*. This information will be useful in devising strategies for crop improvement on enhanced drought tolerance and water use efficiency.

Keywords: microRNAs; dehydration; desiccation; resurrection plants; *Tripogon loliiformis*; post-transcriptional gene silencing; miRNAs

1. Introduction

The majority of higher plants are sensitive to dehydration and lose viability upon the loss of 41–70% of their total water content [1]. However, some plants are well adapted to adverse environments and implement adaptation mechanisms that mitigate the effects of water loss [2]. Although the adaptation mechanisms are effective under diverse environments, enabling plant survival, severe water loss still leads to the death of the plant. Resurrection plants represent a small but diverse group of angiosperms that exhibit unique tolerance mechanisms to cope with desiccation [3]. This unique group of plants can tolerate desiccation in their vegetative tissues for prolonged periods and rapidly recover their metabolic activity within 48 h of watering with minimal or non-existent tissue damage [3,4].

To achieve this phenomenal tolerance, resurrection plants utilize a repertoire of strategies that include the rapid shutdown of photosynthesis; leaf and cell wall structural adjustment; the stabilization of subcellular milieu through the accumulation of sugars, late embryogenesis abundant proteins (LEA), heat shock proteins, and other compatible solutes; and the induction of extensive antioxidant and ROS scavenging systems during desiccation [5–8].

Tripogon loliiformis is a small tufted diploid (2n = 20) annual to short-lived perennial C4 resurrection grass. It is endemic to a wide range of habitats in Australia and "resurrects" from its desiccated state within 72 h [9,10]. It's short life cycle, ploidy level, and ease of propagation makes it a suitable experimental model for desiccation tolerance studies. *T. loliiformis* physiological responses to desiccation are characterized by structural, physiological, and biochemical changes that include leaf folding, cell wall folding and vacuole fragmentation, the early shutdown of photosynthesis, the retention of chlorophyll (homoiochlorophyllous) [9,11], increased anthocyanin accumulation, and the accumulation of sucrose and trehalose [12]. Response to stress is genetically encoded and transcriptional and post-transcriptional reprogramming are important elements in stress tolerance strategies [13–16]. Previous studies have linked miRNAs with many cellular processes [17,18].

MiRNAs are a class of 20–22 nt endogenous non-coding small RNAs that play important roles in the regulation of gene expression at the transcriptional and post-transcriptional level in animals, plants [19,20], and unicellular organisms [21]. In plants, miRNAs are involved in many metabolic and biological processes, where they play crucial regulatory roles in growth and development [22], phytohormone signaling [23], and adaptive responses to abiotic and biotic stress [17,24,25]. In Arabidopsis, using mannitol as a stress-inducing agent, miR396, miR168, miR167, and miR171 were found to be drought-responsive [26]. In rice, 30 differentially expressed miRNAs were identified under drought stress, while in tobacco, miR395 and miR169 were found to be sensitive to drought stress [27]. Using a combination of high throughput sequencing and microarray technology in a genome-wide study of *Populus euphratica*, Li et al. [28] identified 131 differentially expressed miRNAs under drought stress.

Although studies on miRNAs have been conducted in drought sensitive and tolerant plants including wheat [29], sorghum [30], switch grass [31], brachypodium [32], *Arabidopsis* [25], rice [33], and maize [34], their role in stress tolerance in desiccation tolerant plants has not been investigated. While drought sensitive and tolerant plants employ tolerance mechanisms to withstand mild to moderate dehydration stress, it is postulated that resurrection plants utilize a repertoire of unique desiccation tolerance mechanisms to overcome and adapt to extreme conditions. In this study, *T. loliiformis* plants subjected to desiccation stress at strategic dehydration, desiccation, and rehydration stages were analyzed through high throughput sequencing for the identification and enrichment of conserved miRNAs and elucidation of their potential roles in desiccation tolerance. Future crop production is threatened by the effects of global climate change, erratic weather patterns, and population growth. There is an urgent need for the development of climate resilient and water use efficient crops. Understanding the unique desiccation tolerance mechanisms harbored by resurrection plants presents a great potential for the development of robust crops.

2. Results

2.1. Analysis of Small RNAs in Shoot and Root of Tripogon loliiformis

The unique desiccation tolerance capabilities of resurrection plants suggests underlying tolerance mechanisms and effective regulation of transcription and translation. Studies have shown that miRNAs regulate gene expression at the transcriptional and post-transcriptional level. To investigate the role of miRNAs in desiccation tolerance, small RNA libraries from shoots and roots of hydrated, dehydrating (60% and 40% Relative Water Content (RWC)), desiccated (<10% RWC), and rehydrated plants were sequenced and the identification and expression profiling of miRNAs were conducted. High-throughput sequencing generated 142 million small RNAs reads in the range of 13–16 million reads per library, with an average of 1–3 million unique reads (Figure 1A). The raw reads were pre-processed to remove adapter sequences, low quality reads, and <15 nt reads, resulting in 104 (73%) million clean reads that were used for downstream analysis. Consistent with previous studies, the majority of the small RNAs were in the range of 21–24 nt, with 24 nt having the most reads, followed by 21 nt, across all libraries (Figure 1B).

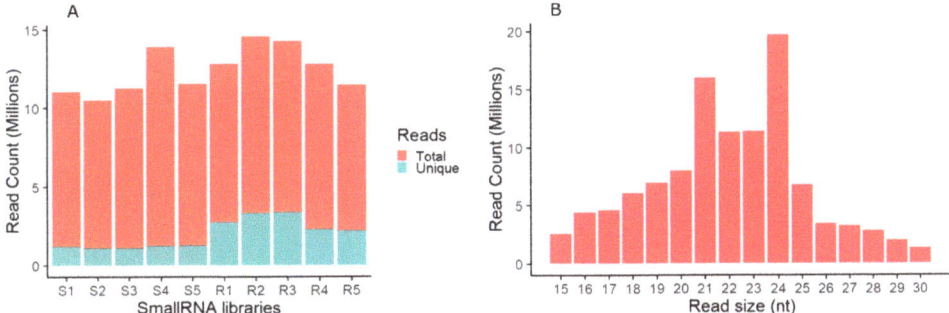

Figure 1. *Tripogon loliiformis* small RNA reads distribution. (**A**) Pre-processed reads distribution in shoot and root small RNA libraries; (**B**) overall reads distribution by size. S1–S5, R1–R5 represent shoot and root libraries at Hydrated (1), 60% RWC (2), 40% RWC (3), <10% RWC (4), and 48 h after rehydration (5), respectively.

2.2. Evolutionary Conservation of MiRNAs in Tripogon loliiformis

Many miRNAs are evolutionary conserved in species within the same kingdom and miRNA genes in one species may exist as orthologs or homologs in other species, presenting a powerful strategy to identify new miRNAs through a homology search [35]. To identify conserved miRNAs in *T. loliiformis*, the small RNA sequences were mapped against known plant miRNAs in the miRBase 21.0 database [36] and a total of 265 unique conserved miRNAs comprising 668 family members and isoforms from 60 MIR families were identified (Supplemental Table S1). Many of the identified miRNAs were family members and isoforms of the nine most evolutionary conserved miRNAs in plants [37]. The majority of the identified miRNAs were from closely related monocot species including *Oryza sativa*, *Sorghum bicolor*, *Zea mays*, *Branchypodium distachyon*, *Hordeum vulgare*, *Triticum aestivum*, and *Aegilops tauschii*, as well as from the model species *Arabidopsis thaliana* (Supplemental Figure S1).

2.3. Spatiotemporal Expression of MiRNAs in Tripogon loliiformis Tissues

Previous studies suggest that resurrection plants are genetically primed to respond to dehydration, even at a hydrated state, through the constitutive expression of stress response mechanisms [38,39]. Do resurrection plants gain their tolerance by post-transcriptionally regulating transcription more tightly than sensitive plants? Differential expression analysis identified 183 conserved miRNAs that differentially accumulated in shoots and roots during dehydration (Supplemental Table S2). A higher number of miRNAs were up-regulated in the shoots compared to the roots (Figure 2A). In the shoots at 60% RWC, 47 miRNAs were up-regulated and 29 down-regulated. At 40% RWC, 37 miRNAs were up-regulated while 40 were down-regulated. Surprisingly, miRNA accumulation was observed in desiccated tissue at <10% RWC, where 41 miRNAs showed an increased abundance, while 40 displayed a decreased accumulation. In rehydrating shoots, 51 miRNAs were up-regulated while 22 were down-regulated. In contrast, a reduced accumulation of miRNAs was observed in the roots where a higher number showed downregulation as dehydration ensued. In summary, 26 miRNAs were up-regulated and 54 down-regulated at 60% RWC, 24 up-regulated and 50 down-regulated at

40% RWC, and 30 up- and 54 down-regulated at <10% RWC, while 23 and 10 miRNAs were up- and down-regulated respectively on rehydration (Figure 2A, Supplemental Figure S2).

Tissue specific expression was observed where some miRNAs exhibited disparate expression between shoots and roots (Figure 2B). For example, miR399b, miR399j, miR167a, and miR393 h:i:j:k were up-regulated, while miR164c:h, miR169r:a, miR528a:b, and miR160d were down-regulated in the shoots, but were not expressed in the roots. The miRNAs miR444f, miR160e, and miR6300 were up-regulated in the shoots and down-regulated in the roots (Figure 2A), while other miRNAs showed a similar expression trend.

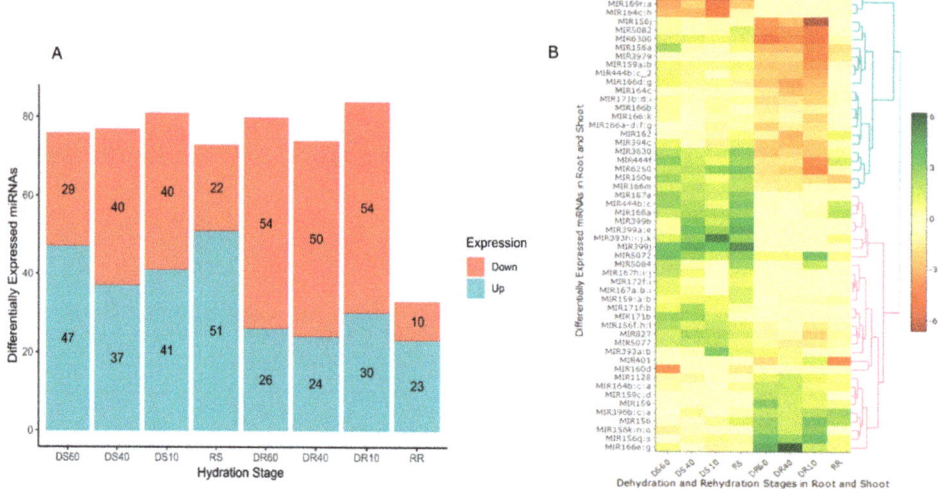

Figure 2. Differentially expressed miRNAs in shoots and roots under dehydration stress. (**A**) Bar graph showing disparate expression between shoot and roots. (**B**) Heatmap showing differentially expressed genes between shoots and roots at the different dehydration, desiccation, and rehydration stages. DS: Dehydrated Shoot, DR: Dehydrated Root, RS: Rehydrated Shoot, RR: Rehydrated Root.

2.4. Tripogon loiiformis Stress-Associated MiRNAs Targets

MicroRNAs exert their post-transcriptional gene silencing role through base complementarity pairing with their cognate mRNA transcripts, leading to cleavage or translation repression. A large number of gene targets were predicted from *T. loliiformis* contigs annotated from RNAseq data generated in the same experiment [40], *S. italica* coding sequences, and *A. thaliana* transcripts preloaded in the psRNATarget genomic library for target prediction. A total of 1236 unique contigs targeted by conserved miRNAs were predicted in *T. loliiformis* (Supplemental Table S3). The majority of the predicted targets were orthologs of known conserved miRNAs targets involved in a broad range of biological processes including metabolism, response to abiotic stress, post-transcriptional gene silencing, regulation of development, and gene expression. The targets included transcription factors (TFs), protein kinases, transporters, chaperones, antioxidants, and carbohydrate metabolism associated genes. The transcription factors and other genes targeted by the miRNAs were interpreted to be differentially expressed based on the accumulation of their cognate miRNAs. The TFs associated with down-regulated miRNAs included members of Auxin response factor (ARF 5, 13, 22) targeted by miR162a, miR160a, and miR164b; *Squamosa* promoter binding-like protein (SPL 17, 18, 19) targeted by miR157d and miR156c; MYB (miR396, miR1128, miR159a), MADs-box (miR157a, 164b, miR1128), NAC (miR390), bHLH (miR408), and WRKY 38, 39, 70 targeted by miR167a, miR395b, and miR390. The TFs associated with up-regulated miRNAs included GATA 12 (miR396a),

Scarecrow (miR171c), WRKY 4, 14 (miR5139, miR2916), ethylene-responsive transcription factor (ERF1) (miR5021), AP2/EREBP (miR166), bZIP (miR5021), and the nuclear factor Y (NFY A, C) as targets of miR169b and miR167a, respectively. However, it is worth noting that some members of the same transcription factor family, for example, WRKY, SPL, and MYB, showed disparate expression (Table S3). The predicted chaperones and heat shock proteins and factors targeted by miRNAs included HSP70, HSP90, and DNAJ. The antioxidants ascorbate peroxidase 4-like, peroxidase, and glutathione-s-transferase were among the targets for down-regulated miRNAs. A number of carbohydrate metabolism associated gene targets with a low accumulation of their cognate miRNAs included sucrose synthase (miR195a), Galactinol-sucrose galactosyltransferase (miR1128), fructokinase-4-like (miR319), hexokinase (miR319g:l, miR319a-d:f:h), and sucrose-phosphate synthase (miR167c) (Table S3). The accumulation of sucrose during dehydration and decline in glucose and fructose levels has been previously reported in a number of desiccation tolerant plants, including *T. loliiformis* [12,41].

The majority of the miRNAs targeting protein kinases were down-regulated during dehydration, indicating the active involvement of kinases in stress response signaling. The predicted kinases included class members of calcium-dependent protein kinase 1, 2, 3 targeted by miR157, miR169p, and miR159a, and CBL-interacting protein kinase as a target of miR408e. The expression of some kinases such as cysteine-rich receptor-like protein kinase (miR156h), MAP kinase (miR166h), serine/threonine-protein kinase (miR529), and phosphatidylinositol 4-kinase (miR408) was enhanced, while family members of some kinases displayed disparate expression. Transport associated members of ABC transporters, aquaporin NIP and H^+ antiporter, were targeted by down-regulated miRNAs, while the K^+ antiporter and aquaporin TIP and SIP were targets of up-regulated miRNAs during dehydration stress (Table S3).

2.5. Functional Roles of MiRNA Targets

MiRNAs control various cellular physiological, biochemical, and molecular processes. To investigate the potential functions and biological relevance of the predicted miRNA target genes, Gene ontology (GO) and enrichment analysis was conducted using Blast2GO [42] and GO terms viewed on Cytoscape [43]. The over-represented GO terms for the up-regulated miRNA gene targets included development associated processes such as cell development, growth and regulation of cell morphogenesis, metabolic and DNA catabolic processes, and regulation of programmed cell death. Other processes were related to vesicle fusion, membrane fusion, and protein transport and localization (Supplemental Figure S3). Over-represented metabolic activities included oxidoreductase, hydrolyase, nuclease, and ligase. The biological processes associated with down-regulated miRNA target genes included stress responses, defense and innate immunity, lipid and fatty acids metabolism, membrane transport, gene expression regulation, and post-transcriptional gene silencing (Supplemental Figure S4).

3. Discussion

MiRNAs research has been extensively conducted and their critical role in many biological processes enumerated [17,24,25,44]. In plants, most miRNA studies have focused on drought sensitive and tolerant species and no studies on desiccation responsive miRNAs in desiccation tolerant plants have been reported. Drought sensitive and tolerant plants employ stress tolerance mechanisms geared towards the restoration of homeostasis through water retention, albeit at different efficacies. Although resurrection plants utilize similar mechanisms at mild and moderate dehydration stress, they are primed for desiccation and utilize additional desiccation tolerance mechanisms to limit cellular damage to repairable levels through maintenance of the cellular structure and physiological integrity [2]. Based on the different response strategies, it is plausible that resurrection plants employ distinct regulatory processes at the post-transcriptional level that may not be observed in other vascular plant species. In this study, we performed a genome-wide profiling of miRNAs and their expression and targets in the desiccation tolerant *T. loliiformis*. High-throughput sequencing of small RNA libraries confirmed the evolutionary conservation of miRNAs through the identification of conserved miRNAs in *T. loliiformis*.

The high number of miRNAs identified suggests the critical role of miRNAs in gene expression regulation and their indispensable position in desiccation tolerance strategies in resurrection plants. The elevated accumulation of miRNAs in desiccated tissues suggested the existence of enhanced regulatory activities as dehydration ensued. However, the presence of miRNAs at a desiccated state could imply the repackaging and storage of miRNAs as the tissues dehydrated below 40% RWC, evidenced by high numbers during rehydration (Figure 2). The high number of miRNAs expressed during rehydration could be attributed to the regulation of cellular processes for protection against rehydration associated damage observed in resurrection plants [45].

Recent transcriptomics studies in resurrection plants indicated transcripts expression at extreme dehydration and attributed signal transduction proteins and retroelements to the instrumental role of gene silencing during desiccation [13,14,46]. In *T. loliiformis*, a similar trend was observed, where miRNAs expression was recorded at a desiccated state. Tissue specific expression was observed where miRNAs were induced in shoots and their expression was suppressed in roots (Figure 2). The expression disparity pointed to an inherent stress adaptation strategy where miRNAs act as master modulators of processes associated with energy metabolism, growth, and development in the shoots, while redirecting resources to promote stress tolerance and enhance protective mechanisms, some of which could be targeted to the roots. In Arabidopsis miR393, miR390/159, and miR159, TIR1, TCP/MYB, and SBP-LIKE, respectively, were induced under biotic and abiotic stress to suppress development-related processes for morphological adaptation to stress [17]. In resurrection plants, transcriptional reprogramming redirecting resources from growth processes towards cellular protection has been previously reported [13].

The majority of plant miRNAs target transcription factors that bind to conserved *cis*-acting promoter elements to affect the gene expression response, particularly those induced by abiotic stress [47]. Among the predicted transcription factors associated with down-regulated miRNAs during desiccation in *T. loliiformis* were nuclear factor Y NF-YA, NF-YB, and NF-YC targeted by miR169m, miR528, and miR167a, respectively. NF-Y is a CCAAT-DNA binding transcription factor that imparts significant tolerance to drought and facilitates increased yields in corn [48]. Members of the squamosa promoter-binding-like proteins (SPL) involved in the temporal regulation of shoot development and phase transition were differentially expressed, as previously reported [49]. The SPL transcription factors SPL17 (miR156a/c) and SPL18 (miR157) were cognates for down-regulated miRNAs and SPL6 (miR166), SPL12 (miR157), and SPL15 (miR5059) were predicted to be down-regulated. Under water deficit conditions, SPL are usually down-regulated as growth and developmental-associated processes including shoot development, vegetative phase transition, flowering, and leaf polarity are normally shutdown [17]. The observed differential expression implies that some members of SPL have stress tolerance-related functions in *T. loliiformis*, as recently reported in wheat [50]. The predicted members of the WRKY transcription factors modulate many processes in plants, including the response to abiotic and biotic stress [51]. The expression of WRKY 4, 38, 39, and 70 showed enhanced expression under extreme dehydration, pointing to their abiotic stress regulatory roles. In Arabidopsis, WRKY 70 was implicated in the positive regulation of defence, negative regulation of senescence, and modulation of osmotic stress tolerance through the regulation of stomatal aperture [52,53]. In the resurrection plant *Haberlea rhodopensis*, WRKY transcription factors were induced during desiccation [13]. The observed delayed senescence and early photosynthesis shutdown in *T. loliiformis* could be attributed to the regulation of WRKY 70 by miR390. Recent studies have associated NAC TFs with an unfolded protein response (UPR) in the ER stress response [54], while the regulation of GAMYB and ARF by miR159 and miR160 has been reported [49].

Protein synthesis and folding machinery in the endoplasmic reticulum (ER) can be compromised under unfavorable conditions, resulting in ER stress that activates the UPR [55,56]. The ER resident molecular chaperones play a critical role in the folding of newly synthesized proteins, maintenance of proteome integrity, and protein homeostasis [57]. In *T. loliiformis*, DNAJ and calnexin chaperones were predicted to be targeted by miR408 and miR5021, respectively. Calnexin (Cnx), an integral membrane protein, coordinates the processing of newly synthesized N-linked glycoproteins. DNAJ, on the other hand, binds directly on the luminal binding protein BiP, an HSP70 molecular chaperone that interacts with the newly synthesized polypeptides [58]. The observed down-regulation of miR408 and miR5021 would enhance the accumulation of DNAJ and calnexin chaperones to counter the ER stress during dehydration stress in *T. loliiformis*. Heterologous expression of *Oryza sativa* calnexin in tobacco conferred dehydration tolerance under mannitol stress [59].

Carbohydrate metabolism is a key process observed in resurrection plants during desiccation [7,60]. The accumulation of sucrose, oligosaccharides, and compatible solutes such as Late Embryogenesis Abundant (LEA) proteins and small heat shock proteins during desiccation leads to cellular stabilization through cytosolic vitrification [61]. Desiccation-induced accumulation of non-reducing trehalose sugar in *T. loliiformis* was recently associated with the induction of the cytoprotective autophagy pathways [12]. Comparative metabolic analysis of sucrose accumulation between desiccation tolerant and sensitive *Eragrostis nindensis* confirmed the crucial role of sucrose in desiccation tolerance, as sucrose only accumulated in the leaves of tolerant species [62]. Sucrose accumulation correlated with the expression of carbohydrate metabolic enzymes during desiccation, as previously reported in the resurrection plants *Craterostigma plantagineum* and *Haberlea rhodopensis* [13]. In *T. loliiformis*, the observed down-regulation of miRNAs targeting sucrose synthesis genes could lead to the enhanced expression of transcripts encoding sucrose synthase, sucrose 6-phosphate synthase, sucrose transporter, and galactinol synthase.

The lack of cellular damage and severe oxidative stress suggests elaborate protective mechanisms involving energy metabolism, growth, and development programs that regulate the induction of stress response mechanisms to provide a cushion against water loss. Differential miRNA expression analysis indicated distinct tissue specific accumulation patterns between the shoots and roots under desiccation stress. The majority of miRNAs were up-regulated in the shoots and down-regulated in the roots. For example, miR528a/b predicted to target LEA proteins and Calmodulin-like proteins (CML21) was down-regulated in shoots and had no expression in the roots. The accumulation of LEA proteins as a stress response mechanism has been previously reported. LEA proteins were reported to accumulate in shoot and scutellar, but not in root tissue of desiccation tolerant wheat seedling [63]. A variety of stress responses are mediated by Ca^{2+} signaling. For example, in rice, the multi-stress responsive gene2 (OsMSR2) was induced by multiple abiotic stress stimuli, and its overexpression in Arabidopsis enhanced tolerance to drought and salinity [64]. Chlorophyll biosynthesis is controlled by the scarecrow SCL27 gene through the regulation of photochlorophyllide oxidoreductase [65]. In *T. loliiformis*, miR171b predicted to target SCL27 was up-regulated in shoots and had zero expression in the roots. Disparate miRNA expression has been previously reported in other plants under diverse stress conditions [17], but not under desiccation tolerance. The observed spatiotemporal miRNAs expression in *T. loliiformis* suggests a unique adaptation strategy where extensive regulatory processes take place in the shoots, including early photosynthesis shutdown, induction of antioxidant systems, stress signaling, and resources mobilized and redirected to the roots to enhance survival.

4. Materials and Methods

4.1. Plant Materials and Stress Treatments

Tripogon loliiformis seeds from a single plant originally collected from Charleville in South western Queensland, Australia, were germinated in 65 mm plastic pots containing 50% native red soil and seed sowing potting mix in a growth chamber at 27 °C and for a 16 h photoperiod. The *Tripogon loliiformis* biological cycle completes in 12–14 weeks after germination. The plants are in their vegetative stage between two to six weeks and reproductive stage from the seventh to eighth week [66]. For this study, the dehydration experiment was conducted at the sixth week when the plants were in the vegetative stage. Twenty four hours prior to dehydration, fifteen pots containing multiple plants were well-watered to saturation. Three replicate samples from the hydrated plants were randomly collected. Since *T. loliiformis* is a small plant of approximately 5 cm in height, the entire aerial part (shoots) and roots, except the corm, were sampled per plant. The remaining plants were dehydrated by withholding water until they were air dry and their RWC dropped below 10% after six days [9]. During dehydration, triplicate shoot and root samples were collected at 60%, 40%, and <10% RWC. The collected samples were snap frozen in liquid nitrogen and stored at −80 °C until RNA extraction. Desiccated plants were watered and rehydrated samples collected after 48 h. The percentage RWC was determined using the leaf tissues and calculated according to Barrs, Weatherley [67]. *T. loliiformis* leaves were weighed upon sampling to get the fresh weight (FW). The leaves were placed in petri dishes containing water in a 4 °C fridge for 4–6 h. The leaves were blotted dry with a paper towel and weighed to get their turgid weight (TW). The leaves were dried overnight in a vacuum oven at 80 °C. The following day, the samples were cooled at room temperature in an aspirator to avoid error due to condensation, after which the dry weight (DW) was weighed. The percentage RWC of the leaf tissue was calculated using the formula

$$RWC\,(\%) = ((\text{Fresh Weight} - \text{Dry Weight})/(\text{Turgid Weight} - \text{Dry Weight})) \times 100.$$

4.2. Total RNA Extraction and High Throughput Sequencing

Total RNA was extracted from shoot and root tissues harvested from hydrated, dehydrating (60%, 40% RWC), desiccated (<10% RWC), and rehydrated plants using the Trizol Reagent (Invitrogen) according to the manufacturer's instructions with modifications. Briefly, 50 mg of sample was ground in liquid nitrogen and transferred into a microfuge tube containing 1 mL of Trizol reagent, mixed thoroughly, and incubated for 5 min at room temperature. A total of 200 µL of chloroform was added, mixed by vortex, and centrifuged for 15 min at 4 °C, 14,000 rpm. Then, 350 µL of the top aqueous phase was homogenized in a Qiagen QIAshredder spin column and centrifuged at maximum speed for 2 min, and the flow-through was transferred into a new 2 mL collection tube for ethanol precipitation by adding 0.5 times 96% ethanol. SmallRNAs enriched total RNA bidding was done by transferring the mixture into a miRNeasy™ minElute™ spin column. Total RNA quality and purity were checked with the NanoDrop® 2000 spectrophotometer (Thermo Scientific, Wilmington, DE, USA) and the integrity and quality were verified using a Bioanalyser (Agilent technologies, Santa Clara, CA, USA) (Supplemental Figure S5). Five triplicate shoot and root small RNAs libraries were prepared using the TruSeq™ Small RNA Sample Preparation protocol from Illumina® and 50 bp single-end reads sequenced at Texas A&M AgriLife Genomics and Bioinformatics service, USA, using an Illumina HiSeq 2500 Sequencer (Illumina Inc., San Diego, CA, USA). The reads have been deposited in the Sequence Read Archive (SRA) at NCBI, Accession number SRP113187 (https://www.ncbi.nlm.nih.gov/Traces/study/?acc=SRP113187).

4.3. Small RNAs Data Analysis and MiRNAs Identification

Small RNA reads were pre-processed for quality and trimmed for removal of the 3′ adapter sequences using CLC genomics workbench 6.52 [68]. The pre-processed reads were mapped against the plant miRNA sequences in the miRBase release 21 [36] using CLC genomics workbench 6.52 [68]. Due to the absence of *T. loliiformis* genomic information at the time of this study, annotated miRNA sequences from 71 plant species in miRBase [36] were used as proxy references against which the miRNAs were annotated. A mismatch allowance of two was used in the conserved miRNA search. The 5′ mature sequences in the size range of 20–22 nt were filtered from the resultant hits as potential conserved miRNAs.

4.4. MiRNAs Expression Analysis

The expression profiles of identified miRNAs were determined using CLC genomics Workbench. The raw small RNA reads were mapped against the identified miRNAs to determine miRNA expression values in the shoot and root tissues across respective dehydration and rehydration states. To determine the differential expression of miRNAs throughout dehydration, desiccation, and rehydration, the expression values were enumerated and normalized against total read counts reported as reads per million (RPM). Using the hydrated stage as the control, all hydration stages were compared to identify the miRNAs that were constitutively expressed, induced, or suppressed. MiRNAs with a fold change of ±2 and a Bonferroni and FDR corrected p-value ≤ 0.05 were considered significant.

4.5. MiRNAs Targets and Their Functional Enrichment

Gene targets of the conserved miRNAs were predicted using the plant small RNA Target Analysis Server, psRNATarget [69], with the following parameters; maximum expectation (0–5):3, length of complementarity (15–30):19, and maximum energy to un-pair the target site (UPE, 0–100):25. The *Tripogon loliiformis* de novo assembled transcriptome dataset [40] and mRNA sequences from the closely related *S. italica*, as well as Arabidopsis, were used for the targets prediction. To identify the functional categories of the differentially expressed miRNA targets (Table 1), *T. loliiformis* transcriptome functional mapping encompassing all the plant metabolic pathways and enzyme functions was developed using the MapMan Mercator tool and visualization of predicted miRNA targets against the mapping was conducted using MapMan [70]. For Gene Ontology (GO) enrichment analysis, a BLAST search of *T. loliiformis* contigs against NCBI non-redundant (nr) Plantae sequences was conducted. The BLAST output was analyzed for GO enrichment with the Fischer exact test and a p-value of 0.05 to identify over-represented GO terms using Blast2GO [42]. The results were visualized using the Cytoscape BiNGO plugin [43].

Table 1. MapMan functional categories of potential targets of differentially expressed conserved miRNAs during dehydration in *Tripogon loliiformis*.

Name	Contig Number	Accession Number	Target Description
ABIOTIC STRESS RESPONSE			
MIR408	Contig_1263	XP_004951273.1	BAG family molecular chaperone regulator 6
MIR172j	Contig_4654	EMS50802.1	DnaJ homolog subfamily C member 7
MIR3630	Contig_195	CAA47948.2	heat shock protein 70
MIR6250	Contig_29066	XP_003578630.1	hydrophobic protein OSR8-like
MIR166d	Contig_10495	XP_003557818.1	probable methyltransferase PMT2-like
CARBOHYDRATE METABOLISM			
MIR319a	Contig_91996	XP_003568782.1	hexokinase-7-like
MIR172h	Contig_914	XP_004984086.1	sucrose synthase 2-like
MIR167c	Contig_12310	XP_004965756.1	sucrose-phosphate synthase 3-like
MIR394c	Contig_1	XP_003566746.1	callose synthase 7-like
MIR166b	Contig_1239	XP_004984579.1	galactinol synthase 2-like
MIR167e	Contig_3855	XP_004958079.1	xylose isomerase-like
DEVELOPMENT ASSOCIATED			
MIR5021	Contig_16313	XP_004966491.1	senescence-associated protein DIN1-like
MIR172b	Contig_2082	XP_004962343.1	serine/threonine-protein kinase TOR-like
MIR528	Contig_16524	XP_004951509.1	serine-threonine kinase receptor protein
MIR156c	Contig_10275	XP_004957197.1	squamosa promoter-binding-like protein 17-like
MIR156	Contig_10275	XP_004957197.1	squamosa promoter-binding-like protein 17-like
HORMONE METABOLISM			
MIR160d	Contig_39233	EMT29346.1	1-aminocyclopropane-1-carboxylate synthase 7
MIR167h	Contig_16664	XP_004961973.1	ABSCISIC ACID-INSENSITIVE 5-like protein
MIR164a	Contig_14909	XP_004985140.1	Allene oxide synthase 2-like
MIR172b	Contig_975	XP_004956401.1	auxin transport protein BIG-like
MIR171e	Contig_10826	XP_004976258.1	IAA-amino acid hydrolase ILR1-like 5-like

Table 1. Cont.

Name	Contig Number	Accession Number	Target Description
DNA REPAIR/SYNTHESIS/CHROMATIN STRUCTURE			
MIR166h	Contig_3556	XP_003576514.1	DNA mismatch repair protein Msh6-1-like
MIR167a	Contig_548	XP_004975516.1	DNA repair helicase XPB1-like
MIR172a	Contig_49068	NP_001151792.1	DNA binding protein
MIR319g:l	Contig_5793	NP_001065884.1	HUA enhancer 2
MIR399a:e	Contig_961	XP_004961835.1	probable histone H2A.4-like

5. Conclusions

Molecular studies have highlighted significant differences between resurrection plants and other dehydration sensitive species at the gene expression level. Some of the observed differences could be attributed to the post-transcriptional regulatory role of miRNAs. In this study, the results obtained suggest that *T. loliiformis*, a desiccation tolerant grass, elicits significant stress responses at the post-transcriptional level. *T. loliiformis* survival under extreme conditions could be attributed to efficient regulatory processes. Our findings in this study will enhance the understanding of *T. loliiformis* miRNA-mediated regulatory mechanisms and provide a foundation for further investigation of the gene pool of resurrection plants that could be holding many answers to agricultural crops.

Supplementary Materials: The following are available online at http://www.mdpi.com/2223-7747/7/3/68/s1, Figure S1: Annotation of *Tripogon loliiformis* conserved miRNAs using miRBase database, Figure S2: Cross-comparison Venn diagram showing the number of differentially expressed genes between shoots and roots, Figure S3: Gene Ontology of overrepresented GO terms associated with the targets of down-regulated miRNAs, Figure S4: Gene Ontology of overrepresented GO terms associated with the targets of up-regulated miRNAs, Figure S5: Representative Electrophoresis assay image of total RNA used for high-throughput sequencing. RNA with RNA Integrity Number (RIN) above 7.5 was used for sequencing library preparation, Table S1: List of conserved miRNAs identified in *Tripogon loliiformis*, Table S2: List of differential expressed conserved miRNAs in shoots and roots during dehydration, Table S3: List of predicted miRNAs targets associated with *T. loliiformis* contigs.

Author Contributions: Conceived and designed the experiments: I.N., B.W., C.C.-G., S.M., M.B.D., and X.Z. Performed the experiments: I.N., B.W., and C.C.-G. Analyzed the data: I.N., B.W., S.M., M.B.D., and X.Z. Wrote the paper: I.N., B.W., and S.M.

Funding: S.M. was supported by a QUT Professional Capacity Grant, B.W. was supported by a QUT Vice Chancellor's Research Fellowship, and I.N. was supported by The Australian Department of Foreign Affairs and Trade (DFAT) under the Australian Awards—Africa PhD scholarship.

Acknowledgments: This work was supported by an Australia Awards - Africa PhD scholarship (I.N.), QUT Vice Chancellor's Research Fellowship (B.W.), and a professional capacity development grant (S.M.). Computational resources and services used in this work were provided by the HPC and Research Support Group, QUT, Brisbane, Australia. Library preparation and Illumina sequencing was performed by Genomics and Bioinformatics services at Texas A&M University. The authors also thank Peraj Karbaschi for assistance with the plant propagation.

Conflicts of Interest: The authors declare no conflict of interest. The funders had no role in the design of the study; in the collection, analyses, or interpretation of data; in the writing of the manuscript, and in the decision to publish the results.

References

1. Höfler, K.; Migsch, H.; Rottenburg, W. Über die Austrocknungresistenz landwirtschaftlicher Kulturpflanzen. *Forschungsdienst* **1941**, *12*, 50–61.
2. Rascio, N.; Rocca, N.L. Resurrection Plants: The Puzzle of Surviving Extreme Vegetative Desiccation. *Crit. Rev. Plant Sci.* **2005**, *24*, 209–225. [CrossRef]
3. Scott, P. Resurrection Plants and the Secrets of Eternal Leaf. *Ann. Bot.* **2000**, *85*, 159–166. [CrossRef]
4. Bernacchia, G.; Salamini, F.; Bartels, D. Molecular characterization of the rehydration process in the resurrection plant Craterostigma plantagineum. *Plant Physiol.* **1996**, *111*, 1043–1050. [CrossRef] [PubMed]
5. Farrant, J.M. A comparison of mechanisms of desicassion tolerance among 3 angiosperms resurrection plant species. *Plant Ecol.* **2000**, *151*, 29–39. [CrossRef]

6. Moore, J.P.; Farrant, J.M. A Systems-Based Molecular Biology Analysis of Resurrection Plants for Crop and Forage Improvement in Arid Environments. In *Improving Crop Resistance to Abiotic Stress*; Wiley-VCH Verlag GmbH & Co. KGaA: Weinheim, Germany, 2012; pp. 399–418. [CrossRef]
7. Peters, S.; Mundree, S.G.; Thomson, J.A.; Farrant, J.M.; Keller, F. Protection mechanisms in the resurrection plant Xerophyta viscosa (Baker): Both sucrose and raffinose family oligosaccharides (RFOs) accumulate in leaves in response to water deficit. *J. Exp. Bot.* **2007**, *58*, 1947–1956. [CrossRef] [PubMed]
8. Asami, P.; Mundree, S.; Williams, B. Saving for a rainy day: Control of energy needs in resurrection plants. *Plant Sci.* **2018**, *271*, 62–66. [CrossRef] [PubMed]
9. Karbaschi, M.R.; Williams, B.; Taji, A.; Mundree, S.G. Tripogon loliiformis elicits a rapid physiological and structural response to dehydration for desiccation tolerance. *Funct. Plant Biol.* **2016**. [CrossRef]
10. Orchard, E.; Wilson, A. Flora of Australia. In *Poaceae Australian Biological Resources Study*; K Mallett, T.M., Ed.; CSIRO: Canberra, Australia, 2005.
11. Gaff, D.; Pate, J.S. *The Biology of Resurrection Plants*; Pate, J.S., McComb, A.J., Eds.; The Biology of Australian Plants, University Western Australia Press: Nedlands, Australia, 1981; pp. 114–146.
12. Williams, B.; Njaci, I.; Moghaddam, L.; Long, H.; Dickman, M.B.; Zhang, X.; Mundree, S. Trehalose Accumulation Triggers Autophagy during Plant Desiccation. *PLoS Genet.* **2015**, *11*, e1005705. [CrossRef] [PubMed]
13. Gechev, T.S.; Benina, M.; Obata, T.; Tohge, T.; Sujeeth, N.; Minkov, I.; Hille, J.; Temanni, M.-R.; Marriott, A.S.; Bergström, E. Molecular mechanisms of desiccation tolerance in the resurrection glacial relic Haberlea rhodopensis. *Cell. Mol. Life Sci.* **2013**, *70*, 689–709. [CrossRef] [PubMed]
14. Rodriguez, M.C.S.; Edsgard, D.; Hussain, S.S.; Alquezar, D.; Rasmussen, M.; Gilbert, T.; Nielsen, B.H.; Bartels, D.; Mundy, J. Transcriptomes of the desiccation-tolerant resurrection plant Craterostigma plantagineum. *Plant J.* **2010**, *63*, 212–228. [CrossRef] [PubMed]
15. Zhou, J.; Wang, X.; Jiao, Y.; Qin, Y.; Liu, X.; He, K.; Chen, C.; Ma, L.; Wang, J.; Xiong, L. Global genome expression analysis of rice in response to drought and high-salinity stresses in shoot, flag leaf, and panicle. *Plant Mol. Biol.* **2007**, *63*, 591–608. [CrossRef] [PubMed]
16. Zhou, L.; Liu, Y.; Liu, Z.; Kong, D.; Duan, M.; Luo, L. Genome-wide identification and analysis of drought-responsive microRNAs in Oryza sativa. *J. Exp. Bot.* **2010**, *61*, 4157–4168. [CrossRef] [PubMed]
17. Khraiwesh, B.; Zhu, J.K.; Zhu, J. Role of miRNAs and siRNAs in biotic and abiotic stress responses of plants. *Biochim. Biophys. Acta* **2012**, *1819*, 137–148. [CrossRef] [PubMed]
18. Sunkar, R. MicroRNAs with macro-effects on plant stress responses. *Semin. Cell Dev. Biol.* **2010**, *21*, 805–811. [CrossRef] [PubMed]
19. Lagos-Quintana, M.; Rauhut, R.; Lendeckel, W.; Tuschl, T. Identification of novel genes coding for small expressed RNAs. *Science* **2001**, *294*, 853–858. [CrossRef] [PubMed]
20. Park, W.; Li, J.; Song, R.; Messing, J.; Chen, X. CARPEL FACTORY, a Dicer Homolog, and HEN1, a Novel Protein, Act in microRNA Metabolism in Arabidopsis thaliana. *Curr. Biol.* **2002**, *12*, 1484–1495. [CrossRef]
21. Zhao, T.; Li, G.; Mi, S.; Li, S.; Hannon, G.J.; Wang, X.-J.; Qi, Y. A complex system of small RNAs in the unicellular green alga Chlamydomonas reinhardtii. *Genes Dev.* **2007**, *21*, 1190–1203. [CrossRef] [PubMed]
22. Yang, T.W.; Xue, L.G.; An, L.Z. Functional diversity of miRNA in plants. *Plant Sci.* **2007**, *172*, 423–432. [CrossRef]
23. Liu, Q.; Chen, Y.Q. Insights into the mechanism of plant development: Interactions of miRNAs pathway with phytohormone response. *Biochem. Biophys. Res. Commun.* **2009**, *384*, 1–5. [CrossRef] [PubMed]
24. Shukla, L.I.; Chinnusamy, V.; Sunkar, R. The role of microRNAs and other endogenous small RNAs in plant stress responses. *Biochim. Biophys. Acta Gene Regul. Mech.* **2008**, *1779*, 743–748. [CrossRef] [PubMed]
25. Sunkar, R.; Zhu, J.K. Novel and stress-regulated microRNAs and other small RNAs from Arabidopsis. *Plant Cell* **2004**, *16*, 2001–2019. [CrossRef] [PubMed]
26. Liu, H.H.; Tian, X.; Li, Y.J.; Wu, C.A.; Zheng, C.C. Microarray-based analysis of stress-regulated microRNAs in Arabidopsis thaliana. *RNA* **2008**, *14*, 836–843. [CrossRef] [PubMed]
27. Frazier, T.P.; Sun, G.; Burklew, C.E.; Zhang, B. Salt and Drought stresses induce the aberrant expression of microRNA genes in tobacco. *Mol. Biotechnol.* **2011**, *49*, 159–165. [CrossRef] [PubMed]
28. Li, B.; Qin, Y.; Duan, H.; Yin, W.; Xia, X. Genome-wide characterization of new and drought stress responsive microRNAs in Populus euphratica. *J. Exp. Bot.* **2011**, *62*, 3765–3779. [CrossRef] [PubMed]

29. Sun, F.; Guo, G.; Du, J.; Guo, W.; Peng, H.; Ni, Z.; Sun, Q.; Yao, Y. Whole-genome discovery of miRNAs and their targets in wheat (*Triticum aestivum* L.). *BMC Plant Biol.* **2014**, *14*, 142. [CrossRef] [PubMed]
30. Zhang, L.; Zheng, Y.; Jagadeeswaran, G.; Li, Y.; Gowdu, K.; Sunkar, R. Identification and temporal expression analysis of conserved and novel microRNAs in Sorghum. *Genomics* **2011**, *98*, 460–468. [CrossRef] [PubMed]
31. Matts, J.; Jagadeeswaran, G.; Roe, B.A.; Sunkar, R. Identification of microRNAs and their targets in switchgrass, a model biofuel plant species. *J. Plant Physiol.* **2010**, *167*, 896–904. [CrossRef] [PubMed]
32. Budak, H.; Akpinar, A. Dehydration stress-responsive miRNA in Brachypodium distachyon: Evident by genome-wide screening of microRNAs expression. *OMICS* **2011**, *15*, 791–799. [CrossRef] [PubMed]
33. Zhao, B.; Liang, R.; Ge, L.; Li, W.; Xiao, H.; Lin, H.; Ruan, K.; Jin, Y. Identification of drought-induced microRNAs in rice. *Biochem. Biophys. Res. Commun.* **2007**, *354*, 585–590. [CrossRef] [PubMed]
34. Ding, D.; Zhang, L.; Wang, H.; Liu, Z.; Zhang, Z.; Zheng, Y. Differential expression of miRNAs in response to salt stress in maize roots. *Ann. Bot.* **2009**, *103*, 29–38. [CrossRef] [PubMed]
35. Zhang, B.; Pan, X.; Cannon, C.H.; Cobb, G.P.; Anderson, T.A. Conservation and divergence of plant microRNA genes. *Plant J.* **2006**, *46*, 243–259. [CrossRef] [PubMed]
36. Griffiths-Jones, S.; Grocock, R.J.; Van Dongen, S.; Bateman, A.; Enright, A.J. miRBase: MicroRNA sequences, targets and gene nomenclature. *Nucleic Acids Res.* **2006**, *34*, D140–D144. [CrossRef] [PubMed]
37. Luo, Y.; Guo, Z.; Li, L. Evolutionary conservation of microRNA regulatory programs in plant flower development. *Dev. Biol.* **2013**, *380*, 133–144. [CrossRef] [PubMed]
38. Oliver, M.J.; Guo, L.; Alexander, D.C.; Ryals, J.A.; Wone, B.W.; Cushman, J.C. A sister group contrast using untargeted global metabolomic analysis delineates the biochemical regulation underlying desiccation tolerance in Sporobolus stapfianus. *Plant Cell Online* **2011**, *23*, 1231–1248. [CrossRef] [PubMed]
39. Gechev, T.; Dinakar, C.; Benina, M.; Toneva, V.; Bartels, D. Molecular mechanisms of desiccation tolerance in resurrection plants. *Cell. Mol. Life Sci.* **2012**, *69*, 3175–3186. [CrossRef] [PubMed]
40. Sra, N. Bioproject: Transcriptome Analysis of Tripogon Loliiformis throughout Dehydration, Desiccation and Rehydration. Available online: https://www.ncbi.nlm.nih.gov/bioproject/?term=PRJNA288839 (accessed on 22 April 2018).
41. Whittaker, A.; Bochicchio, A.; Vazzana, C.; Lindsey, G.; Farrant, J. Changes in leaf hexokinase activity and metabolite levels in response to drying in the desiccation-tolerant species Sporobolus stapfianus and Xerophyta viscosa. *J. Exp. Bot.* **2001**, *52*, 961–969. [CrossRef] [PubMed]
42. Conesa, A.; Götz, S. Blast2GO: A Comprehensive Suite for Functional Analysis in Plant Genomics. *Int. J. Plant Genom.* **2008**, *2008*, 619832. [CrossRef] [PubMed]
43. Maere, S.; Heymans, K.; Kuiper, M. BiNGO: A Cytoscape plugin to assess overrepresentation of gene ontology categories in biological networks. *Bioinformatics* **2005**, *21*, 3448–3449. [CrossRef] [PubMed]
44. Mallory, A.C.; Vaucheret, H. Functions of microRNAs and related small RNAs in plants. *Nat. Genet.* **2006**, *38*, S31–S36. [CrossRef] [PubMed]
45. Alpert, P.; Oliver, M.J. 1 Drying Without Dying. In *Desiccation and Survival in Plants: Drying without Dying*; CABI: Wallingford, UK, 2002; p. 3.
46. Farrant, J.; Cooper, K.; Hilgart, A.; Abdalla, K.; Bentley, J.; Thomson, J.; Dace, H.W.; Peton, N.; Mundree, S.; Rafudeen, M. A molecular physiological review of vegetative desiccation tolerance in the resurrection plant Xerophyta viscosa (Baker). *Planta* **2015**, *242*, 407–426. [CrossRef] [PubMed]
47. Zhu, J.K. Salt and drought stress signal transduction in plants. *Annu. Rev. Plant Biol.* **2002**, *53*, 247–273. [CrossRef] [PubMed]
48. Nelson, D.E.; Repetti, P.P.; Adams, T.R.; Creelman, R.A.; Wu, J.; Warner, D.C.; Anstrom, D.C.; Bensen, R.J.; Castiglioni, P.P.; Donnarummo, M.G. Plant nuclear factor Y (NF-Y) B subunits confer drought tolerance and lead to improved corn yields on water-limited acres. *Proc. Natl. Acad. Sci. USA* **2007**, *104*, 16450–16455. [CrossRef] [PubMed]
49. Ferdous, J.; Hussain, S.S.; Shi, B.J. Role of microRNAs in plant drought tolerance. *Plant Biotechnol. J.* **2015**, *13*, 293–305. [CrossRef] [PubMed]
50. Wang, B.; Geng, S.; Wang, D.; Feng, N.; Zhang, D.; Wu, L.; Hao, C.; Zhang, X.; Li, A.; Mao, L. Characterization of Squamosa Promoter Binding Protein-LIKE genes in wheat. *J. Plant Biol.* **2015**, *58*, 220–229. [CrossRef]
51. Wang, C.; Deng, P.; Chen, L.; Wang, X.; Ma, H.; Hu, W.; Yao, N.; Feng, Y.; Chai, R.; Yang, G.; et al. A Wheat WRKY Transcription Factor TaWRKY10 Confers Tolerance to Multiple Abiotic Stresses in Transgenic Tobacco. *PLoS ONE* **2013**, *8*, e65120. [CrossRef] [PubMed]

52. Li, J.; Besseau, S.; Törönen, P.; Sipari, N.; Kollist, H.; Holm, L.; Palva, E.T. Defense-related transcription factors WRKY70 and WRKY54 modulate osmotic stress tolerance by regulating stomatal aperture in Arabidopsis. *New Phytol.* **2013**, *200*, 457–472. [CrossRef] [PubMed]
53. Besseau, S.; Li, J.; Palva, E.T. WRKY54 and WRKY70 co-operate as negative regulators of leaf senescence in Arabidopsis thaliana. *J. Exp. Bot.* **2012**, *63*, 2667–2679. [CrossRef] [PubMed]
54. Ruberti, C.; Kim, S.-J.; Stefano, G.; Brandizzi, F. Unfolded protein response in plants: One master, many questions. *Curr. Opin. Plant Biol.* **2015**, *27*, 59–66. [CrossRef] [PubMed]
55. Howell, S.H. Endoplasmic reticulum stress responses in plants. *Annu. Rev. Plant Biol.* **2013**, *64*, 477–499. [CrossRef] [PubMed]
56. Williams, B.; Verchot, J.; Dickman, M. When Supply Does Not Meet Demand-ER Stress and Plant Programmed Cell Death. *Front. Plant Sci.* **2014**, *5*, 211. [CrossRef] [PubMed]
57. Kim, Y.E.; Hipp, M.S.; Bracher, A.; Hayer-Hartl, M.; Ulrich Hartl, F. Molecular chaperone functions in protein folding and proteostasis. *Annu. Rev. Biochem.* **2013**, *82*, 323–355. [CrossRef] [PubMed]
58. Gupta, D.; Tuteja, N. Chaperones and foldases in endoplasmic reticulum stress signaling in plants. *Plant Signal. Behav.* **2011**, *6*, 232–236. [CrossRef] [PubMed]
59. Sarwat, M.; Naqvi, A.R. Heterologous expression of rice calnexin (OsCNX) confers drought tolerance in Nicotiana tabacum. *Mol. Biol. Rep.* **2013**, *40*, 5451–5464. [CrossRef] [PubMed]
60. Ghasempour, H.; Gaff, D.; Williams, R.; Gianello, R. Contents of sugars in leaves of drying desiccation tolerant flowering plants, particularly grasses. *Plant Growth Regul.* **1998**, *24*, 185–191. [CrossRef]
61. Farrant, J.M.; Moore, J.P. Programming desiccation-tolerance: From plants to seeds to resurrection plants. *Curr. Opin. Plant Biol.* **2011**, *14*, 340–345. [CrossRef] [PubMed]
62. Illing, N.; Denby, K.J.; Collett, H.; Shen, A.; Farrant, J.M. The signature of seeds in resurrection plants: A molecular and physiological comparison of desiccation tolerance in seeds and vegetative tissues. *Integr. Comp. Biol.* **2005**, *45*, 771–787. [CrossRef] [PubMed]
63. Ried, J.L.; Walker-Simmons, M. Group 3 late embryogenesis abundant proteins in desiccation-tolerant seedlings of wheat (*Triticum aestivum* L.). *Plant Physiol.* **1993**, *102*, 125–131. [CrossRef] [PubMed]
64. Bender, K.W.; Snedden, W.A. Calmodulin-Related Proteins Step Out from the Shadow of Their Namesake. *Plant Physiol.* **2013**, *163*, 486–495. [CrossRef] [PubMed]
65. Ma, Z.; Hu, X.; Cai, W.; Huang, W.; Zhou, X.; Luo, Q.; Yang, H.; Wang, J.; Huang, J. Arabidopsis miR171-Targeted Scarecrow-Like Proteins Bind to GT cis-Elements and Mediate Gibberellin-Regulated Chlorophyll Biosynthesis under Light Conditions. *PLOS Genet.* **2014**, *10*, e1004519. [CrossRef] [PubMed]
66. Le, T.T.T. Molecular and Functional Characterisation of an Osmotin Gene from the Resurrection Plant Tripogon Loliiformis. Ph.D. Thesis, Queensland University of Technology, Brisbane, Australia, 2018. Available online: https://eprints.qut.edu.au/115835 (accessed on 30 August 2018).
67. Barrs, H.; Weatherley, P. A re-examination of the relative turgidity technique for estimating water deficits in leaves. *Aust. J. Biol. Sci.* **1962**, *15*, 413–428. [CrossRef]
68. Qiagen. CLC Genomics Workbench. Available online: https://www.qiagenbioinformatics.com/products/clc-genomics-workbench/ (accessed on 22 April 2018).
69. Dai, X.; Zhao, P.X. psRNATarget: A plant small RNA target analysis server. *Nucleic Acids Res.* **2011**, *39*, W155–W159. [CrossRef] [PubMed]
70. Lohse, M.; Nagel, A.; Herter, T.; May, P.; Schroda, M.; Zrenner, R.; Tohge, T.; Fernie, A.R.; Stitt, M.; Usadel, B. Mercator: A fast and simple web server for genome scale functional annotation of plant sequence data. *Plant Cell Environ.* **2014**, *37*, 1250–1258. [CrossRef] [PubMed]

 © 2018 by the authors. Licensee MDPI, Basel, Switzerland. This article is an open access article distributed under the terms and conditions of the Creative Commons Attribution (CC BY) license (http://creativecommons.org/licenses/by/4.0/).

Review

Biology and Function of miR159 in Plants

Anthony A. Millar *, Allan Lohe and Gigi Wong

Division of Plant Science, Research School of Biology, The Australian National University,
Canberra ACT 2601, Australia
* Correspondence: tony.millar@anu.edu.au

Received: 10 May 2019; Accepted: 23 July 2019; Published: 30 July 2019

Abstract: MicroR159 (miR159) is ancient, being present in the majority of land plants where it targets a class of regulatory genes called *GAMYB* or *GAMYB-like* via highly conserved miR159-binding sites. These *GAMYB* genes encode R2R3 MYB domain transcription factors that transduce the gibberellin (GA) signal in the seed aleurone and the anther tapetum. Here, *GAMYB* plays a conserved role in promoting the programmed cell death of these tissues, where miR159 function appears weak. By contrast, *GAMYB* is not involved in GA-signaling in vegetative tissues, but rather its expression is deleterious, leading to the inhibition of growth and development. Here, the major function of miR159 is to mediate strong silencing of *GAMYB* to enable normal growth. Highlighting this requirement of strong silencing are conserved RNA secondary structures associated with the miR159-binding site in *GAMYB* mRNA that promotes miR159-mediated repression. Although the miR159-*GAMYB* pathway in vegetative tissues has been implicated in a number of different functions, presently no conserved role for this pathway has emerged. We will review the current knowledge of the different proposed functions of miR159, and how this ancient pathway has been used as a model to help form our understanding of miRNA biology in plants.

Keywords: miR159; *GAMYB*; programmed cell death; aleurone; tapetum; vegetative growth; flowering

1. Introduction

Associated with the emergence and diversification of land plants, is a core set of conserved microRNA (miRNA) families that arose early in terrestrial plant evolution and which are conserved in modern day plant species [1]. This conservation implies these endogenous gene regulators are fundamental to plant biology and have been indispensable for the conquest of plant life on land. One such core family is microR159 (miR159), which has now been extensively studied in multiple, diverse plant species. In this review, we will highlight the major functions identified for miR159, its use as a model for gaining greater insights into miRNA biology in general, and finally highlight the many outstanding questions surrounding this ancient gene regulator.

2. MiR159 is Strongly Conserved and Highly Abundant Throughout the Plant Kingdom

Surveys of the many deep sequencing experiments on the small RNA fractions of plants, find miR159 ubiquitously present as a 21 nucleotide (nt) miRNA in all eudicots and monocotyledonous plants examined [2], and present in the majority of basal angiosperms, gymnosperms, ferns, and lycopods examined [3]. It has either been classified as a Class I (ubiquitous) or a Class II (present in most taxonomic groups) miRNA [2,3]. There is some uncertainty regarding whether miR159 is present in Bryophytes, where it is generally regarded as being absent [4], but it has been reported in a liverwort [5]. However, the reported miR159 sequence was only 18 nt long, suggesting it may not be a genuine miR159 homologue, so further analyses will be required to resolve this. Nevertheless, it is apparent that miR159 arose early in basal land plants and has been strongly conserved henceforth.

MiR159 fits the curious observation that the stronger the miRNA is conserved, the greater its expression or abundance [2]. This was derived from a multitude of small RNA-sequencing experiments from a wide diversity of plant species, where miR159 is often among the most abundant small RNA species (e.g., [6–10]). Additionally, highly similar miR159 isoforms are present in most land plants (Figure 1; [11]), so the sequence of this canonical miR159 appears to have remained fixed for 100s of millions of years. Nevertheless, like most miRNA families, considerable variation exists within small RNAs defined as miR159, with most plant species containing multiple family members that encode identical or highly similar isoforms, or "isomiRs", that differ by one to several nucleotides. For example, maize has 11 different *MIR159* loci, encoding four different miR159 isoforms [12]. For the most part, nucleotide variation occurs at the extremities of the miRNA, at positions considered less important for its specificity [13]. This is the case for the three different miR159 isoforms found in Arabidopsis that vary by 1–2 nucleotides; however, as these isoforms appear functionally redundant, this variation unlikely impacts which genes they target for repression [14,15]. Some species have even more variant miR159 isoforms (e.g., poplar; grape, soybean, and maize with 3–5 sequence variations [16]), so whether these miR159 variants have sub-functionalized to regulate different targets remains a possibility. Indeed, the ancient miRNA miR319 is closely related to miR159. In Arabidopsis, these two families are identical at 17 of 21 nucleotide positions, but have distinct target genes, demonstrating their sub-functionalization [17]. Their similarity extends to their primary-*MIRNA* precursors, where *pri-MIR159* and *pri-MIR319* are both unusually long fold-back structures that are processed in a non-canonical loop-to-base direction [18]. Phylogenetic analysis of primary-*MIRNA* precursor sequences of these families supports a common origin of miR319 and miR159, with the likelihood that miR159 has arisen and specialized from miR319 in basal land plants [19].

3. GAMYB and GAMYB-like Genes are the Only Conserved Targets of miR159

Core to understanding the function of a miRNA is the identification of the genes that they target. A clear and recurrent theme is that miR159 targets a family of genes encoding R2R3 MYB transcription factors referred to as "GAMYB" or "GAMYB-like". Similar to the conservation of miR159, *GAMYB*-homologues with a highly conserved miR159 binding site are found in most lineages of land plants (Figure 1). This extends to basal plants such as lycopods (e.g., *Selaginella moellendorffii*), moss (e.g., *Physcomitrella patens*), and the liverwort *Marchantia polymorpha* [20,21]. However, in *Marchantia* it appears that the *GAMYB* homologue is not regulated by miR159, but rather by miR319 [21,22]. Even in Arabidopsis, miR319 can regulate the *GAMYB* targets [17]. However, as miR319 in Arabidopsis is narrowly and weakly expressed compared to the widely and abundantly expressed miR159, miR319-mediated regulation of the *GAMYB-like* genes is insignificant relative to miR159-mediated regulation [17]. This makes miR159 functionally specific for the *GAMYB-like* targets, whereas miR319 is functionally specific for genes encoding another class of transcription factors, the *TCP* family, which miR159 is unable to regulate [17,23]. Therefore, it appears likely the more specific miR159 has arisen from miR319 in basal land plants, sub-functionalizing to become specific for the *GAMYB-like* genes. Although there is sequence variation in both the miR159 and its binding site within the *GAMYB-like* genes, both have appeared to have become fixed, arguing that this ancient miR159-*GAMYB* target relationship is critical for the life of land plants [24].

Strong experimental evidence supports the prediction of conserved miR159-mediated regulation of *GAMYB*. Firstly, degradome analysis from multiple diverse species confidently identifies *GAMYB* homologues are being regulated via a miR159-mediated cleavage mechanism. Although this analysis only detects targets regulated by the miRNA-guided cleavage mechanism (and not the translational-repression mechanism), functionally important targets appear to be preferentially detected [25]. Degradome experiments have been mainly performed on higher plants, including eudicots such as Arabidopsis [25], soybean [26], cotton [27], tomato [28], orchids [29], and peach [30]; also monocots such as wheat [31], rice [32], and barley [33], among many others, all of which experimentally validate *GAMYB* homologues as targets of miR159. Although many of these degradome experiments also pick up other target genes (e.g., [27,31]), these other targets are diverse in their identity and do not appear to be broadly conserved miR159 targets; i.e., they are not identified in degradomes from multiple diverse species. This argues that although miR159 may regulate additional targets, this does not appear to be at the expense of its main target, *GAMYB*. For instance, in tomato, miR159 has acquired a novel target, a gene that encodes a protein with a NOZZLE-like domain, and this miR159-mediated regulation is important for tomato development [34]. However, miR159 still regulates *GAMYB-like* genes in tomato, which is important for fruit development [35].

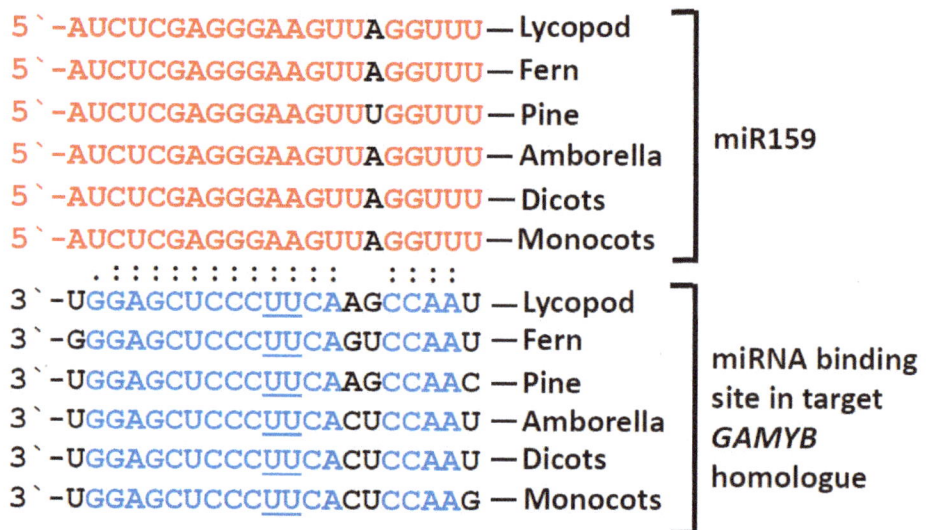

Figure 1. The microR159 (miR159)-*GAMYB* regulatory pathway appears highly conserved in land plants. Similar/identical miR159 isomiRs (shown in red) are found in most plant linages, including Lycopods (*Selaginella uncinata;* [3]) ferns (*Salvinia cucullata;* [3]), pine (*Pinus densata;* [36]), Amborella [37], dicots, and monocots (miRbase; [11]). Highly similar and complementary miR159 binding sites (shown in blue) are found in *GAMYB* homologues from lycopods (*Selaginella moellendorffii;* [20]), ferns (*Salvinia cucullata;* [11]), pine (*Larix kaempferi;* [38]), Amborella and many different monocots and dicots. Variant nucleotide positions are shown in black. However, throughout the plant kingdom, variation is not limited to these positions; for example, see [39].

4. The miR159-*GAMYB* Pathway in Arabidopsis

The Arabidopsis miR159 family has been extensively studied as a model for plant miRNA-mediated gene regulation (Figure 2). Arabidopsis has three *MIR159* genes (*MIR159a*, *MIR159b*, and *MIR159c*), each encoding a distinct isoform that differ from one another by 1–2 nucleotides [8]. Examination of their expression domains with promoter: GUS constructs found *MIR159a: GUS* and *MIR159b: GUS* had highly similar expression patterns, being broadly expressed throughout the plant, but strongest in shoot and root meristematic regions [14]. By contrast, the expression domain of a *MIR159c: GUS* reporter gene was much narrower, being restricted mainly to anthers and the shoot apical region [15], suggesting sub-functionalization. Regarding their level of expression, both deep sequencing and qPCR has found miR159a to be the most abundant family member, with miR159c being very weakly expressed [8,15]. To investigate their function, T-DNA loss-of-function mutant alleles were generated for each gene, however, none of the single *mir159* mutant plants displayed any phenotypic defects [14,15]. However, consistent with the highly similar expression domains, miR159a and miR159b were demonstrated to be functionally redundant, as a double *mir159a.mir159b* (*mir159ab*) mutant displayed severe growth and developmental defects, most notably a smaller rosette with upwardly curled leaves (Figure 2; [14]).

As a triple *mir159abc* mutant appeared indistinguishable from *mir159ab*, this and other data suggested miR159c in Arabidopsis has little to no activity and possibly corresponds to a pseudogene [15,17]. This is one of the few instances in Arabidopsis where T-DNA mutants have been identified and combined for all members of a miRNA family, and the *mir159ab* and *mir159abc* mutants have been used extensively in the functional characterization of miR159.

A bioinformatic search of miR159 targets in Arabidopsis using the standard target prediction program psRNATarget, identifies almost 100 potential miR159 targets with four or less mismatches [40]. The top twenty targets are shown in Table 1, which includes eight *MYB* genes with highly conserved miR159 binding sites [15]. By contrast, the non-*MYB* genes are highly diverse and their miR159-binding sites do not appear conserved [15]. Of the conserved *MYB* targets, seven are *GAMYB-like* genes (*MYB33*, *MYB65*, *MYB81*, *MYB97*, *MYB101*, *MYB104*, and *MYB120*), and the other is non-*GAMYB-like* gene, *DUO1* (*DUO POLLEN1*), which has a conserved miR159 binding site at a position distinct from the *GAMYB-like* genes [17]. Despite the fact that miR159-mediated cleavage products can be isolated for many of these predicted targets (Table 1), transcript profiling of the *mir159ab* mutant only identified two genes that appeared strongly de-regulated, the *GAMYB-like* targets, *MYB33* and *MYB65* (Table 1). This de-regulation resulted in *MYB33* and *MYB65* being strongly expressed throughout the plant [14,41]. Consistently, the only genes detected in multiple degradome analyses were *MYB33* and *MYB65* (Table 1) [25,42]. Eliminating the expression of these genes via the introduction of *myb33* and *myb65* loss-of-function alleles, suppressed all vegetative phenotypic defects of *mir159ab*, as a *mir159ab.myb33.myb65* quadruple mutant appeared indistinguishable from wild-type, other than male sterility [14]. The phenotype of male sterility is the only apparent defect of *myb33.myb65* plants, as *MYB33* and *MYB65* are two redundant genes that facilitate anther development (Figure 2) [43].

These genetic experiments demonstrated the major role of miR159 in Arabidopsis as being the widespread suppression of *MYB33* and *MYB65* expression, whose activity has severe deleterious impacts on plant growth and development, including stunted growth and curled leaves (Figure 2). The experiments also defined the functional specificity of miR159 in Arabidopsis as being *MYB33* and *MYB65* [14]. Supporting this is the expression of either a miR159-resistant *MYB33* or miR159-resistant *MYB65* transgene, both of which can phenocopy the *mir159ab* mutant [14,23,44]. This much narrower functional specificity compared to the bioinformatic prediction of many more targets is a common theme in miRNA biology, where both in animal and plants, pleiotropic defects of miRNA mutants can be suppressed via the repression of one-two target genes, despite bioinformatic programs predicting many targets with conserved miRNA binding sites [24,45]. Partially explaining this phenomenon for miR159, many of the bioinformatically predicted miR159 targets appear to have transcriptional domains that are mutually exclusive to that of miR159. Hence, the miRNA and targets are physically separated spatially and/or temporally preventing interaction (Figure 2) [14,15].

Despite their deleterious impact on vegetative growth, *MYB33* and *MYB65* appear ubiquitously transcribed throughout the plant, but only to be strongly and ubiquitously silenced, other than in seeds and anthers (Figure 2) [41,46]. There are multiple lines of evidence supporting this claim; (1) the vegetative phenotype of *myb33.myb65* appears indistinguishable from wild-type; (2) the transcriptome profiles of shoot apical regions of wild-type versus *myb33.myb65* appear indistinguishable; (3) the expression of a *MYB33: GUS* transgene is undetectable in GUS-stained vegetative tissues, but a miR159 resistant version of the reporter gene, *mMYB33: GUS*, is widely and strongly expressed (Figure 2) [41]. The efficiency of this silencing is highlighted by the *mir159a* single mutant; although deep sequencing demonstrates miR159a is the predominant isoform (e.g., miR159a–6621 reads, miR159b–982 reads [8]), the *mir159a* mutant appears indistinguishable from wild-type [14], implying strong reductions in miR159 levels do not impact the silencing of *MYB33/MYB65*. Conversely, overexpression of a wild-type *MYB33* gene fails to result in any phenotypic defects [47]. Although these *MYB33* overexpressing Arabidopsis plants have high *MYB33* mRNA levels, they do not exhibit any phenotypic defects, indicating miR159 also represses expression of *MYB33/MYB65* mRNA via a translational repression mechanism [47]. The importance of this mechanism was shown via the complementation of *mir159ab* with a mutated miR159 variant that had two mismatches to *MYB33/MYB65* at the cleavage site; although this attenuated cleavage, this miR159 variant could still potently silence *MYB33/MYB65* [47]. Therefore, these combined silencing mechanisms ensure *MYB33/MYB65* are strongly repressed in vegetative tissues.

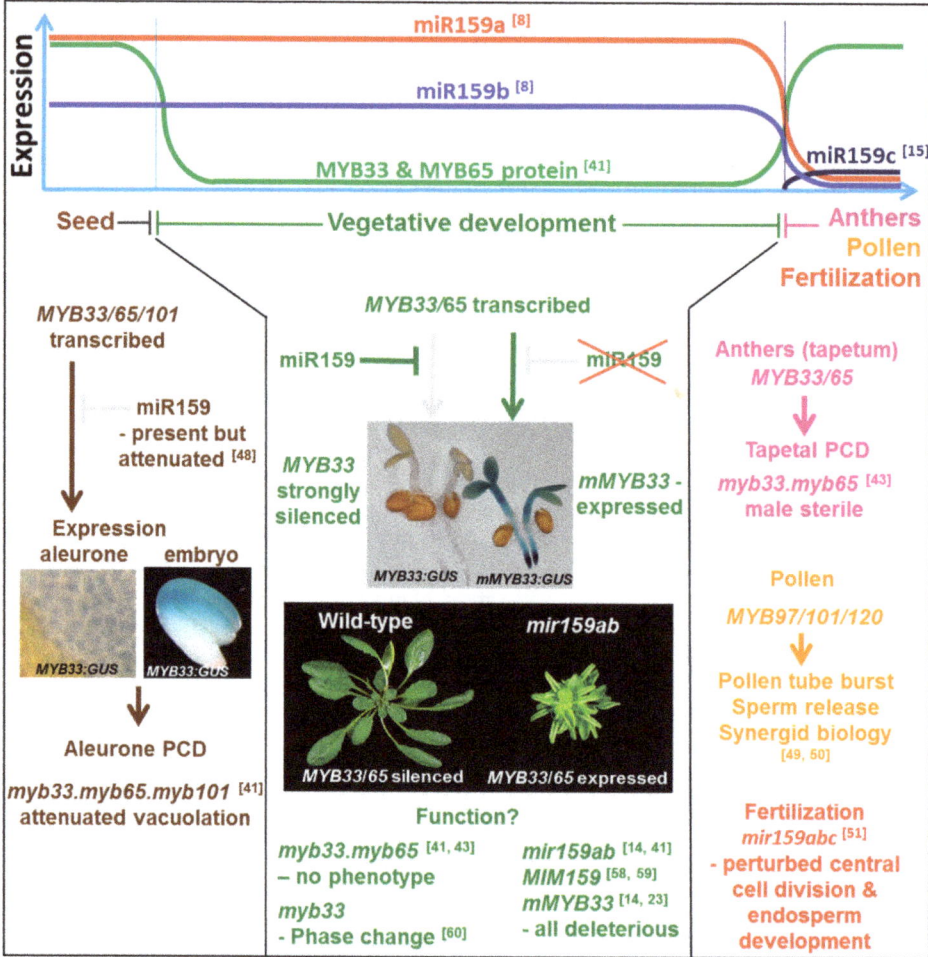

Figure 2. The *miR159-GAMYB-like* pathway in Arabidopsis. miR159a is the predominant family member, being expressed in seed and throughout plant development at a constantly high level, but it is absent in anthers [8,14,46]. miR159b is expressed at a lower level than miR159a [8,15,46], but its expression pattern appears highly similar to miR159a [14]. miR159c, is weakly expressed and appears mainly confined to anthers [15]. In seeds, miR159 efficacy appears attenuated [48], enabling *GAMYB-like* gene expression which promotes PCD of the aleurone [41]. In contrast, throughout vegetative development, miR159 efficacy is strong, and *MYB33/65* expression is strongly silenced. Only via inhibition of miR159, or mutation of the miR159 binding site within *MYB33* or *MYB65*, will expression occur, which leads to strong deleterious outcomes, such as stunted growth and curled leaves [14,41]. Although the function of the pathway has been suggested to be involved in flowering-time and phase change, the purpose of this pathway in vegetative development is still unclear. In anthers, miR159 activity is low. Here, *MYB33* and *MYB65* are expressed to promote PCD in the tapetum [43]. *MYB97/101/120* expression is required for pollen function [49,50]. Finally, miR159 is required for fertilization [51].

Table 1. miR159 targets in Arabidopsis as determined by different approaches. The top 20 miR159a targets in Arabidopsis as identified by the bioinformatic program psRNATarget with standard search parameters [40]; the number of mismatches is indicated by the score. Confirming this prediction, 5'-RACE analysis can detect miR159-guided cleavage products for at least nine of these genes. In contrast, the more quantitative degradome analysis only identifies three of these genes, with only *MYB33* and *MYB65* being frequently detected in multiple degradome analyses [25,42]. Overexpression of miR159 could detect down-regulation of multiple targets [13,52]. However, genetic analysis using a loss-of-function *mir159ab* mutant identify *MYB33* and *MYB65* as the major important targets [14,41].

	At Number	Score	Name	5'-RACE	Degradome	miR159 OE	*miR159ab*
1	AT4G37770	1.5	ACS8	[53]		[13]	
2	AT2G32460	2	MYB101	[15,17,53]		[13]	
3	AT3G60460	2	DUO1	[15,17,53]			
4	AT2G26950	2	MYB104				
5	AT4G26930	2	MYB97				
6	AT5G06100	2.5	MYB33	[15,23]	[25,42]	[52]	[14,41]
7	AT3G11440	2.5	MYB65	[23]	[25,42]		[14,41]
8	AT2G34010	2.5	MRG1	[53]	[42]		
9	AT2G21600	2.5	RER1B				
10	AT5G55020	2.5	MYB120	[15]		[13]	
11	AT4G27330	2.5	SPL				
12	AT5G27395	2.5	Tim44-related				
13	AT3G61740	3	SDG14, ATX3				
14	AT1G29010	3	MRG-LIKE				
15	AT4G31240	3	NRX2				
16	AT2G26960	3	MYB81	[15]			
17	AT2G22810	3	ACS4				
18	AT3G08850	3	RAPTOR1B				
19	AT5G55930	3.5	OPT1	[13]		[13]	[41]
20	AT2G44450	3.5	beta gluc 15				

5. Conserved RNA Secondary Structures in *MYB33/65* Promote miR159-Mediated Silencing

Highlighting this efficient silencing were miR159 efficacy assays performed on the various Arabidopsis *MYB* targets [44]. Here, it was demonstrated that *MYB33* and *MYB65* were very sensitive targets of miR159, being strongly silenced. In contrast, the other *MYB* genes (*MYB81*, *MYB97*, *MYB101*, *MYB104*, and *DUO1*), were poorly silenced by miR159. As all these *MYB* targets had highly complementary miR159 binding sites, it implies factors other than complementarity must be contributing to this differential miR159-mediated silencing [44]. Correlating with this difference, is a predicted RNA secondary structure that abuts the miR159 binding site of *MYB33* and *MYB65*, but which is absent in the poorly regulated targets (Figure 3; also see [44] for RNA secondary structures of the various Arabidopsis *GAMYB-like* genes). To determine the significance of this in silico predicted RNA structure, a structure/function analysis was performed. Mutation of this structure within the *MYB33* context attenuates silencing, whereas the restoration of the structure, although with a different primary nucleotide sequence, restores strong silencing of *MYB33* [44]. Therefore, this demonstrates that this RNA secondary structure facilitates *MYB33* and *MYB65* silencing, earmarking them as functional targets of miR159. It argues that a fully functional miR159 target site of *MYB33/MYB65* encompasses nucleotides beyond that of the binding site.

Further evidence of the importance of this RNA secondary structure is its strong conservation in *GAMYB-like* homologues throughout the plant kingdom (Figure 3), as the nucleotides that correspond to the stems of the RNA secondary structures are conserved in *GAMYB* homologues of eudicots, monocots, and basal angiosperms, such as Amborella [44]. This indicates this structure is part of the miR159-*GAMYB* regulatory relationship and that the mechanism of regulation is likely more complex than miRNA-binding site complementarity alone. Given that so many miRNA-target relationships are ancient, it will be interesting to investigate how many other miRNA targets have conserved RNA elements associated with their miRNA binding sites, as these ancient regulatory relationships have had 100s of millions of years to evolve greater regulatory complexity.

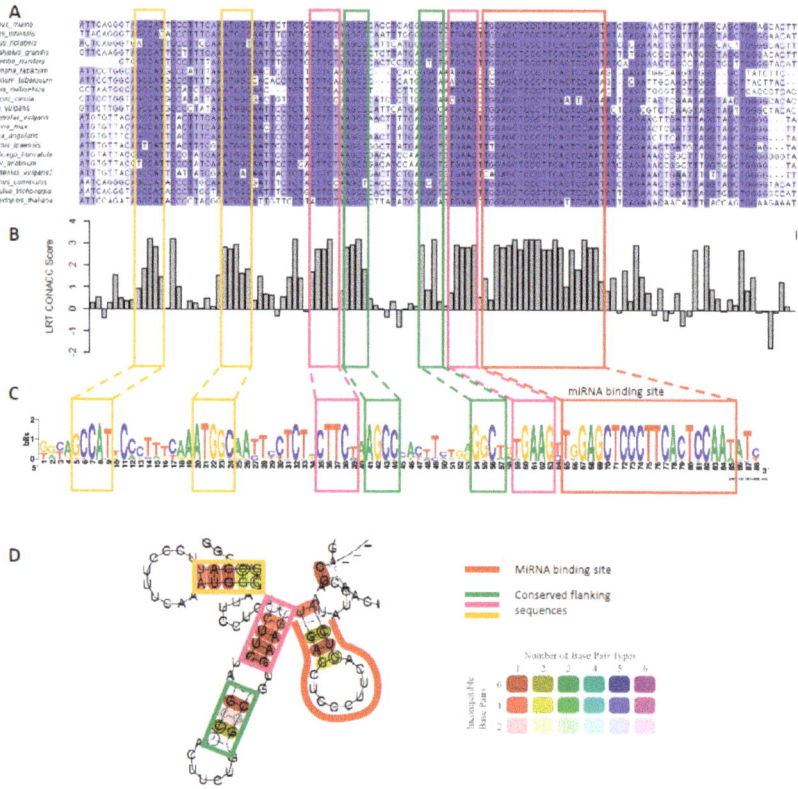

Figure 3. (A) Multiple alignment of *MYB33* homologues from different plants species. The binding site is boxed in red, and the conserved flanking sequences in yellow, pink, and green throughout the Figure. (B) phyloP score of the multiple sequence alignment of *MYB33* sequences. A positive score denotes evolutionary conservation, whereas, a negative score denotes acceleration [54]. A likelihood ratio test (LRT) was used as the method to detect non-neutral substitution rates. Scores were generated using rPHAST [55]. (C) Sequence logo of the binding site and conserved flanking sequences generated using WebLogo [56]. (D) RNA secondary structure prediction of the consensus sequence from the multiple alignment in A. generated using RNAalifold [57] at 22 °C and default parameters. Colours represent the number of base pairs types (i.e., AU, UA, CG, GC, UG, GU), and hue the number of non-conserved nucleotides at that position.

In Arabidopsis, not only do the poorly regulated *GAMYB-like* genes lack this conserved RNA structure, but they have highly specific transcriptional domains, predominantly in seeds and anthers, where miR159 activity appears attenuated or absent [14,15,49,50]. Therefore, strong selection of this RNA secondary structural element may have occurred for *GAMYB* homologues that are transcribed in vegetative tissues and require strong miR159-mediated silencing to prevent deleterious outcomes [44]. Investigating whether this also applies to other species with multiple *GAMYB* homologues will be interesting to follow up. Finally, the efficacy of miR159-mediated silencing of *MYB33* in Arabidopsis varies between tissues, being strong in the rosette, but weak in the seed [48]. As RNA secondary structures are dynamic in vivo, they may be operating as a riboswitch, with certain formations facilitating silencing, and others attenuating silencing. It will be interesting to determine whether the conformation of the RNA secondary structure changes between tissues, controlling the ability of miR159 to silence *MYB33*.

6. The Function of miR159-*MYB* Pathway in Plant Development

The functional role of the miR159-*GAMYB* pathway has been studied in numerous plant species, and this is summarized in Table 2.

Table 2. Functional analyses of the miR159-*GAMYB* pathway in plants.

Species	Approach	Phenotype	Ref.
Arabidopsis	T-DNA *mir159ab* mutant	Pleiotropic defects, stunted growth, curled leaves, reduced apical dominance	[14]
Arabidopsis	T-DNA *mir159c* mutant	none	[15]
Arabidopsis	T-DNA *mir159abc* mutant	Perturbed fertilization	[51]
Arabidopsis	MIM159 mimic–loss-of-function.	Pleiotropic defects, stunted growth, curled leaves, defective sepals, petals and anthers	[58,59]
Arabidopsis (Col-0)	miR159a overexpression	Male sterility	[13]
Arabidopsis (Ler)	miR159a overexpression	Male sterility, delayed flowering-time	[52]
Arabidopsis	T-DNA *myb33.myb65* mutant	Male sterile	[43]
Arabidopsis	T-DNA *myb33* mutant	Altered phase change	[60]
Rice	STTM159 mimic–loss-of-function	Stunted growth, curled leaves, smaller seeds	[61,62]
Rice	miR159 overexpression	Delayed heading, shorten internode I, malformed flowers, male sterility	[63]
Rice	*gamyb-1* insertion mutant	Male sterility	[64]
Barley	miR159 overexpression	Male sterility	[65]
Wheat	miR159 overexpression	Delayed heading, male sterility, increased tillering	[66]
Gloxinia	MIM159 mimic loss-of-function, miR159 over-expression (OE)	Accelerated flowering (MIM159) or delayed flowering (miR159 OE)	[67]
Tomato	miR159 overexpression	Fruit set, parthenocarpy, ovule development, seedless fruits	[35]
Cucumber	RNAi against *GAMYB*	Altered ratio of male to female flowers	[68]
Strawberry	RNAi against *GAMYB*	Inhibition of receptacle ripening	[69]

6.1. A Role in Male Reproductive Development

The GAMYB/GAMYB-like family of transcription factors is found throughout the plant kingdom, where they share high sequence similarity in their R2R3 DNA-binding domains located towards the N-terminal region, but are much more diverse in their C-terminal regions [20]. Nevertheless, the functions of these *GAMYB* homologues appear to be highly conserved, as *GAMYB* homologues from *Lycopods* or Bryophytes can partially complement a *gamyb-2* rice mutant [20], or a cucumber *GAMYB* can complement the male sterile phenotype of the Arabidopsis *myb33.myb65* mutant [68]. Hence, despite the sequence diversity of the C-terminal regions, this complementation of distant species argues the biochemical function of GAMYB has been strongly conserved.

To date, a role in male reproductive development appears the clearest function for *GAMYB* [70]. Inhibition of its activity perturbs male development, whether in basal plants such as lycopods (*Selaginella moellendorffii*) or bryophytes (moss-*Physcomitrella patens*) [20], or in higher flowering plants, such as rice [64] or Arabidopsis [43]. Moreover, GAMYB was shown to positively regulate the *CYP703* gene, which is required for male development in both basal and higher plants [20,71]. It appears this *GAMYB-CYP703* pathway arose very early in land plant development, and then has come under the control of gibberellin (GA) in lycopods, likely explaining why male reproductive development in plants is under the control of GA [20,71]. It was speculated that the GA regulation of the *GAMYB-CYP703* pathway was a step in the evolution of the sporophyte-dominated life cycle, which requires greater regulatory control for its more complex reproductive system [20].

Consistently, there have been many reports of plants with multiple *GAMYB* homologues for which at least one is strongly transcribed in anthers (e.g., *CsGAMYB1* in cucumber, [68]; *TaGAMYB1* in wheat, [66]; *HvGAMYB* in barley, [65]), or is anther-specific (e.g., *PtrMYB012* in poplar, [16]; *MYB97*, *MYB120* in Arabidopsis, [49]), many of which are positively regulated by GA. Inhibition of GAMYB activity perturbs programmed cell death (PCD) in the anther tapetum [71], where in both a rice *gamyb* mutant, and the Arabidopsis *myb33.myb65* mutant, the tapetum fails to degenerate, resulting in hypertrophy, leading to male sterility [43,64,71]. Additionally, *MYB33* and *MYB65* in Arabidopsis are also required for the formation of the radial microtubule array surrounding nuclei immediately following meiosis II [72]. In the *myb33.myb65* mutant, the resulting defects in male meiotic cytokinesis produce diploid pollen with a defective pollen wall morphology [72]. However, the role that miR159 plays in regulating GAMYB expression in male development is unclear, where it may be fine-tuning expression or preventing expression occurring in particular cell layers, but for most species, this is yet to be resolved. In rice anthers, miR159 and *GAMYB* are co-expressed, suggesting potential fine-tuning of *GAMYB* expression [63].

In Arabidopsis, the role of the miR159-*GAMYB* pathway and its interaction with the miR319-*TCP* pathway in flower maturation has been investigated using miRNA loss-of-function *MIM159* and *MIM319* transgenic plants, both of which display multiple pleiotropic defects [59]. In sepals, petals and anthers of these plants, it was found that the GAMYB and TCP proteins are expressed and directly interact to regulate another miRNA, miR167, which creates a miR159-miR319-miR167 network. It is proposed that the function of miR159/miR319 is to dampen MYB/TCP expression, resulting in low miR167, and hence enabling strong ARF6/8 expression, which in turn regulates many genes required for floral development including that of auxin signaling [59]. However, in wild-type plants, it appears *MYB33/65* expression in flowers is restricted to anthers, and the *myb33.myb65* mutant only displays anther defects [43]. Therefore, in wild-type, it appears the role of miR159 in flowers is to strongly repress the *MYB* genes in sepals and petals to prevent strong expression of miR167 to ultimately enable strong ARF6/8 expression.

MiR159 is present in pollen where it has a crucial role in fertility [51]. It has been known for some time that sperm cell entry alone triggers central cell division, suggesting that the male genome and/or unknown factors transmitted by the sperm control the initiation of endosperm development. Unexpectedly, the central cell usually fails to initiate division after pollination by *mir159abc* mutants, or stops dividing after one or two divisions, resulting in reduced seed set. It was found that both *MYB33* and *MYB65* are highly expressed in the central cell of the embryo sac before fertilization, but after fertilization, both transcripts are rapidly cleared from the central cell and the endosperm initiates development. It was observed that *MYB33* and *MYB65* transcripts are not cleared in pollinations with *mir159abc* pollen, suggesting that miR159 in pollen is transmitted to the central cell by fertilization where it degrades *MYB33* and *MYB65* transcripts [51]. Thus, miR159 has a paternal effect on seed development: miR159 carried in pollen abolishes central cell repression after fertilization permitting endosperm nuclear divisions [51]. Loss of maternal miR159 also results in seed defects but these defects are less severe on seed set than loss of paternal miR159, and the mechanism of this maternal effect is unknown.

6.2. A Role in Seed Development

GAMYB was first identified as a GA signaling component in the barley aleurone, hence giving these MYB genes their name "GA"MYB [73]. Here, GAMYB positively transduces the GA signal to activate expression of α-amylase and other hydrolytic enzymes [74], as well as promote PCD in the aleurone [75]. This latter function appears conserved in Arabidopsis, as a *myb33.myb65.myb101* triple mutant has attenuated vacuolation in aleurone cells, a PCD-mediated process that is positively regulated by GA [41]. Therefore, a conserved role in PCD in the aleurone and tapetum in both monocots and dicots is currently a unifying function for these GAMYB transcription factors. Curiously, both these tissues are single cell layers that provide nutrients upon death to the embryo (aleurone) or pollen (tapetum). It is possible other *GAMYB-like* genes may play similar roles in terms of inhibiting growth and promoting cell death. For example, *MYB97*, *MYB101*, and *MYB120* are all expressed in the pollen tube, and in a *myb97.myb101.myb120* mutant, the pollen tube fails to undergo growth arrest and then fails to degenerate in order to release the sperm cells to the ovules [49,50].

Many downstream targets of the miR159-*GAMYB* pathway in Arabidopsis support a role in PCD. Micro-array analysis on the shoot apical region of *mir159ab* plants found that of the 166 up-regulated genes, many appeared aleurone related [41]. Many of these aleurone related genes were also down-regulated in *myb33.myb65.myb101* seeds, making them strong candidates of being downstream of *GAMYB* activity [41]. This includes the most up-regulated gene in *mir159ab*, *CYSTEINE PROTEINASE 1* (*CP1*), whose expression appears tightly correlated with *GAMYB* expression [41,46], and corresponds to a class of enzymes which have been associated with PCD and cell lysis. Similarly, inhibition of miR159 in transgenic rice results in up-regulation of pathways associated with PCD, suggesting that *GAMYB* promotes these pathways in the rice grain [61]. Again, what role miR159 plays in regulating GAMYB activity in the seed is unclear. In Arabidopsis germinating seeds, miR159 and *MYB33* are co-transcribed in the aleurone and embryo, however, MYB33 protein is expressed, which suggests miR159 may only be fine-tuning the expression of *MYB33* in this tissue [48]. Nevertheless, *mir159ab* plants produce malformed seeds [14], implying miR159 is required for proper seed development.

6.3. The Role of miR159-GAMYB Pathway in Vegetative Tissues

In Arabidopsis, the widespread transcription of *MYB33/MYB65*, only to be strongly silenced by miR159 raises the question of what is the purpose of this seemingly futile regulatory pathway. Although miR159 is sometimes associated with leaf development due to the smaller, upwardly curled leaves of the *mir159ab* mutant (Figure 2), this phenotype appears more a consequence of the deleterious impact of *MYB33/MYB65* expression rather than the alteration of a developmental program [41]. In general, de-regulated expression of GAMYB in leaves results in strong perturbation of growth. This was shown in *mir159ab*, as well as transgenic Arabidopsis expressing miRNA decoys to inhibit miR159 function, either *MIMIC159* (*MIM159*), Short target tandem *MIMIC159* (*STTM159*), or *SPONGE159* (*SP159*) [58,76], or with *STTM159* rice [61,62], which all result in the similar phenotypic defect of stunted growth. For instance, *STTM159* rice plants are shorter than wild-type rice, with decreased cell numbers, and the most down-regulated genes in *STTM159* rice are associated with cell division. Therefore, the main role of rice miR159 is to suppress *GAMYB* expression to enable cell proliferation [61]. Likewise, the expression of miR159-resistant *GAMYB* transgenes in Arabidopsis [14,16,23,44], lead to the same phenotypic defects. Therefore, it is clear that these *GAMYB* genes encode a class of transcription factors that when expressed inhibit growth, a phenotype contrary to a role in promoting the GA signal for which they were originally identified. Supporting this, GA treatments do not alter the RNA levels of *MYB33*, *MYB65* or miR159 in Arabidopsis rosettes, and the response of *myb33.myb65* plants to GA is not perturbed in vegetative tissues [41]. Therefore, the role the *GAMYB* in transducing the GA signal appears to be tissue dependent, where it is involved in transducing the GA signal in seeds and anthers [70,73,77], but not in vegetative tissues [41]. Supporting this, the Arabidopsis *myb33.myb65* or rice *gamyb* mutants do not appear to have any obvious phenotypic defects at the vegetative stage [43,64].

Contrary to the growth inhibition phenotype of the leaves, the roots of *mir159ab* Arabidopsis are longer than wild-type and have a larger apical meristem zone. Thus, in roots, *GAMYB* expression appears to enhance cell cycle progression, leading to extended roots. However, the root lengths of *myb33.myb65* or *myb33.myb65.myb101* plants were unchanged compared to wild-type, again indicating that these *GAMYB-like* genes are likely fully silenced in roots, again raising the question of what is the role of this pathway in this vegetative tissue [78].

6.4. A Role of miR159 in Controlling GA-Mediated Flowering-Time and Growth?

This clear role in inhibiting growth appears at odds with a role often ascribed to *GAMYB* in promoting flowering-time [79,80]. This idea arose from the fact that GA promotes flowering-time, and that the *GAMYB* or *GAMYB-like* genes were thought to be positive regulators of GA throughout the plant [81]. Supporting this idea was the finding that the *LEAFY* gene, a central regulator of flowering, contained a MYB-binding site within its promoter, and this binding site appeared critical in transducing the GA-signal [82]. Subsequently, it was shown that *MYB33* transcription was induced at the shoot apical region upon the induction of flowering, either through GA-application or exposure to long-day conditions, and that the MYB33 protein could bind the *LEAFY* promoter in in vitro gel shift assays [81]. Then, overexpression of miR159 in Arabidopsis [ecotype Landsberg *erecta*, (L*er*)] resulted in down-regulation of *MYB33* expression, which correlated with a decrease in *LEAFY* expression and a delayed flowering-time under short-day conditions [52]. Supporting this is the manipulation of miR159 levels in other plant species that result in altered flowering-times. This includes overexpression of miR159 in rice and wheat which lead to a reduced heading-time [63,66]. Additionally, in the ornamental flowering plant *Gloxinia*, the over-expression of miR159 delayed flowering, whereas the inhibition of miR159 with a *MIM159* transgene accelerated flowering-time [67]. Unlike *MIM159* Arabidopsis or *STTM159* rice [58,59,61,62,76], *MIM159 Gloxinia* did not exhibit any defects in vegetative growth or development [67].

Such evidence argues for a clear and conserved role for miR159 in flowering-time, and that *GAMYB* is likely promoting the GA-signal with regard to flowering. However, overexpression of miR159 in Arabidopsis (ecotype Columbia) did not affect flowering-time [13]. So, although overexpression of miR159 in both ecotypes (L*er* and Columbia) resulted in male sterility due the requirement of GAMYB activity in anthers, there was a differential response with regard to flowering-time. Moreover, a *myb33.myb65* mutant (ecotype Columbia) did not have a delayed flowering-time, and the *mir159ab* mutant (greater GAMYB activity) displayed a late flowering-time under short-day conditions, implying greater GAMYB activity was inhibiting flowering [41]. Given the severe pleiotropic defects of *mir159ab*, it is uncertain whether delayed flowering is a direct result of greater GAMYB activity, or a secondary effect of the severe growth and developmental defects [41].

In addition to delayed flowering, the *mir159ab* Arabidopsis mutant was found to have a strong delay in vegetative phase change (VPC), with the first leaf with abaxial trichomes being *leaf* 16.0 as opposed to *leaf* 7.9 for wild-type, and this was tightly correlated with the increased levels of miR156, one of the key determinants of VPC [60]. In a complex regulatory mechanism, it was found that MYB33 activated transcription of both the *MIR156* gene and its target, *SPL9*, via direct interaction with their promoters [60]. Conversely, Arabidopsis plants overexpressing miR159 (*leaf* 7.3) or the *myb33* mutant (*leaf* 7.1) only had slight increases in VPC compared to wild-type (*leaf* 7.9). This argued that MYB33 protein is expressed to some extent in the Arabidopsis rosette. However, the VPC of a *myb65* mutant (*leaf* 8.1) was unchanged from wild-type, implying *MYB65* did not appear to impact this pathway [60]. Given the subtle changes to vegetative phase change in the *myb33* mutant, the miR159-*GAMYB* pathway was considered a modifier of VPC, where miR159 promotes VPC by preventing MYB33 expression which negatively regulates VPC [60]. It will be interesting to see what role the miR159-*GAMYB* pathway is found to have in this process in other plant species.

Therefore, regarding the miR159-*GAMYB* pathway in growth and flowering, there is strong conflicting evidence. Although the difference in Arabidopsis is possibly due to ecotype variation, a role for GAMYB in either promoting flowering, or alternatively, deleteriously inhibiting growth, will need further experimentation for clarification of how such diametrically opposed outcomes can arise.

6.5. Fruit and Reproductive Development

There is growing evidence that the miR159-*GAMYB* pathway plays a role in fruit development. In strawberries, fruit development is GA-regulated, and miR159 is strongly expressed in the fruit's receptacle tissue and appears to regulate *GAMYB*, as miR159 and *GAMYB* expression is reciprocal [83]. *GAMYB* is a key regulator of strawberry fruit development, as repression of *GAMYB* via RNAi inhibits receptacle ripening and color formation [69]. In tomato, the miR159-*GAMYB* pathway is present in ovules, and overexpression of miR159 resulted in abnormal ovule development, precocious fruit initiation and seedless fruits [35]. Similarly, in grapes, the pathway appears active in the fruits, and under the control of GA [84]. In the monoecious plant cucumber, inhibition of *GAMYB* activity via RNAi altered the ratio of male to female flowers, decreasing the number of nodes with male flowers [68]. Therefore, it appears this pathway is involved in many different functions of the reproductive process in different plant species.

7. The Function of the miR159-*MYB* Pathway in Plant Stress

7.1. Abiotic Stress

Given the ubiquity and abundance of miR159 throughout the plant kingdom, it is not surprising that numerous studies have implicated miR159 in a wide range of stresses from many different plant species (for review see [85]). In Arabidopsis, miR159 levels increase under salinity [86], and in germinating seeds, miR159 has been found to accumulate in response to the stress hormone ABA as well as to drought [87]. MiR159 also accumulates to higher levels in response to drought in maize, wheat and barley [85]. Such results suggests that increased levels of miR159 may result in greater stress tolerance. However, in some species, miR159 levels decrease in response to drought or salinity [85], and overexpression of miR159 in rice resulted in increased sensitivity to heat-stress [66]. In potato, in which the drought tolerant gene *cap-binding 80* protein has been down-regulated, miR159 levels were decreased and mRNA levels of *GAMYB-like* homologues were higher [88]. Therefore, these studies have found no consistent or unified role for miR159 in plant stress response.

The functional role of the Arabidopsis miR159-*GAMYB* pathway to abiotic stress was investigated by comparing the response a mutant lacking this entire pathway, the *mir159ab.myb33.myb65* quadruple mutant, to that of wild-type plants [46]. Two-week old plants were exposed to three weeks of treatments with either ABA, high temperature, high light, drought or cold. However, no differential response between the *mir159ab.myb33.myb65* mutant and wild-type plants were identified. As it was demonstrated that miR159 fully represses *MYB33* and *MYB65* in vegetative tissues of Arabidopsis plants [41], it was rationalized that miR159 levels would need to decrease in Arabidopsis to enable activation of these two *GAMYB-like* genes [46]. However, none of the treatments appeared to repress miR159 to levels in which would allow *MYB33* and *MYB65* expression, and this was supported by assaying the downstream marker gene *CP1*, whose levels appeared completely repressed [46]. Based on this, no clear role for this pathway was identified, and it remains uncertain what role it plays in stress response in Arabidopsis.

7.2. Biotic Stress

Similarly, the levels of miR159 respond to many different biotic stresses. Recently it was shown that cotton and Arabidopsis accumulate elevated levels of miR159 in response to the fungus, *Verticillium dahlia* [89]. MiR159 was exported into the fungal hyphae, where it targeted the gene encoding isotrichodermin C-15 hydroxylase (HiC-15), which is critical for hyphal growth. As expression of a miR159-resistant HiC-15 gene in *V. dahlia* resulted in greater virulence, it was concluded that exporting miR159 from the plant was conferring greater pathogenic resistance. Given that the miR159-binding site is highly conserved in HiC-15, it was hypothesized that this has evolved to dampen HiC-15 expression as to avoid rapid death of the host, which then enables establishment of the fungus on the plant [89]. Currently, this is the only clear role for miR159 in pathogen response.

MiR159 also accumulates to higher levels in Arabidopsis root galls that form in response to root knot nematodes (RKN). The *MYB33* gene appears dynamically expressed during gall formation, as a *MYB33: GUS* reporter was expressed during early gall development, but not at later stages. Functional evidence for the involvement of the miR159-*GAMYB* pathway is that an Arabidopsis *mir159abc* triple mutant has greater resistance to root knot nematodes (RKN) [90]. Further investigation will be needed to understand the precise role of the pathway in gall formation and the response pathway to RKN infection.

8. Conclusions and Some Unresolved Questions

The miR159-*GAMYB* pathway appears nearly ubiquitous in terrestrial plants, implying it has played an important role in plant's conquest of the land. Although its role in some tissues now appear to be relatively clear, this is far from the case in others. Below are some of the questions we believe still need to be resolved.

1. Why are *MYB33* and *MYB65* transcribed in vegetative tissues where failure to fully repress them results in a detrimental effect? What selective advantage does this give the plant?

a. One hypothesis is that if miR159 is inhibited by a certain trigger, and strong *MYB33/65* expression occurs, growth inhibition (or another unknown process) may result in a beneficial outcome (e.g., drought conditions to slow growth). However, currently, no triggers to inhibit miR159 to enable strong *MYB* expression are known.
  b. A second hypothesis would be that *MYB33/65* are not silenced in all vegetative tissues, but in certain cells they are expressed where they confer a selective advantage. Some evidence suggests *GAMYB* is involved in the transition to flowering, and VPC in Arabidopsis. But currently there is much conflicting data. For instance in Arabidopsis, overexpressing miR159 represses flowering-time, and inhibition of miR159 represses VPC. Other studies have found no role for miR159 in flowering. More work is needed here to clarify these roles, and how conserved they are across species.
2. Why is expression of *GAMYB* in vegetative tissues deleterious and how does it inhibit growth? What down-stream events are these genes triggering? Although some studies have started to address this, more work is needed for a clearer understanding.
3. Is *GAMYB* function related to the way it is regulated, i.e., strongly transcribed, only to then be strongly silenced by miR159? Does miR159 have a role in stress response? Again, many studies have identified changes to miR159 levels in response to a host of different biotic/abiotic stresses, but currently there is no clearly defined role for this miR159 concerning stress tolerance/response.
4. How does the conserved RNA secondary structure associated with the miR159-binding sites of *GAMYB* genes promote their silencing by miR159? Can this structure facilitate a complex regulatory mechanism, enabling strong silencing in some tissues, but poor silencing in others, depending on a dynamic secondary structure configuration? i.e., acting like a riboswitch concerning silencing.
5. What is the role of miR159-mediate regulation on non-*GAMYB* targets? For example, *DUO1* has a conserved miR159-binding site, but the role of miR159 in controlling the expression of this gene remains unclear.
6. What is the role of miR159 in female fertility? Why are Arabidopsis *mir159ab* seeds small and misshapen (likewise rice STTM159 grains are small)? Why does the central cell still divide in some *mir159abc* ovules? How can a seed still form (from *mir159abc* pollen) with a viable embryo when the endosperm divisions stop apparently so early?

Author Contributions: All authors contributed to the writing of the review.

Funding: This research received no external funding.

Acknowledgments: G.W. was supported by an Australian Government Research Training Program RTP Scholarship.

Conflicts of Interest: The authors declare no conflict of interest.

References

1. Axtell, M.J.; Bartel, D.P. Antiquity of microRNAs and their targets in land plants. *Plant Cell* **2005**, *17*, 1658–1673. [CrossRef] [PubMed]
2. Chávez Montes, R.A.; de Fátima Rosas-Cárdenas, F.; De Paoli, E.; Accerbi, M.; Rymarquis, L.A.; Mahalingam, G.; Marsch-Martínez, N.; Meyers, B.C.; Green, P.J.; de Folter, S. Sample sequencing of vascular plants demonstrates widespread conservation and divergence of microRNAs. *Nat. Commun.* **2014**, *5*, 3722. [CrossRef] [PubMed]
3. You, C.; Cui, J.; Wang, H.; Qi, X.; Kuo, L.Y.; Ma, H.; Gao, L.; Mo, B.; Chen, X. Conservation and divergence of small RNA pathways and microRNAs in land plants. *Genome Biol.* **2017**, *18*, 158. [CrossRef] [PubMed]
4. Axtell, M.J.; Meyers, B.C. Revisiting criteria for plant microRNA annotation in the era of big data. *Plant Cell* **2018**, *30*, 272–284. [CrossRef] [PubMed]
5. Alaba, S.; Piszczalka, P.; Pietrykowska, H.; Pacak, A.M.; Sierocka, I.; Nuc, P.W.; Singh, K.; Plewka, P.; Sulkowska, A.; Jarmolowski, A.; et al. The liverwort Pellia endiviifolia shares microtranscriptomic traits that are common to green algae and land plants. *Plant J.* **2014**, *80*, 331–344.

6. Fahlgren, N.; Howell, M.D.; Kasschau, K.D.; Chapman, E.J.; Sullivan, C.M.; Cumbie, J.S.; Givan, S.A.; Law, T.F.; Grant, S.R.; Dang, J.L.; et al. High-throughput sequencing of Arabidopsis microRNAs: Evidence for frequent birth and death of *MIRNA* genes. *PLoS ONE* **2007**, *2*, e219. [CrossRef] [PubMed]
7. Jeong, D.H.; Park, S.; Zhai, J.; Gurazada, S.G.R.; De Paoli, E.; Meyers, B.C.; Green, P.J. Massive analysis of rice small RNAs: Mechanistic implications of regulated microRNAs and variants for differential target RNA cleavage. *Plant Cell* **2011**, *23*, 4185–4207. [CrossRef]
8. Rajagopalan, R.; Vaucheret, H.; Trejo, J.; Bartel, D.P. A diverse and evolutionarily fluid set of microRNAs in *Arabidopsis thaliana*. *Genes Dev.* **2006**, *20*, 3407–3425. [CrossRef] [PubMed]
9. Szittya, G.; Moxon, S.; Santos, D.M.; Jing, R.; Fevereiro, M.P.; Moulton, V.; Dalmay, T. High-throughput sequencing of Medicago truncatula short RNAs identifies eight new miRNA families. *BMC Genom.* **2008**, *9*, 593. [CrossRef]
10. Mao, W.; Li, Z.; Xia, X.; Li, Y.; Yu, J. A combined approach of high-throughput sequencing and degradome analysis reveals tissue specific expression of microRNAs and their targets in cucumber. *PLoS ONE* **2012**, *7*, e33040. [CrossRef]
11. Kozomara, A.; Birgaoanu, M.; Griffiths-Jones, S. miRBase: From microRNA sequences to function. *Nucleic Acids Res.* **2019**, *47*, D155–D162. [CrossRef] [PubMed]
12. Zhang, L.; Chia, J.M.; Kumari, S.; Stein, J.C.; Liu, Z.; Narechania, A.; Maher, C.A.; Guill, K.; McMullen, M.D.; Ware, D. A genome-wide characterization of microRNA genes in maize. *PLoS Genet.* **2009**, *5*, e1000716. [CrossRef] [PubMed]
13. Schwab, R.; Palatnik, J.F.; Riester, M.; Schommer, C.; Schmid, M.; Weigel, D. Specific effects of microRNAs on the plant transcriptome. *Dev. Cell* **2005**, *8*, 517–527. [CrossRef] [PubMed]
14. Allen, R.S.; Li, J.; Stahle, M.I.; Dubroué, A.; Gubler, F.; Millar, A.A. Genetic analysis reveals functional redundancy and the major target genes of the Arabidopsis miR159 family. *Proc. Natl. Acad. Sci. USA* **2007**, *104*, 16371–16376. [CrossRef] [PubMed]
15. Allen, R.S.; Li, J.; Alonso-Peral, M.M.; White, R.G.; Gubler, F.; Millar, A.A. MicroR159 regulation of most conserved targets in Arabidopsis has negligible phenotypic effects. *Silence* **2010**, *1*, 18. [CrossRef] [PubMed]
16. Kim, M.H.; Cho, J.S.; Lee, J.H.; Bae, S.Y.; Choi, Y.I.; Park, E.J.; Lee, H.; Ko, J.H. Poplar MYB transcription factor PtrMYB012 and its Arabidopsis AtGAMYB orthologs are differentially repressed by the Arabidopsis miR159 family. *Tree Physiol.* **2018**, *38*, 801–812. [CrossRef] [PubMed]
17. Palatnik, J.F.; Wollmann, H.; Schommer, C.; Schwab, R.; Boisbouvier, J.; Rodriguez, R.; Warthmann, N.; Allen, E.; Dezulian, T.; Huson, D.; et al. Sequence and expression differences underlie functional specialization of Arabidopsis microRNAs miR159 and miR319. *Dev. Cell* **2007**, *13*, 115–125. [CrossRef]
18. Bologna, N.G.; Mateos, J.L.; Bresso, E.G.; Palatnik, J.F. A loop-to-base processing mechanism underlies the biogenesis of plant microRNAs miR319 and miR159. *EMBO J.* **2009**, *28*, 3646–3656. [CrossRef]
19. Li, Y.; Li, C.; Ding, G.; Jin, Y. Evolution of *MIR159/319* microRNA genes and their post-transcriptional regulatory link to siRNA pathways. *BMC Evol. Biol.* **2011**, *11*, 122. [CrossRef]
20. Aya, K.; Hiwatashi, Y.; Kojima, M.; Sakakibara, H.; Ueguchi-Tanaka, M.; Hasebe, M.; Matsuoka, M. The Gibberellin perception system evolved to regulate a pre-existing GAMYB-mediated system during land plant evolution. *Nat. Commun.* **2011**, *2*, 544. [CrossRef]
21. Tsuzuki, M.; Nishihama, R.; Ishizaki, K.; Kurihara, Y.; Matsui, M.; Bowman, J.L.; Kohchi, T.; Hamada, T.; Watanabe, Y. Profiling and characterization of small RNAs in the Liverwort, *Marchantia polymorpha*, belonging to the first diverged land plants. *Plant Cell Physiol.* **2016**, *57*, 359–372. [CrossRef] [PubMed]
22. Lin, S.S.; Bowman, J.L. MicroRNAs in *Marchantia polymorpha*. *New Phytol.* **2018**, *220*, 409–416. [CrossRef] [PubMed]
23. Palatnik, J.F.; Allen, E.; Wu, X.; Schommer, C.; Schwab, R.; Carrington, J.C.; Weigel, D. Control of leaf morphogenesis by microRNAs. *Nature* **2003**, *425*, 257–263. [CrossRef] [PubMed]
24. Li, J.; Reichel, M.; Li, Y.; Millar, A.A. The functional scope of plant microRNA-mediated silencing. *Trends Plant Sci.* **2014**, *19*, 750–756. [CrossRef] [PubMed]
25. Addo-Quaye, C.; Eshoo, T.W.; Bartel, D.P.; Axtell, M.J. Endogenous siRNA and miRNA targets identified by sequencing of the Arabidopsis degradome. *Curr. Biol.* **2008**, *18*, 758–762. [CrossRef] [PubMed]
26. Song, Q.X.; Liu, Y.F.; Hu, X.Y.; Zhang, W.K.; Ma, B.; Chen, S.Y.; Zhang, J.S. Identification of miRNAs and their target genes in developing soybean seeds by deep sequencing. *BMC Plant Biol.* **2011**, *11*, 5. [CrossRef] [PubMed]

27. Liu, N.; Tu, L.; Tang, W.; Gao, W.; Lindsey, K.; Zhang, X. Small RNA and degradome profiling reveals a role for miRNAs and their targets in the developing fibers of *Gossypium barbadense*. *New Phytol.* **2015**, *206*, 352–367.
28. Zhang, J.; Zeng, R.; Chen, J.; Liu, X.; Liao, Q. Identification of conserved microRNAs and their targets from *Solanum lycopersicum* Mill. *Gene* **2008**, *423*, 1–7. [CrossRef]
29. An, F.M.; Chan, M.T. Transcriptome-wide characterization of miRNA-directed and non-miRNA-directed endonucleolytic cleavage using degradome analysis under low ambient temperature in *Phalaenopsis aphrodite* subsp. formosana. *Plant Cell Physiol.* **2012**, *53*, 1737–1750. [CrossRef]
30. Luo, X.; Gao, Z.; Shi, T.; Cheng, Z.; Zhang, Z.; Ni, Z. Identification of miRNAs and their target genes in peach (*Prunus persica* L.) using high-throughput sequencing and degradome analysis. *PLoS ONE* **2013**, *8*, e79090. [CrossRef]
31. Sun, F.; Guo, G.; Du, J.; Guo, W.; Peng, H.; Ni, Z.; Sun, Q.; Yao, Y. Whole-genome discovery of miRNAs and their targets in wheat (*Triticum aestivum* L.). *BMC Plant Biol.* **2014**, *14*, 142. [CrossRef] [PubMed]
32. Li, Y.F.; Zheng, Y.; Addo-Quaye, C.; Zhang, L.; Saini, A.; Jagadeeswaran, G.; Axtell, M.J.; Zhang, W.; Sunkar, R. Transcriptome-wide identification of microRNA targets in rice. *Plant J.* **2010**, *62*, 742–759. [CrossRef]
33. Curaba, J.; Spriggs, A.; Taylor, J.; Li, Z.; Helliwell, C. miRNA regulation in the early development of barley seed. *BMC Plant Biol.* **2012**, *12*, 120. [CrossRef] [PubMed]
34. Buxdorf, K.; Hendelman, A.; Stav, R.; Lapidot, M.; Ori, N.; Arazi, T. Identification and characterization of a novel miR159 target not related to *MYB* in tomato. *Planta* **2010**, *232*, 1009–1022. [CrossRef] [PubMed]
35. Da Silva, E.M.; Silva, G.F.F.E.; Bidoia, D.B.; da Silva Azevedo, M.; de Jesus, F.A.; Pino, L.E.; Peres, L.E.P.; Carrera, E.; López-Díaz, I.; Nogueira, F.T.S. microRNA159-targeted *SlGAMYB* transcription factors are required for fruit set in tomato. *Plant J.* **2017**, *92*, 95–109. [CrossRef] [PubMed]
36. Wan, L.C.; Zhang, H.; Lu, S.; Zhang, L.; Qiu, Z.; Zhao, Y.; Zeng, Q.Y.; Lin, J. Transcriptome-wide identification and characterization of miRNAs from *Pinus densata*. *BMC Genom.* **2012**, *13*, 132. [CrossRef] [PubMed]
37. Amborella Genome Project. The Amborella genome and the evolution of flowering plants. *Science* **2013**, *342*, 1241089. [CrossRef]
38. Li, W.F.; Zhang, S.G.; Han, S.Y.; Wu, T.; Zhang, J.H.; Qi, L.W. Regulation of *LaMYB33* by miR159 during maintenance of embryogenic potential and somatic embryo maturation in *Larix kaempferi* (Lamb.) Carr. *Plant Cell Tissue Organ Cult.* **2013**, *113*, 131–136. [CrossRef]
39. Pappas, M.D.C.R.; Pappas, G.J.; Grattapaglia, D. Genome-wide discovery and validation of *Eucalyptus* small RNAs reveals variable patterns of conservation and diversity across species of *Myrtaceae*. *BMC Genom.* **2015**, *16*, 1113. [CrossRef]
40. Dai, X.; Zhuang, Z.; Zhao, P.X. psRNATarget: A plant small RNA target analysis server (2017 release). *Nucleic Acids Res.* **2018**, *46*, W49–W54. [CrossRef]
41. Alonso-Peral, M.M.; Li, J.; Li, Y.; Allen, R.S.; Schnippenkoetter, W.; Ohms, S.; White, R.G.; Millar, A.A. The microRNA159-regulated *GAMYB-like* genes inhibit growth and programmed cell death in Arabidopsis. *Plant Physiol.* **2010**, *154*, 757–771. [CrossRef] [PubMed]
42. German, M.A.; Pillay, M.; Jeong, D.H.; Hetawal, A.; Luo, S.; Janardhanan, P.; Kannan, V.; Rymarquis, L.A.; Nobuta, K.; German, R.; et al. Global identification of microRNA–target RNA pairs by parallel analysis of RNA ends. *Nat. Biotech.* **2008**, *26*, 941. [CrossRef] [PubMed]
43. Millar, A.A.; Gubler, F. The Arabidopsis *GAMYB-like* genes, *MYB33* and *MYB65*, are microRNA-regulated genes that redundantly facilitate anther development. *Plant Cell* **2005**, *17*, 705–721. [CrossRef] [PubMed]
44. Zheng, Z.; Reichel, M.; Deveson, I.; Wong, G.; Li, J.; Millar, A.A. Target RNA secondary structure is a major determinant of miR159 efficacy. *Plant Physiol.* **2017**, *174*, 1764–1778. [CrossRef] [PubMed]
45. Seitz, H. Redefining microRNA targets. *Curr. Biol.* **2009**, *19*, 870–873. [CrossRef] [PubMed]
46. Li, Y.; Alonso-Peral, M.; Wong, G.; Wang, M.B.; Millar, A.A. Ubiquitous miR159 repression of *MYB33/65* in Arabidopsis rosettes is robust and is not perturbed by a wide range of stresses. *BMC Plant Biol.* **2016**, *16*, 179. [CrossRef] [PubMed]
47. Li, J.; Reichel, M.; Millar, A.A. Determinants beyond both complementarity and cleavage govern microR159 efficacy in Arabidopsis. *PLoS Genet.* **2014**, *10*, e1004232. [CrossRef] [PubMed]
48. Alonso-Peral, M.M.; Sun, C.; Millar, A.A. MicroRNA159 can act as a switch or tuning microRNA independently of its abundance in Arabidopsis. *PLoS ONE* **2012**, *7*, e34751. [CrossRef] [PubMed]

49. Leydon, A.R.; Beale, K.M.; Woroniecka, K.; Castner, E.; Chen, J.; Horgan, C.; Palanivelu, R.; Johnson, M.A. Three MYB transcription factors control pollen tube differentiation required for sperm release. *Curr. Biol.* **2013**, *23*, 1209–1214. [CrossRef] [PubMed]
50. Liang, Y.; Tan, Z.M.; Zhu, L.; Niu, Q.K.; Zhou, J.J.; Li, M.; Chen, L.Q.; Zhang, X.Q.; Ye, D. *MYB97*, *MYB101* and *MYB120* function as male factors that control pollen tube-synergid interaction in *Arabidopsis thaliana* fertilization. *PLoS Genet.* **2013**, *9*, e1003933. [CrossRef]
51. Zhao, Y.; Wang, S.; Wu, W.; Li, L.; Jiang, T.; Zheng, B. Clearance of maternal barriers by paternal miR159 to initiate endosperm nuclear division in Arabidopsis. *Nat. Commun.* **2018**, *9*, 5011. [CrossRef] [PubMed]
52. Achard, P.; Herr, A.; Baulcombe, D.C.; Harberd, N.P. Modulation of floral development by a gibberellin-regulated microRNA. *Development* **2004**, *131*, 3357–3365. [CrossRef] [PubMed]
53. Alves-Junior, L.; Niemeier, S.; Hauenschild, A.; Rehmsmeier, M.; Merkle, T. Comprehensive prediction of novel microRNA targets in *Arabidopsis thaliana*. *Nucleic Acids Res.* **2009**, *37*, 4010–4021. [CrossRef] [PubMed]
54. Pollard, K.S.; Hubisz, M.J.; Rosenbloom, K.R.; Siepel, A. Detection of nonneutral substitution rates on mammalian phylogenies. *Genome Res.* **2010**, *20*, 110–121. [CrossRef]
55. Hubisz, M.J.; Pollard, K.S.; Siepel, A. PHAST and RPHAST: Phylogenetic analysis with space/time models. *Brief Bioinform.* **2011**, *12*, 41–51. [CrossRef] [PubMed]
56. Crooks, G.; Hon, G.; Chandonia, J.; Brenner, S. WebLogo: A sequence logo generator. *Genome Res.* **2004**, *14*, 1188–1190. [CrossRef] [PubMed]
57. Bernhart, S.H.; Hofacker, I.L.; Will, S.; Gruber, A.R.; Stadler, P.F. RNAalifold: Improved consensus structure prediction for RNA alignments. *BMC Bioinform.* **2008**, *9*, 1–13. [CrossRef]
58. Todesco, M.; Rubio-Somoza, I.; Paz-Ares, J.; Weigel, D. A collection of target mimics for comprehensive analysis of microRNA function in *Arabidopsis thaliana*. *PLoS Genet.* **2010**, *6*, e1001031. [CrossRef]
59. Rubio-Somoza, I.; Weigel, D. Coordination of flower maturation by a regulatory circuit of three microRNAs. *PLoS Genet.* **2013**, *9*, e1003374. [CrossRef]
60. Guo, C.; Xu, Y.; Shi, M.; Lai, Y.; Wu, X.; Wang, H.; Zhu, Z.; Poethig, R.S.; Wu, G. Repression of miR156 by miR159 regulates the timing of the juvenile-to-adult transition in Arabidopsis. *Plant Cell* **2017**, *29*, 1293–1304. [CrossRef]
61. Zhao, Y.; Wen, H.; Teotia, S.; Du, Y.; Zhang, J.; Li, J.; Sun, H.; Tang, G.; Peng, T.; Zhao, Q. Suppression of microRNA159 impacts multiple agronomic traits in rice (*Oryza sativa* L.). *BMC Plant Biol.* **2017**, *17*, 215. [CrossRef] [PubMed]
62. Zhang, H.; Zhang, J.; Yan, J.; Gou, F.; Mao, Y.; Tang, G.; Botella, J.R.; Zhu, J.K. Short tandem target mimic rice lines uncover functions of miRNAs in regulating important agronomic traits. *Proc. Natl. Acad. Sci. USA* **2017**, *114*, 5277–5282. [CrossRef] [PubMed]
63. Tsuji, H.; Aya, K.; Ueguchi-Tanaka, M.; Shimada, Y.; Nakazono, M.; Watanabe, R.; Nishizawa, N.K.; Gomi, K.; Shimada, A.; Kitano, H.; et al. *GAMYB* controls different sets of genes and is differentially regulated by microRNA in aleurone cells and anthers. *Plant J.* **2006**, *47*, 427–444. [CrossRef] [PubMed]
64. Kaneko, M.; Inukai, Y.; Ueguchi-Tanaka, M.; Itoh, H.; Izawa, T.; Kobayashi, Y.; Hattori, T.; Miyao, A.; Hirochika, H.; Ashikari, M.; et al. Loss-of-function mutations of the rice *GAMYB* gene impair α-amylase expression in aleurone and flower development. *Plant Cell* **2004**, *16*, 33–44. [CrossRef] [PubMed]
65. Murray, F.; Kalla, R.; Jacobsen, J.; Gubler, F. A role for *HvGAMYB* in anther development. *Plant J.* **2003**, *33*, 481–491. [CrossRef] [PubMed]
66. Wang, Y.; Sun, F.; Cao, H.; Peng, H.; Ni, Z.; Sun, Q.; Yao, Y. TamiR159 directed wheat *TaGAMYB* cleavage and its involvement in anther development and heat response. *PLoS ONE* **2012**, *7*, e48445. [CrossRef]
67. Li, X.; Bian, H.; Song, D.; Ma, S.; Han, N.; Wang, J.; Zhu, M. Flowering time control in ornamental gloxinia (*Sinningia speciosa*) by manipulation of miR159 expression. *Ann. Bot.* **2013**, *111*, 791–799. [CrossRef]
68. Zhang, Y.; Zhang, X.; Liu, B.; Wang, W.; Liu, X.; Chen, C.; Liu, X.; Yang, S.; Ren, H. A *GAMYB* homologue *CsGAMYB1* regulates sex expression of cucumber via an ethylene-independent pathway. *J. Exp. Bot.* **2014**, *65*, 3201–3213. [CrossRef]
69. Vallarino, J.G.; Osorio, S.; Bombarely, A.; Casañal, A.; Cruz-Rus, E.; Sánchez-Sevilla, J.F.; Amaya, I.; Giavalisco, P.; Fernie, A.R.; Botella, M.A.; et al. Central role of FaGAMYB in the transition of the strawberry receptacle from development to ripening. *New Phytol.* **2015**, *208*, 482–496. [CrossRef]
70. Plackett, A.R.; Thomas, S.G.; Wilson, Z.A.; Hedden, P. Gibberellin control of stamen development: A fertile field. *Trends Plant Sci.* **2011**, *16*, 568–578. [CrossRef]

71. Aya, K.; Ueguchi-Tanaka, M.; Kondo, M.; Hamada, K.; Yano, K.; Nishimura, M.; Matsuoka, M. Gibberellin modulates anther development in rice via the transcriptional regulation of *GAMYB*. *Plant Cell* **2009**, *21*, 1453–1472. [CrossRef] [PubMed]
72. Liu, B.; De Storme, N.; Geelen, D. Gibberellin induces diploid pollen formation by interfering with meiotic cytokinesis. *Plant Physiol.* **2017**, *173*, 338–353. [CrossRef] [PubMed]
73. Gubler, F.; Kalla, R.; Roberts, J.K.; Jacobsen, J.V. Gibberellin-regulated expression of a myb gene in barley aleurone cells: Evidence for Myb transactivation of a high-pI alpha-amylase gene promoter. *Plant Cell* **1995**, *7*, 1879–1891. [CrossRef] [PubMed]
74. Gubler, F.; Raventos, D.; Keys, M.; Watts, R.; Mundy, J.; Jacobsen, J.V. Target genes and regulatory domains of the GAMYB transcriptional activator in cereal aleurone. *Plant J.* **1999**, *17*, 1–9. [CrossRef] [PubMed]
75. Guo, W.J.; Ho, T.H.D. An abscisic acid-induced protein; HVA22, inhibits gibberellin-mediated programmed cell death in cereal aleurone cells. *Plant Physiol.* **2008**, *147*, 1710–1722. [CrossRef] [PubMed]
76. Reichel, M.; Li, Y.; Li, J.; Millar, A.A. Inhibiting plant microRNA activity: Molecular *SPONGE*s, target *MIMIC*s and STTMs all display variable efficacies against target microRNAs. *Plant Biotech. J.* **2015**, *13*, 915–926. [CrossRef] [PubMed]
77. Gong, X.; Bewley, D.J. A *GAMYB-like* gene in tomato and its expression during seed germination. *Planta* **2008**, *228*, 563–572. [CrossRef] [PubMed]
78. Xue, T.; Liu, Z.; Dai, X.; Xiang, F. Primary root growth in Arabidopsis thaliana is inhibited by the miR159 mediated repression of *MYB33*, *MYB65* and *MYB101*. *Plant Sci.* **2017**, *262*, 182–189. [CrossRef]
79. Spanudakis, E.; Jackson, S. The role of microRNAs in the control of flowering time. *J. Exp. Bot.* **2014**, *65*, 365–380. [CrossRef]
80. Conti, L. Hormonal control of the floral transition: Can one catch them all? *Dev. Biol.* **2017**, *430*, 288–301. [CrossRef]
81. Gocal, G.F.; Sheldon, C.C.; Gubler, F.; Moritz, T.; Bagnall, D.J.; MacMillan, C.P.; Li, S.F.; Parish, R.W.; Dennis, E.S.; Weigel, D.; et al. *GAMYB-like* genes, flowering, and gibberellin signaling in Arabidopsis. *Plant Physiol.* **2001**, *127*, 1682–1693. [CrossRef] [PubMed]
82. Blazquez, M.A.; Green, R.; Nilsson, O.; Sussman, M.R.; Weigel, D. Gibberellins promote flowering of arabidopsis by activating the *LEAFY* promoter. *Plant Cell* **1998**, *10*, 791–800. [CrossRef]
83. Csukasi, F.; Donaire, L.; Casañal, A.; Martínez-Priego, L.; Botella, M.A.; Medina-Escobar, N.; Llave, C.; Valpuesta, V. Two strawberry miR159 family members display developmental-specific expression patterns in the fruit receptacle and cooperatively regulate *Fa-GAMYB*. *New Phytol.* **2012**, *195*, 47–57. [CrossRef] [PubMed]
84. Wang, C.; Jogaiah, S.; Zhang, W.; Abdelrahman, M.; Fang, J.G. Spatio-temporal expression of miRNA159 family members and their *GAMYB* target gene during the modulation of gibberellin-induced grapevine parthenocarpy. *J. Exp. Bot.* **2018**, *69*, 3639–3650. [CrossRef] [PubMed]
85. Zhang, B. MicroRNA: A new target for improving plant tolerance to abiotic stress. *J. Exp. Bot.* **2015**, *66*, 1749–1761. [CrossRef] [PubMed]
86. Liu, H.H.; Tian, X.; Li, Y.J.; Wu, C.A.; Zheng, C.C. Microarray-based analysis of stress-regulated microRNAs in Arabidopsis thaliana. *RNA* **2008**, *14*, 836–843. [CrossRef] [PubMed]
87. Reyes, J.L.; Chua, N.H. ABA induction of miR159 controls transcript levels of two *MYB* factors during Arabidopsis seed germination. *Plant J.* **2007**, *49*, 592–606. [CrossRef]
88. Pieczynski, M.; Marczewski, W.; Hennig, J.; Dolata, J.; Bielewicz, D.; Piontek, P.; Wyrzykowska, A.; Krusiewicz, D.; Strzelczyk-Zyta, D.; Konopka-Postupolska, D.; et al. Down-regulation of *CBP80* gene expression as a strategy to engineer a drought-tolerant potato. *Plant Biotechnol. J.* **2013**, *11*, 459–469. [CrossRef]
89. Zhang, T.; Zhao, Y.L.; Zhao, J.H.; Wang, S.; Jin, Y.; Chen, Z.Q.; Fang, Y.Y.; Hua, C.L.; Ding, S.W.; Guo, H.S. Cotton plants export microRNAs to inhibit virulence gene expression in a fungal pathogen. *Nat. Plants* **2016**, *2*, 16153. [CrossRef]
90. Medina, C.; da Rocha, M.; Magliano, M.; Ratpopoulo, A.; Revel, B.; Marteu, N.; Magnone, V.; Lebrigand, K.; Cabrera, J.; Barcala, M.; et al. Characterization of microRNAs from Arabidopsis galls highlights a role for miR159 in the plant response to the root-knot nematode *Meloidogyne incognita*. *New Phytol.* **2017**, *216*, 882–896. [CrossRef]

© 2019 by the authors. Licensee MDPI, Basel, Switzerland. This article is an open access article distributed under the terms and conditions of the Creative Commons Attribution (CC BY) license (http://creativecommons.org/licenses/by/4.0/).

Review

Perspectives on microRNAs and Phased Small Interfering RNAs in Maize (*Zea mays* L.): Functions and Big Impact on Agronomic Traits Enhancement

Zhanhui Zhang [1,*], Sachin Teotia [1,2,3], Jihua Tang [1] and Guiliang Tang [1,2,*]

[1] State Key Laboratory of Wheat and Maize Crop Science, Henan Agricultural University, Zhengzhou 450002, China; steotia@mtu.edu (S.T.); tangjihua1@163.com (J.T.)
[2] Department of Biological Sciences, Michigan Technological University, Houghton, MI 49931, USA
[3] Department of Biotechnology, Sharda University, Greater Noida 201306, India
* Correspondence: zhanhuiz15@icloud.com (Z.Z.); gtang1@mtu.edu (G.T.);
Tel.: +86-0371-56990188 (Z.Z.); +1-906-487-2174 (G.T.)

Received: 24 March 2019; Accepted: 11 June 2019; Published: 12 June 2019

Abstract: Small RNA (sRNA) population in plants comprises of primarily micro RNAs (miRNAs) and small interfering RNAs (siRNAs). MiRNAs play important roles in plant growth and development. The miRNA-derived secondary siRNAs are usually known as phased siRNAs, including phasiRNAs and tasiRNAs. The miRNA and phased siRNA biogenesis mechanisms are highly conserved in plants. However, their functional conservation and diversification may differ in maize. In the past two decades, lots of miRNAs and phased siRNAs have been functionally identified for curbing important maize agronomic traits, such as those related to developmental timing, plant architecture, sex determination, reproductive development, leaf morphogenesis, root development and nutrition, kernel development and tolerance to abiotic stresses. In contrast to *Arabidopsis* and rice, studies on maize miRNA and phased siRNA biogenesis and functions are limited, which restricts the small RNA-based fundamental and applied studies in maize. This review updates the current status of maize miRNA and phased siRNA mechanisms and provides a survey of our knowledge on miRNA and phased siRNA functions in controlling agronomic traits. Furthermore, improvement of those traits through manipulating the expression of sRNAs or their targets is discussed.

Keywords: maize (*Zea mays* L.); miRNA; phasiRNA; tasiRNA; agronomic traits; crop improvement

---

1. Introduction

Plant and animal small RNAs (sRNAs) are short noncoding regulatory RNAs in the size range of ~20 to 30 nucleotides (nt) [1,2]. These sRNAs play crucial roles in various biological regulatory processes through mediating gene silencing at both transcriptional and posttranscriptional levels [1,3]. According to the origin and biogenesis, plant sRNAs can be categorized into several major classes, micro RNAs (miRNAs), heterochromatic small interfering RNAs (hc-siRNAs), phased small interfering RNAs (phased siRNAs), and natural antisense transcript small interfering RNAs (NAT-siRNAs) [4].

Plants miRNAs are processed from long *MIRNA* transcripts by a microprocessor and dicing complexes [5–7]. Compared to animal miRNAs, plant miRNAs tend to have fewer targets that mainly encode transcription factors and F-box proteins [8]. This indicates that miRNA is at the central position of gene expression regulatory networks of plant growth and development. The accumulating studies proved miRNAs to be key regulators of various biological regulatory processes in plants, including developmental timing, plant architecture, organ polarity, inflorescence development and responses to biotic and abiotic stresses [9,10]. Additionally, miRNAs also drive secondary siRNA generation that are defined as phased siRNAs. Such secondary siRNAs, including canonical phased siRNAs (phasiRNAs) and phased trans-acting siRNAs (tasiRNAs), also play key roles in plant development [11–13]. Moreover, manipulation of mRNA transcript abundance via miRNA control provides a unique strategy for the improvement of the complex agronomic traits of crops [14,15]. Thus, understanding the functions of miRNAs and related secondary siRNAs in various plant species, especially in crops like maize, is essential for crop improvement.

Maize is not only a model plant genetic system, but is also an important crop species for food, fuel and feed [16]. Like *Drosophila* and the worm *Caenorhabditis elegans*, maize has been a significant contributor to a number of important discoveries, including the so-called "jumping genes" (transposons), activator/dissociation (Ac/Ds) and Mutator, as well as the epigenetic phenomenon termed paramutation [17,18]. However, there have been limited studies on the roles of miRNAs and miRNA-derived secondary siRNAs in maize metabolism, development and stress responses [12,19–34], making it far from utilized in agronomic traits improvement through genetic engineering. Compared to *Arabidopsis* and rice, maize sRNA and RNAi mechanisms remain only partially resolved.

This review examines the current status of our understanding of the biogenesis and functions of miRNA and phased siRNA in maize, with a focus on their key components and the missing links of the pathways. Such study can help evaluate the potential roles of maize sRNAs in the enhancement of agronomic traits. First, we survey the recent findings regarding miRNA and phased siRNA working mechanisms in maize. We further compare the mechanistic differences for those mechanisms between maize and the model plants *Arabidopsis* and rice highlighting the missing links in maize. Furthermore, we review the identified miRNA and phased siRNA functions in regulating important agronomic traits in maize. Finally, we discuss the potential applications of these small regulatory RNAs or of their target genes in agronomic traits enhancement.

2. MiRNA and phasiRNA Biogenesis in Maize

2.1. Core Components of sRNA Biogenesis in Plants

In plants, the sRNAs biogenesis and gene silencing mainly depends on the activities of three kinds of proteins, dicer or dicer-like proteins (DCLs), argonautes (AGOs) and RNA-dependent RNA polymerases (RDRs) [1,32,35,36]. The sRNA-mediated gene silencing is initiated by double stranded RNA (dsRNA) generation by RDRs or the folding of *MIRNA* gene transcripts [1,36]. The dsRNA is processed into sRNAs, 20–30 nt in length, by the cleavages of microprocessors, DCLs [1,5,36]. Different classes of sRNAs are recognized by specific AGOs to assemble RNA-induced silencing complex (RISC) [1,36]. In *Arabidopsis*, 4 DCLs, 10 AGOs and six RDRs are encoded, while in rice, eight DCLs, 19 AGOs and five RDRs are encoded [35]. In maize, these major components have also been identified mainly based on their orthologs in *Arabidopsis* and rice genomes, which include five ZmDCL, 17 ZmAGO and five ZmRDR genes (Table 1, Figure 1A) [32,37]. Among them, only a few ones have been experimentally verified so far, including *fuzzy tassel* (*fzt*, *ZmDCL1*) [22], *ragged seedling2* (*rgd2*, *ZmAGO7*) [38], *ZmRDR1* [37], *mediator of paramutation 1* (*mop1*, *ZmRDR2*) [39].

2.2. MiRNA-Mediated Gene Silencing in Maize

In plants, 21-nt miRNA biogenesis includes four steps (Figure 1): (1) *MIRNAs* transcription; (2) precursor miRNA (pre-miRNA) generation by dicer-like RNase III protein I (DCL1) cleavage; (3) miRNA duplexes release; (4) miRNA duplexes methylation and export to cytoplasm and miRNA-RISC assembly [6,7,40–42]. In the cytoplasm, the miRNA-RISCs mediate their target mRNA degradation [43,44], or translational inhibition in plants [45]. In contrast to the 21 nt miRNAs, a class of 24 nt miRNAs was discovered in plants (Figure 1). These 24 nt miRNAs are processed by DCL3 during their biogenesis [46,47]. In RNA-directed DNA methylation (RdDM), these 24 nt miRNAs are sorted into AGO4 to direct DNA methylation at the loci of their origin, thus regulating their target genes in trans [46,47]. After their biogenesis, miRNAs are also subjected to catabolism [48], in which demethylated or uridylated miRNAs are degraded by small RNA degrading nucleases (SDNs) [40,49].

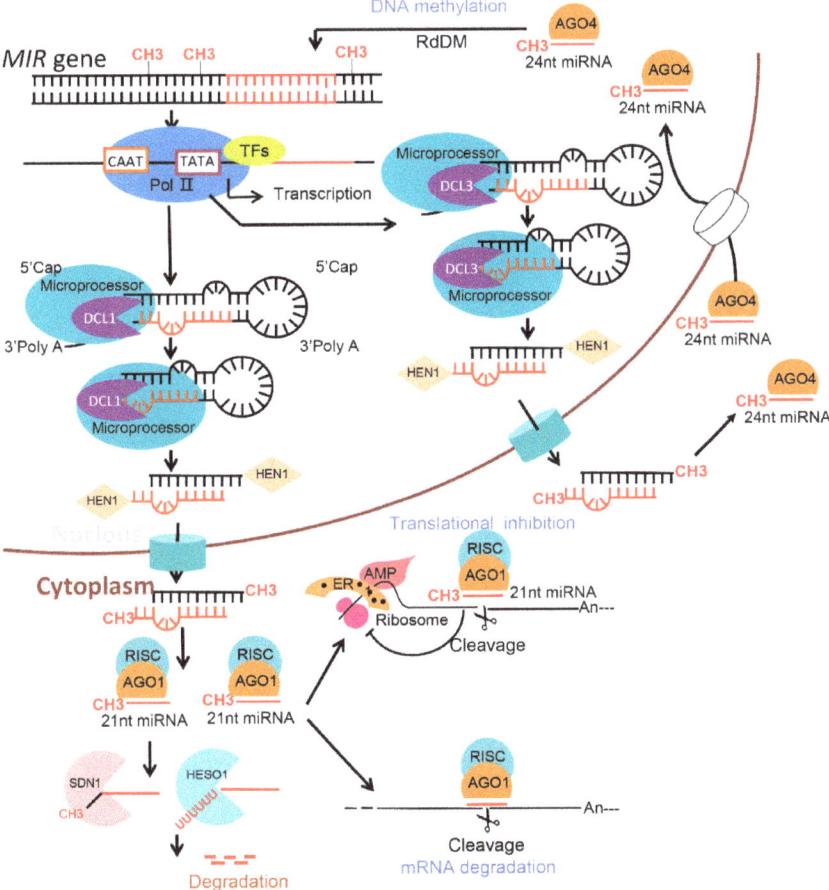

Figure 1. An overview showing micro RNA (miRNA) biogenesis and functioning in plants. *MIRNA* genes are transcribed to form primary miRNAs, from which 21 and 24 nt miRNAs are processed by DCL1 and DCL3, respectively. Their 3' ends are methylated by HEN1. While the 21 nt species are involved in cleavage or translational inhibition of the target mRNAs, the 24 nt miRNAs are involved in DNA methylation.

2.3. Origin and Biogenesis of phasiRNAs in Maize

Generally, phasiRNA biogenesis is initiated by cleavage of single-stranded *PHAS* loci transcripts by 22 nt miRNAs. Then, those cleaved single-stranded RNAs are used to generate dsRNAs by RDRs. DCLs further phase dsRNAs to produce 21 or 24 nt phasiRNAs. PhasiRNAs are subsequently loaded to AGOs to regulate gene expression network [11,50] (Figure 2A,B). In grasses, including maize, phasiRNA precursors, *PHAS* loci transcripts, are transcribed by RNA polymerase II. These long noncoding precursor transcripts are internally cleaved, guided by 22 nt miR2118 to generate the 21 nt phasiRNAs or by miR2275 for the 24 nt phasiRNA. Such special class of small RNAs are specifically expressed in reproductive organs, conferring male fertility [11,12,51]. In maize, the biogenesis of 21 and 24 nt phasiRNAs are regulated by DCL4 and DCL5, respectively (Figure 2A,B). Next, 21 and 24 nt phasiRNAs are recruited by AGO5c and AGO18b, respectively, to assemble RISC and regulate gene expression [12].

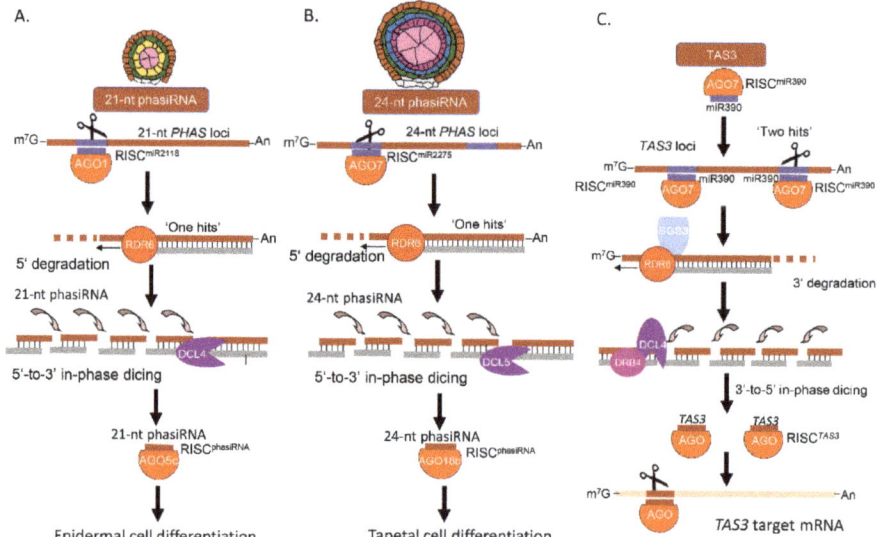

Figure 2. Phased small interfering RNA (phasiRNA) and trans-acting simple interfering RNA (tasiRNA) biogenesis pathways in maize. (**A**) The 21 nt phasiRNA biogenesis pathway. (**B**) The 24 nt phasiRNA biogenesis pathway. The regulatory mechanism of 24 nt has not been fully uncovered. (**C**) TAS3-tasiRNA biogenesis pathway.

The production of the 21-nt tasiRNAs is initiated by a miRNA through RDR6 and DCL4 [52]. In *Arabidopsis*, miR173, miR390 and miR828 trigger the production of *TAS1a-c/TAS2*, *TAS3*, and *TAS4* siRNAs, respectively [53,54]. The maize *TAS3* pathway has been identified through the mutations, *leafbladeless1* (*ldl1*) and *ragged seedling2* (*rgd2*), which encode the orthologs of SGS3 and AGO7 of *Arabidopsis* (Figure 2C) [55]. After *TAS3* siRNAs is generated, they are recruited by AGO7 to assemble RISC and induce ARF3 gene silencing by targeting mRNA transcripts.

2.4. Functional Redundancy and Divergence of the Key Components in Maize sRNA Biogenesis Pathways

2.4.1. DCLs

Based on the phylogenetic analysis, different DCLs from *Arabidopsis*, rice and maize were classified into four subgroups (Table 1, Figure 3A,B). ZmDCL1 showed high similarity with *Arabidopsis* AtDCL1 and rice OsDCL1a–1c; ZmDCL3a and ZmDCL5/3b are similar to *Arabidopsis* DCL3 and rice OsDCL3a–3b; and ZmDCL2 and ZmDCL4 are most similar to AtDCL2 and AtDCL4, respectively [23,32]. In *Arabidopsis*, AtDCL1 produces mature miRNAs [56]; AtDCL2 is involved in virus defense-related siRNA generation and has functional redundancy with AtDCL4 [57]; while AtDCL3 catalyzes the production of 24-nt siRNAs [58]; and AtDCL4 is mainly for the production of tasiRNAs [59]. Although the DCL family proteins are largely functionally conserved among the three plant species, DCL3a and DCL3b are considered specific to monocots and predate the divergence of rice and maize [60].

Table 1. Known and putative components of the micro RNA (miRNA) and simple interfering (siRNA) pathways in maize.

Gene	Accession Number	Chromosomal Location (5'-3')	Type
1. ZmDCLs			
ZmDCL1	GRMZM2G040762_P01	Chr. 1: 4,600,841–4,608,248	DCL1
ZmDCL2	GRMZM2G301405_P01	Chr. 5: 19,916,753–19,927,967	DCL2
ZmDCL3a	GRMZM5G814985_P01	Chr. 3: 164,415,209–164,418,189	DCL3
ZmDCL5/3b	GRMZM2G413853_P01	Chr. 1: 229,801,762–229,819,069	DCL3
ZmDCL4	GRMZM2G160473_P01	Chr. 10: 129,990,456–129,992,917	DCL4
2. ZmAGOs			
ZmAGO1a	GRMZM2G441583_P01	Chr. 6: 43,253,105–43,261,555	AGO1
ZmAGO1b	AC209206.3_FGP011	Chr. 10: 137,506,877–137,513,415	AGO1
ZmAGO1c	GRMZM2G039455_P01	Chr. 2: 17,563,301–17,573,156	AGO1
ZmAGO1d	GRMZM2G361518_P01	Chr. 5: 64,791,077–64,796,881	AGO1
ZmAGO2a	GRMZM2G007791_P01	Chr. 2: 9,973,816–9,981,340	ZIPPY
ZmAGO2b	GRMZM2G354867_P01	Chr. 1:142,397,812–142,403,450	ZIPPY
ZmAGO4	GRMZM2G589579_P01	Chr. 8: 2,511,663–2,519,008	AGO4
ZmAGO5a	GRMZM2G461936_P02	Chr. 5: 13,611,800–13,618,698	MEL1
ZmAGO5b	GRMZM2G059033_P01	Chr. 2: 233,385,077–233,392,000	MEL1
ZmAGO5c	GRMZM2G347402_P01	Chr. 7: 72,044,775–72,053,779	MEL1
ZmAGO5d	GRMZM2G123063_P01	Chr. 5:4,000,995–4,009,425	MEL1
ZmAGO7	GRMZM2G354867_P01	Chr. 10: 141,823,070–141,828,449	ZIPPY
ZmAGO9	GRMZM2G141818_P03	Chr. 6: 168,642,369–168,650,358	AGO4
ZmAGO10a	AC189879.3_FG003	Chr. 9: 87,408,375–87,414,276	AGO1
ZmAGO10b	GRMZM2G079080_P02	Chr. 6: 103,286,236–103,293,200	AGO1
ZmAGO18a	GRMZM2G105250_P01	Chr. 2: 199,510,528–199,516,085	OsAGO18
ZmAGO18b	GRMZM2G457370_P01	Chr. 1: 250,132,189–250,137,737	OsAGO18
ZmAGO18c	GRMZM2G457370_P02	Chr. 1: 250,132,189–250,137,737	OsAGO18
3. ZmRDRs			
ZmRDR1	GRMZM2G481730_P01	Chr. 5: 205,385,818–205,389,710	RDR1
ZmMOP1	GRMZM2G042443_P01	Chr. 2: 41,131,324–41,136,928	RDR2
ZmRDR6a	GRMZM2G357825_P01	Chr. 9: 109,055,576–109,093,885	RDR6
ZmRDR6b	GRMZM2G145201_P01	Chr. 3: 102,532,883–102,536,036	RDR6
ZmRDR6c	GRMZM2G347931_P01	Chr. 9: 106,302,354–106,306,175	RDR6

Note: this information for maize dicer-like (DCL), argonaute (AGO), and RNA-dependent RNA polymerase (RDR), including accession number, chromosomal location and ORFs, was retrieved from the B73 maize sequence database (http://www.maizesequence.org/index.html).

2.4.2. AGOs

In *Arabidopsis*, AtAGO1 is associated with miRNA-mediated gene silencing [61]; AtAGO7 is preferentially associated with a single miRNA, miR390, to trigger production of *TAS3* [52]; and AtAGO5 is a putative germline-specific Argonaute complex associated with miRNAs in mature *Arabidopsis* pollen [62]. In addition, AtAGO2 was identified to have a stand-in role for AtAGO1 in antivirus defense when AGO1-targeted silencing is overcome by viral suppressors [63], AtAGO4 is associated with endogenous siRNAs that direct DNA methylation [64].

In maize, 17 genes encoding 18 AGO family proteins were identified, almost double the number reported in *Arabidopsis* (Table 1, Figure 3A,C) [24,32]. These ZmAGOs were divided phylogenetically into five subgroups: AGO1 (ZmAGO1a-1d and ZmAGO10a, b), MEL1/AGO5 (ZmAGO5a-5d), AGO7 (ZmAGO2 and ZmAGO7), AGO4 (ZmAGO4), and finally the ZmAGO18 (ZmAGO18a-c) [32]. The ZmAGO18 subgroup, ZmAGO18a, ZmAGO18b and ZmAGO18c, are encoded by two genes (*GRMZM2G105250* encodes ZmAGO18a, and ZmAGO18b and ZmAGO18c are encoded by two transcripts of *GRMZM2G457370*) [32]. They displayed high structural similarity to OsAGO18, whose expression is strongly induced by viral infection in rice and confers broad-spectrum virus resistance by sequestering the OsmiR168 from targeting OsAGO1 [65]. Nonetheless, ZmAGO18a is highly expressed in ears, while ZmAGO18b is mostly enriched in tassels, suggesting that ZmAGO18 family may have functional diversities from the OsAGO18 [66]. In fact, ZmAGO18b was proposed to bind the 24 nt phasiRNAs that are suggested to be the products of ZmDCL5/3b in the phasiRNA pathway, based on their concurrent spatial and temporal expression in developing maize ear/tassel development [12]. The mutant *ragged seedling2* (*rgd2*) has been identified to encode an AGO7-like protein required to produce *TAS3* [38], and its functions are highly conserved among *Arabidopsis*, rice and maize [67–69].

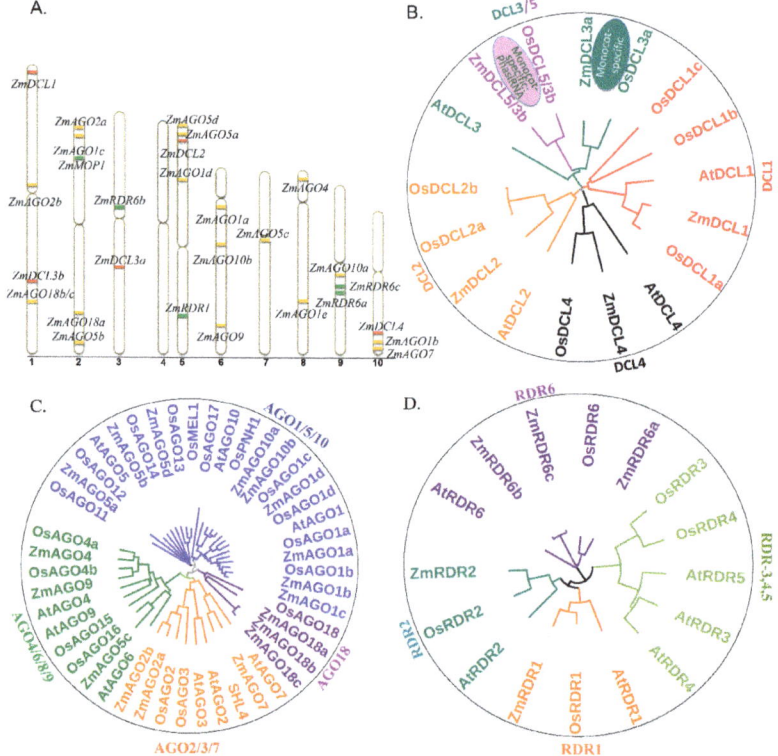

Figure 3. Chromosomal locations and phylogenetic analysis of known and putative components of the RNAi and miRNA pathways in maize. The protein sequences of dicer-like (DCL), argonautes (AGOs) and RNA-dependent RNA polymerase (RDR) in *Arabidopsis*, rice and maize AGOs were obtained from protein database (http://www.ncbi.nlm.nih.gov/protein). The neighbor-joining tree was constructed using Clustal omega [70] and iTol online software [71]. (**A**) The abbreviation of *At* represents *Arabidopsis thaliana*, *Os* for *Oryza sativa*, and *Zm* for *Zea mays*. Red bars indicate the chromosomal locations of *ZmDCLs*, yellow bars for *ZmAGOs*, and green bars for *ZmRDRs*. (**B**) Plants have four types of DCL proteins. There are 4 DCLs encoded in *Arabidopsis* genome, 4 DCL family members in rice, and 4 DCLs in maize. Of these DCL proteins, DCL3a and DCL3b are considered specific to monocots and predate the divergence of rice and maize. (**C**) 10, 19 and 18 AGOs are encoded by *Arabidopsis*, rice and maize, respectively, that can be divided phylogenetically into five subgroups in maize: AGO1, MEL1/AGO5, AGO7, AGO4, and AGO18. AGO18 subgroup has three members in maize. ZmAGO18a-c are considered specific to monocots along with OsAGO18. (**D**) Plants have six types of functionally distinct RDRs. While *Arabidopsis* has all the six types, rice lacks RDR5 and maize lacks RDR3, 4, and 5. In contrast to *Arabidopsis* and rice which have only single member RDR6 family, maize has a multiple member RDR6 family, which is composed of ZmRDR6a, ZmRDR6b, and ZmRDR6c.

2.4.3. RDRs

Six, five and five RDRs have been identified in *Arabidopsis*, rice and maize, respectively (Table 1, Figure 1D) [32,35]. These RDRs were divided phylogenetically into four subgroups: RDR1, RDR2, RDR3/4/5, and RDR6. AtRDR1 and its homolog in maize, ZmRDR1, have been reported to be involved in antiviral defense [37]. AtRDR2 plays a crucial role in RNA-directed DNA methylation and repressive chromatin modifications of certain transgenes, endogenous genes and centromeric repeats that correlate with the production of 24 nt interfering sRNAs [72]. In maize, MOP1 (a homolog of AtRDR2) has proven to be essential for a siRNA-directed gene-silencing pathway, and is also involved in the maintenance of transposon silencing and paramutation [39]. The remaining three RDR homologs of *Arabidopsis* RDR6 in maize, ZmRDR6a-c, are involved in tasiRNA biogenesis [67,73]. We tentatively renamed these three RDRs, previously known as ZmRDR3 and ZmRDR4 [32], to be ZmRDR6a, ZmRDR6b, and ZmRDR6c. ZmRDR6, such a multiple membered family, can be better revealed by identifying their double/triple mutants. Similar to how RDR6 was identified to be important in production of tasiRNAs, an unidentified RDR is expected to play a key role in production of phasiRNAs in maize [12].

3. Functions of miRNAs and phasiRNAs in Maize

3.1. The Interaction of miR156 and miR172 Fine Tunes Plant Developmental Timing

In maize, the transition from juvenile to adult leaves is marked by changes in cell shape, the production of epidermal wax deposits and of specialized cell types like leaf hairs, and a change in the identity of organs that grow from their axillary meristems. In maize and *Arabidopsis*, the roles of miR156 and miR172 interaction in developmental transitions have been widely explored [25,27,74–76]. MiR156 expression levels decrease with leaf age, while that of miR172 increase (Figure 5A). Their targets, encoding squamosa promoter binding protein-like (SBP-Like) and Apetala 2 (AP2) transcription factors, respectively, are expressed in complementary patterns. The mutant *Corngrass1* (*Cg1*) with increased levels of miR156 and reduced miR172 activity, displays restrained developmental transitions, prolonged juvenile features and delayed flowering (Figure 4) [25,77]. In turn, releasing *SPLs* from miR156 regulation leads to premature acquisition of adult leaf features and early flowering, resembling phenotypes of *glossy15* (*gl15*) plants, with reduced activity of miR172 targets (Figure 4) [27,29].

3.2. Plant Architecture Modulated by miR156 and miR319

In maize, plant architecture is mainly determined by tillers, plant height, leaf number, leaf angle and tassel branches. Compared with its ancestor, teosinte (*Zea mays* ssp. parviglumis), maize exhibits a profound increase in apical dominance with a single tiller [78]. Previous researches have proved *teosinte branched1* (*tb1*) gene, encoding a TCP transcription factor that is targeted by miR319, as a major contributor to this domestication change in maize (Figures 4 and 5B) [79,80]. By increasing JA levels, the *tb1* mutant of maize causes a complete loss of apical dominance, allowing the unrestrained outgrowth of axillary buds and inflorescent architectural alterations [79,81]. MiR156 has been proved to be the important regulator in maize and rice plant architecture formation [25,82]. The dominant *Corngrass1* (*Cg1*) mutant of maize has phenotypic changes that are present in the grass-like ancestors of maize, exhibiting numerous tillers, inflorescent architectural alterations and erect leaves (Figure 5B) [25]. The research by Lu et al. [83] in rice revealed that the *ideal plant architecture1* (*IPA1*, *OsSPL14*) could directly bind to the promoter of rice *teosinte branched1* (*Ostb1*), to suppress rice tillering. Likewise, the maize tillering related *ZmSPL* (miR156 target) gene is possible at the upstream of *tb1* in related regulatory pathway. The roles of miR156 in leaf angle and inflorescent architectural modulation have been identified in the corresponding *ZmSPL* mutants, such as *LIGULELESS1* (*LG1*), *tasselsheath4* (*tsh4*, *ZmSBP2*), *UNBRANCHED 2* (*UB2*) and *UB3* [30,33,84,85](Figure 4; Figure 5B).

3.3. Roles of miR172, miR156 and miR159 in Sex Determination

In maize, inflorescence development and sex determination are key factors for grain yield. MiR172 has been identified to play important roles in inflorescence development and sex determination (Figure 4) [86]. Especially, the interplay of miR156 and miR172 contributes largely in maize sex determination and meristem cell fate. In *Cg1* mutant, increased levels of miR156 cause similar phenotypic alterations as seen in *ts4* mutants [25]. Moreover, *STTMmiR172* and *ts4* mutants have reduced expression of miR172 and increased expression of at least two of its targets, *ids1* (*indeterminate spikelet1*) and *sid1* (*sister of indeterminate spikelet1*). These mutants displayed irregular branching within the inflorescence and feminization of the tassel caused by a lack of pistil abortion [86–88]. Decreased levels of miR156 have been detected in feminized tassels of maize *mop1* and *ts1* (*tasselseed1*), implying the missing link of miR156-SPLs with sex-determination genes *ts1*, *ts2*, *ts4*, *Ts6*, and *mop1* [34,89]. Additionally, the mutants of *fuzzy tassel* (encoding dicer-like1 protein) exhibit indeterminate meristems, fasciation, and alterations in sex determination [22]. Such reproductive development alterations are possibly associated with miR159-*GAMYB* pathway, with miR159 and its targets playing the important roles in another development [21,90].

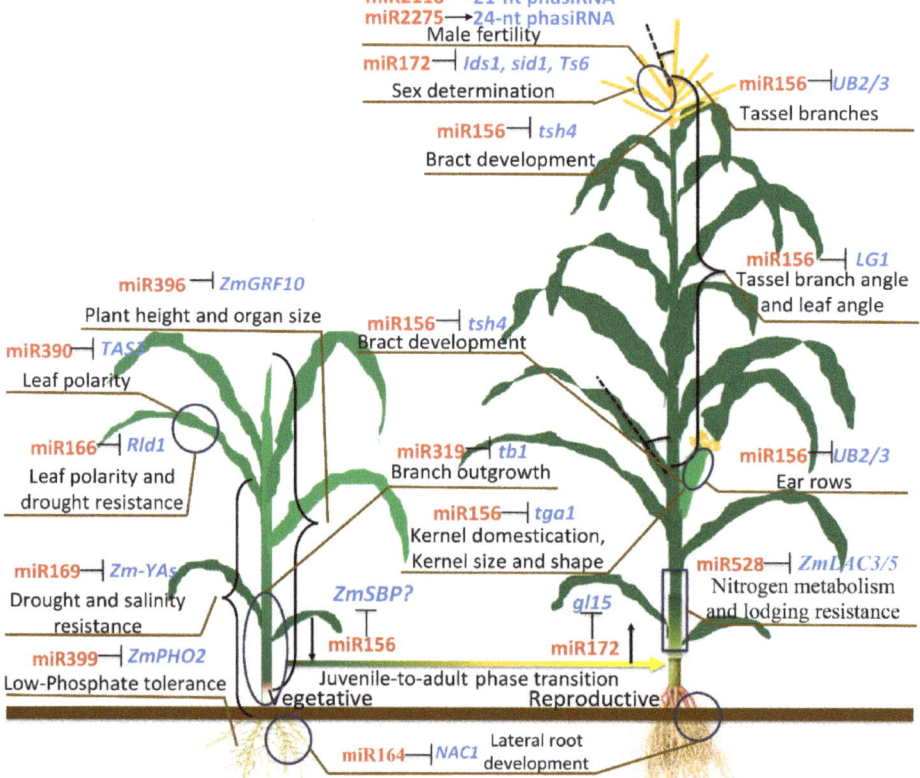

Figure 4. Summary of functionally validated miRNAs and their targets in maize. Nine miRNAs (in red font) and three phasiRNAs (in red font) that regulate specific agronomic traits (in black font) by inducing their targets (in blue font) gene silencing. Question marks indicate that specific *ZmSBP* taking part in juvenile-to-adult phase transition is not known.

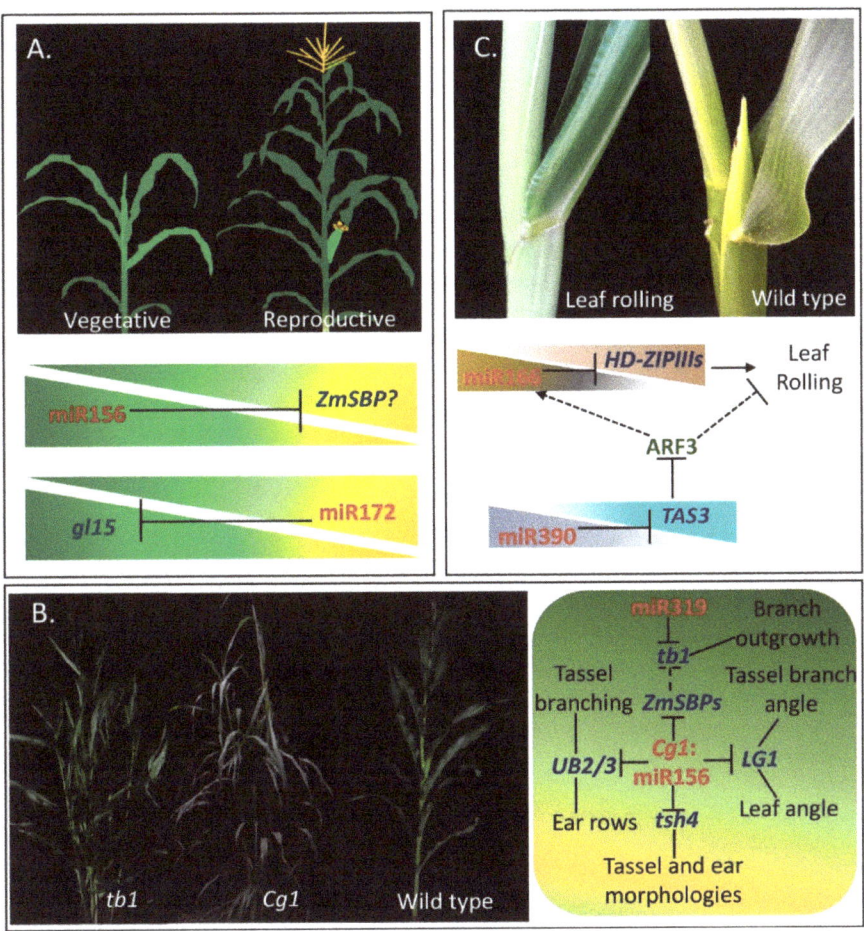

Figure 5. miRNAs or miRNA-phasiRNA interactions in agronomic traits. (**A**) Maize plant developmental timing is fine-tuned by the interaction between miR156 and miR172. However, the misslink between the two miRNAs still need to be addressed; (**B**) plant architectural modulation by the interaction of miR156 and miR319. Representative plants, *tb1* (leaf), *Cg1* (**middle**), and wild type (**right**) (all in the background of Chinese inbred line Zheng58) (Unpublished data), are shown. The potential *ZmSBP* gene probably connects the phenotype of apical dominance loss between maize mutants *tb1* and *Cg1*. In this context, the connection between *ZmSBPs* and *tb1* still need to be experimentally identified; (**C**) leaf shapes are being regulated by miR166 and miR390-TAS3 regulatory networks. *STTMmiR166* mutants have rolling leaf phenotype (**left**), the wild type is ZZC01 (**right**). In this context, the connection between miR166 and ARF3 is still unclear.

3.4. Leaf Patterns Are Shaped by miR166 and miR390-TAS3

Leaves are the most important photosynthetic organs in land plants, which are nearly flat organs designed to efficiently capture light and perform photosynthesis. In maize the specification of abaxial/adaxial polarity was found to be intimately associated with sRNAs, such as miR166, miR390 and *TAS3* (Figure 4; Figure 5C) [26,38,91,92]. The miR166 targets belong to class III homeodomain/leucine zipper (*HD-ZIPIII*) genes. The maize miR166 knockdown and miR166 target over-expression mutants, *STTMmiR166* and *rolled leaf1* (*rld1*), displays an upward curling of the leaf blade that causes adaxialization or partial reversal of leaf polarity [26,88]. The roles of miR166 and *HD-ZIPIII* in leaf polarity are conserved between *Arabidopsis* and maize [93,94]. In plants, miR390 triggers *TAS3*-tasiRNA biogenesis, which interplay with ARF3 to take part in plant development regulation [54,67]. In maize, the mutants of tasiRNA biogenesis pathway components exhibit leaf polarity alterations, *ragged seedling2* (*rgd2*) or *leaf bladeless1* (*lbl1*) [13]. Moreover, several researches proposed that miR390-TAS3 define the adaxial side of the leaf by restricting the expression domain of miR166, which in turn demarcates the abaxial side of leaves by restricting the expression of adaxial determinants [38,92,95].

3.5. PhasiRNAs and Maize Male Fertility

In hybrid maize, male sterility has been widely studied due to both its biological significance and commercial use in hybrid seed production [96]. Maize male fertility is determined by dozens of genes and sRNAs, especially phasiRNAs (Figures 2A,B and 4) [11,12,96]. Indeed, a study reported that two classes of phasiRNAs, 21 and 24 nt in length, were detected to be highly expressed in maize anthers and confer male fertility [12]. The mutant lacking 21 nt phasiRNA, *ocl4*, showed male sterility due to defects in epidermal signaling. Meanwhile, the mutant lacking 24 nt phasiRNA lacking mutants also showed male sterility for due to defective anther subepidermis. This indicated that two types of phasiRNAs regulate anther development independently, with 21 nt premeiotic phasiRNAs regulating epidermal and 24 nt meiotic phasiRNAs regulating tapetal cell differentiation [12].

3.6. Other miRNA Functions in Maize

Several miRNAs have been identified to regulate important maize agronomic traits, such as kernel development, plant growth, abiotic stress tolerance, root development, and nutrition metabolism (Figure 4). The miR156 target, *tga1*, not only confers the domestication of maize naked grains, but also determines the maize kernel shape and size [97,98]. A report on *ZmGRF10*, a miR396 target, indicated that this miRNA is a potential regulator for maize leaf size and plant height [99]. The overexpression *ZmGRF10* mutant displayed reduction in leaf size and plant height by decreasing cell proliferation. Other studies have shown that drought and salinity stresses induce aberrant expression of many miRNAs in maize, for example miR166 and miR169. In maize, miR169 plays a critical role during plant drought, salt and ABA stress response by targeting *NUCLEAR FACTOR-Y subunit A* (*NF-YA*) genes [28]. In *Arabidopsis* and rice, miR166-*HD-ZIP IIIs* have been proven to be associated with drought and ABA stress resistance through maintaining ABA homeostasis [100,101]. Based on our unpublished data, the maize miR166 probably affects tolerance to drought and salinity stresses like in rice and *Arabidopsis*. In maize, miR164 was experimentally identified to be an important regulator in lateral root development by targeting *ZmNAC1* [31,102].

A recent research identified miR528, a monocot-specific miRNA, to be an important regulator for maize nitrogen metabolism in maize. In the miR528 knock-down mutant of maize, under nitrogen-luxury conditions, targets of miR528 are upregulated and mediate increase in lignin content along with superior lodging resistance [20]. The miR399 was identified to regulate the low-phosphate responses in maize [103]. The transgenic plant with miR399 over-expression showed significant phosphorus-toxicity phenotypes, indicating that miR399 is functionally conserved in monocots and dicots.

4. Exploiting the Roles of Maize Small RNAs in Important Agronomic Traits Improvement

Most of agronomic traits are quantitative traits, which are controlled by multiple loci and complex regulatory networks. MiRNAs and phased siRNAs are important participants in these complex regulatory networks. Manipulating the expression levels of miRNAs, miRNA targets, and phased siRNAs is a possible way for agronomic traits improvement. With grain yield increasing, the agronomic traits of maize have been improved through genetic selection [104], which is probably consistent with the elite allele selection of miRNAs and their targets in breeding. Compared with old maize varieties, modern varieties usually have reduced stature, more upright leaves, decreased tassel size, rolling leaf, superior staygreen, less tillers, shorter anthesis-silking interval, less ears per plant and superior stress resistance [104]. Based upon the knowledge about miRNA and phased siRNA functions, manipulating the expression of these small regulatory RNAs and their targets is a possible approach for agronomic traits improvement.

Flowering time represents the developmental transition from vegetative to reproductive phase. Maize spread from its origin to worldwide places with the gradually adaption of flowering time to the local climate [105]. Flowering time determines the length of vegetative phase, biomass and grain yield in maize. The interplay between miR156 and miR172 fine tunes the maize developmental timing and tillering [25,86,88]. Increasing the expression levels of miR156 can elongate the vegetative phase and tillering in maize, which is important to achieve high biomass for silage feed. MiR156 silencing or miR172 over-expression is able to impel maize flowering and precocity, which is in favor of maize mechanized harvest in special regions.

Ideal plant architecture is highly associated with maize planting density and lodging resistance, thereby achieving higher yield. Maize miR156 also regulates plant architectural traits through binding its target genes, such as *tsh4*, *LG1*, *UB2* and *UB3* [30,33,84,85,106,107]. Manipulating the expression of these *ZmSBPs* at optimal levels is needed for idea plant architectural traits. For instance, decreased expression of *LG1* can promote the leaf angle and reduce tassel branches. Furthermore, manipulating the expression of *UB2* and *UB3* in tassel branches and ear rows by using tissue-specific promotor is helpful to get ideal tassel and ear architecture. Additionally, repressing the expression of miR166 or increasing the expression of its targets will increase the leaf rolling, which can be helpful for improving the leaf shapes [88].

In global maize production, lodging and drought are two main abiotic stresses that accounts for large yield loss annually. In a recent research, miR528 has been proved to affect lodging resistance through regulating lignin biosynthesis [20]. Gene silencing of miR528 or overexpression of its targets is helpful for enhancing maize lodging resistance. Knock-down of miR164 promotes maize lateral root development, which can help toward drought and lodging resistance [31]. In the response toward abiotic stress, such as drought, ABA and salinity, miR169 and its targets (NF-YAs) contribute the major regulatory roles through ABA signaling [108]. Lowering the expression of miR169, or increasing that of NF-YAs, can facilitate maize resistance to drought. MiR166 silencing confers resistance against drought in rice and *Arabidopsis*, which is likely conserved in maize too [88,100,101].

5. Future Perspectives

As discussed above, enhanced knowledge on miRNA and phased siRNA functions will be helpful for improving some agronomic traits, including developmental timing, plant architecture, and abiotic stress resistance. Genetic engineering for elite maize germplasms and hybrids still face several hurdles. First, only a small proportion of miRNAs and phased siRNAs have been studied in maize, their complex regulatory networks remain largely unknown. The functional identification of sRNAs is largely dependent upon creating mutants. In maize, the abundant genetic variations or mutations in germplasm pools can provide useful raw materials for the study of these regulatory sRNAs or their targets [109]. Creating new mutants for specific sRNA using artificial miRNA, Short tandem target mimic (STTM) or target mimic (TM) techniques, are efficient strategies for uncovering the functions of these regulatory sRNAs in maize [110–112]. Second, plant miRNA and phased siRNA usually express in spatial and temporal manner. Thus, manipulating the expression of miRNA and phased siRNA in specific tissues and developmental stages can precisely target the traits for improvement. This can be achieved by expressing the transgene expression using tissue- or development-specific promoters, or inducible promoters. Fine genome editing of miRNAs, phased siRNAs and target genes by the CRISPR/Cas9 system can facilitate more subtle manipulations for the target agronomic traits, which is an alternative strategy. Third, for maize hybrids worldwide planted, screening elite hybrid is the most important mask in maize breeding. Usually, the ideal phenotypes in parental inbred lines do not always transfer to the corresponding hybrid. Screening of an elite hybrid is bit of an art and magic, which requires all the yield related traits to reach a balance, and with high heterosis and stress resistance. The current theory of heterosis model facilitate the breeders to make hybrid crosses with high heterosis. Screening the inbred lines with elite genotype/haplotype of miRNAs, phased siRNAs and their targets is fundamental in breeding. Introgressing the elite genotype or haplotype into inbred lines based on heterosis model/heterotic groups will enable the parental elite phenotypes get transferred to their hybrids.

Author Contributions: G.T. and J.T. provided the guideline, Z.Z. collected related references and wrote the manuscript, Z.Z. drew the Figures, J.T., G.T. and S.T. modified the Figures and revised the manuscript.

Funding: This study received the support of funds from National Key Research and Development Program of China (No. 2017YFD0101203), and NSFC (No. 31571679). G.T. is supported by NSF grants (IOS-1048216 and IOS-1340001).

Acknowledgments: We are thankful to anonymous reviewers for their valuable suggestions to improve this article.

Conflicts of Interest: No conflict of interest declared.

References

1. Chen, X. Small RNAs and their roles in plant development. *Annu. Rev. Cell Dev. Biol.* **2009**, *25*, 21–44. [CrossRef] [PubMed]
2. Kim, V.N.; Han, J.; Siomi, M.C. Biogenesis of small RNAs in animals. *Nat. Rev. Mol. Cell Biol.* **2009**, *10*, 126–139. [CrossRef] [PubMed]
3. Fei, Q.; Xia, R.; Meyers, B.C. Phased, secondary, small interfering RNAs in posttranscriptional regulatory networks. *Plant Cell* **2013**, *25*, 2400–2415. [CrossRef] [PubMed]
4. Axtell, M.J. Classification and comparison of small RNAs from plants. *Annu. Rev. Plant Biol.* **2013**, *64*, 137–159. [CrossRef] [PubMed]
5. Wang, Z.; Ma, Z.; Castillo-González, C.; Sun, D.; Li, Y.; Yu, B.; Zhao, B.; Li, P.; Zhang, X. SWI2/SNF2 ATPase CHR2 remodels pri-miRNAs via Serrate to impede miRNA production. *Nature* **2018**, *557*, 516–521. [CrossRef]
6. Rogers, K.; Chen, X. Biogenesis, turnover, and mode of action of plant microRNAs. *Plant Cell* **2013**, *25*, 2383–2399. [CrossRef] [PubMed]
7. Voinnet, O. Origin, biogenesis, and activity of plant microRNAs. *Cell* **2009**, *136*, 669–687. [CrossRef]
8. Rhoades, M.W.; Reinhart, B.J.; Lim, L.P.; Burge, C.B.; Bartel, B.; Bartel, D.P. Prediction of plant microRNA targets. *Cell* **2002**, *110*, 513–520. [CrossRef]
9. Cuperus, J.T.; Fahlgren, N.; Carrington, J.C. Evolution and functional diversification of *MIRNA* Genes. *Plant Cell* **2011**, *23*, 431–442. [CrossRef]
10. Jones-Rhoades, M.W.; Bartel, D.P.; Bartel, B. MicroRNAs and their regulatory roles in plants. *Annu. Rev. Plant Biol.* **2006**, *57*, 19–53. [CrossRef]
11. Yu, Y.; Zhou, Y.; Zhang, Y.; Chen, Y. Grass phasiRNAs and male fertility. *Sci. China Life Sci.* **2018**, *61*, 148–154. [CrossRef] [PubMed]
12. Zhai, J.; Zhang, H.; Arikit, S.; Huang, K.; Nan, G.L.; Walbot, V.; Meyers, B.C. Spatiotemporally dynamic, cell-type-dependent premeiotic and meiotic phasiRNAs in maize anthers. *Proc. Natl. Acad. Sci. USA* **2015**, *112*, 3146–3151. [CrossRef]
13. Juarez, M.T.; Twigg, R.W.; Timmermans, M.C. Specification of adaxial cell fate during maize leaf development. *Development* **2004**, *131*, 4533–4544. [CrossRef] [PubMed]
14. Wang, H.; Wang, H. The miR156/SPL module, a regulatory hub and versatile toolbox, gears up crops for enhanced agronomic traits. *Mol. Plant* **2015**. [CrossRef] [PubMed]
15. Tang, J.; Chu, C. MicroRNAs in crop improvement: Fine-tuners for complex traits. *Nat. Plants* **2017**, *3*, 17077. [CrossRef] [PubMed]
16. Gore, M.A.; Chia, J.M.; Elshire, R.J.; Sun, Q.; Ersoz, E.S.; Hurwitz, B.L.; Peiffer, J.A.; McMullen, M.D.; Grills, G.S.; Ross-Ibarra, J.; et al. A first-generation haplotype map of maize. *Science* **2009**, *326*, 1115–1117. [CrossRef] [PubMed]
17. Brutnell, T.P. Transposon tagging in maize. *Funct. Integr. Genom.* **2002**, *2*, 4–12. [CrossRef]
18. Arteaga-Vazquez, M.A.; Chandler, V.L. Paramutation in maize: RNA mediated trans-generational gene silencing. *Curr. Opin. Genet. Dev.* **2010**, *20*, 156–163. [CrossRef] [PubMed]

19. Sun, W.; Xiang, X.; Zhai, L.; Zhang, D.; Cao, Z.; Liu, L.; Zhang, Z. AGO18b negatively regulates determinacy of spikelet meristems on the tassel central spike in maize. *J. Integr. Plant Biol.* **2018**, *60*, 65–78. [CrossRef] [PubMed]
20. Sun, Q.; Liu, X.; Yang, J.; Liu, W.; Du, Q.; Wang, H.; Fu, C.; Li, W.X. MicroRNA528 affects lodging resistance of maize by regulating lignin biosynthesis under nitrogen-luxury conditions. *Mol. Plant* **2018**, *11*, 806–814. [CrossRef]
21. Field, S.; Thompson, B. Analysis of the Maize dicer-like1 Mutant, fuzzy tassel, Implicates MicroRNAs in Anther Maturation and Dehiscence. *PLoS ONE* **2016**, *11*, e0146534. [CrossRef] [PubMed]
22. Thompson, B.E.; Basham, C.; Hammond, R.; Ding, Q.; Kakrana, A.; Lee, T.F.; Simon, S.A.; Meeley, R.; Meyers, B.C.; Hake, S. The dicer-like1 homolog fuzzy tassel is required for the regulation of meristem determinacy in the inflorescence and vegetative growth in maize. *Plant Cell* **2014**, *26*, 4702–4717. [CrossRef] [PubMed]
23. Petsch, K.; Manzotti, P.S.; Tam, O.H.; Meeley, R.; Hammell, M.; Consonni, G.; Timmermans, M.C. Novel DICER-LIKE1 siRNAs bypass the requirement for DICER-LIKE4 in maize development. *Plant Cell* **2015**, *27*, 2163–2177. [CrossRef] [PubMed]
24. Xu, D.; Yang, H.; Zou, C.; Li, W.X.; Xu, Y.; Xie, C. Identification and functional characterization of the *AGO1* ortholog in maize. *J. Integr. Plant Biol.* **2016**, *58*, 749–758. [CrossRef] [PubMed]
25. Chuck, G.; Cigan, A.M.; Saeteurn, K.; Hake, S. The heterochronic maize mutant *Corngrass1* results from overexpression of a tandem microRNA. *Nat. Genet.* **2007**, *39*, 544–549. [CrossRef] [PubMed]
26. Juarez, M.T.; Kui, J.S.; Thomas, J.; Heller, B.A.; Timmermans, M.C. MicroRNA-mediated repression of *rolled leaf1* specifies maize leaf polarity. *Nature* **2004**, *428*, 84–88. [CrossRef] [PubMed]
27. Lauter, N.; Kampani, A.; Carlson, S.; Goebel, M.; Moose, S.P. MicroRNA172 down-regulates *glossy15* to promote vegetative phase change in maize. *Proc. Natl. Acad. Sci. USA* **2005**, *102*, 9412–9417. [CrossRef] [PubMed]
28. Luan, M.; Xu, M.; Lu, Y.; Zhang, L.; Fan, Y.; Wang, L. Expression of zma-miR169 miRNAs and their target *ZmNF-YA* genes in response to abiotic stress in maize leaves. *Gene* **2015**, *555*, 178–185. [CrossRef] [PubMed]
29. Xu, D.; Wang, X.; Huang, C.; Xu, G.; Liang, Y.; Chen, Q.; Wang, C.; Li, D.; Tian, J.; Wu, L.; et al. *Glossy15* plays an important role in the divergence of the vegetative transition between maize and its progenitor, teosinte. *Mol. Plant* **2017**, *10*, 1579–1583. [CrossRef] [PubMed]
30. Chuck, G.S.; Brown, P.J.; Meeley, R.; Hake, S. Maize SBP-box transcription factors *unbranched2* and *unbranched3* affect yield traits by regulating the rate of lateral primordia initiation. *Proc. Natl. Acad. Sci. USA* **2014**, *111*, 18775–18780. [CrossRef] [PubMed]
31. Li, J.; Guo, G.; Guo, W.; Guo, G.; Tong, D.; Ni, Z.; Sun, Q.; Yao, Y. miRNA164-directed cleavage of *ZmNAC1* confers lateral root development in maize (*Zea mays* L.). *BMC Plant Biol.* **2012**, *12*, 220. [CrossRef] [PubMed]
32. Qian, Y.; Cheng, Y.; Cheng, X.; Jiang, H.; Zhu, S.; Cheng, B. Identification and characterization of Dicer-like, Argonaute and RNA-dependent RNA polymerase gene families in maize. *Plant Cell Rep.* **2011**, *30*, 1347–1363. [CrossRef] [PubMed]
33. Chuck, G.; Whipple, C.; Jackson, D.; Hake, S. The maize SBP-box transcription factor encoded by *tasselsheath4* regulates bract development and the establishment of meristem boundaries. *Development* **2010**, *137*, 1243–1250. [CrossRef] [PubMed]
34. Hultquist, J.F.; Dorweiler, J.E. Feminized tassels of maize *mop1* and *ts1* mutants exhibit altered levels of miR156 and specific SBP-box genes. *Planta* **2008**, *229*, 99–113. [CrossRef] [PubMed]
35. Kapoor, M.; Arora, R.; Lama, T.; Nijhawan, A.; Khurana, J.P.; Tyagi, A.K.; Kapoor, S. Genome-wide identification, organization and phylogenetic analysis of Dicer-like, Argonaute and RNA-dependent RNA Polymerase gene families and their expression analysis during reproductive development and stress in rice. *BMC Genom.* **2008**, *9*, 451. [CrossRef] [PubMed]
36. Matzke, M.A.; Birchler, J.A. RNAi-mediated pathways in the nucleus. *Nat. Rev. Genet.* **2005**, *6*, 24–35. [CrossRef]
37. He, J.; Dong, Z.; Jia, Z.; Wang, J.; Wang, G. Isolation, expression and functional analysis of a putative RNA-dependent RNA polymerase gene from maize (*Zea mays* L.). *Mol. Biol. Rep.* **2010**, *37*, 865–874. [CrossRef]
38. Douglas, R.N.; Wiley, D.; Sarkar, A.; Springer, N.; Timmermans, M.C.; Scanlon, M.J. *Ragged seedling2* Encodes an ARGONAUTE7-like protein required for mediolateral expansion, but not dorsiventrality, of maize leaves. *Plant Cell* **2010**, *22*, 1441–1451. [CrossRef]

39. Dorweiler, J.E.; Carey, C.C.; Kubo, K.M.; Hollick, J.B.; Kermicle, J.L.; Chandler, V.L. Mediator of paramutation1 is required for establishment and maintenance of paramutation at multiple maize loci. *Plant Cell* **2000**, *12*, 2101–2118. [CrossRef]
40. Ramachandran, V.; Chen, X. Small RNA metabolism in *Arabidopsis*. *Trends Plant Sci.* **2008**, *13*, 368–374. [CrossRef]
41. Yu, B.; Yang, Z.; Li, J.; Minakhina, S.; Yang, M.; Padgett, R.W.; Steward, R.; Chen, X. Methylation as a crucial step in plant microRNA biogenesis. *Science* **2005**, *307*, 932–935. [CrossRef] [PubMed]
42. Reinhart, B.J.; Weinstein, E.G.; Rhoades, M.W.; Bartel, B.; Bartel, D.P. MicroRNAs in plants. *Genes Dev.* **2002**, *16*, 1616–1626. [CrossRef] [PubMed]
43. Llave, C.; Xie, Z.; Kasschau, K.D.; Carrington, J.C. Cleavage of Scarecrow-like mRNA targets directed by a class of *Arabidopsis* miRNA. *Science* **2002**, *297*, 2053–2056. [CrossRef]
44. Zhang, Z.; Hu, F.; Sung, M.W.; Shu, C.; Castillo-Gonzalez, C.; Koiwa, H.; Tang, G.; Dickman, M.; Li, P.; Zhang, X. RISC-interacting clearing 3′-5′ exoribonucleases (RICEs) degrade uridylated cleavage fragments to maintain functional RISC in *Arabidopsis thaliana*. *eLife* **2017**, *6*, e24466. [CrossRef] [PubMed]
45. Huntzinger, E.; Izaurralde, E. Gene silencing by microRNAs: Contributions of translational repression and mRNA decay. *Nat. Rev. Genet.* **2011**, *12*, 99–110. [CrossRef]
46. Pontes, O.; Costa-Nunes, P.; Vithayathil, P.; Pikaard, C.S. RNA polymerase V functions in *Arabidopsis* interphase heterochromatin organization independently of the 24-nt siRNA-directed DNA methylation pathway. *Mol. Plant* **2009**, *2*, 700–710. [CrossRef] [PubMed]
47. Teotia, S.; Singh, D.; Tang, G. DNA Methylation in plants by microRNAs. In *Plant Epigenetics*; Rajewsky, N., Jurga, S., Barciszewski, J., Eds.; Springer International Publishing: Cham, Switzerland, 2017; pp. 247–262.
48. Axtell, M.J. Lost in translation? microRNAs at the rough ER. *Trends Plant Sci.* **2017**, *22*, 273–274. [CrossRef] [PubMed]
49. Yu, Y.; Ji, L.; Le, B.H.; Zhai, J.; Chen, J.; Luscher, E.; Gao, L.; Liu, C.; Cao, X.; Mo, B.; et al. ARGONAUTE10 promotes the degradation of miR165/6 through the SDN1 and SDN2 exonucleases in Arabidopsis. *PLoS Biol.* **2017**, *15*, e2001272. [CrossRef] [PubMed]
50. Borges, F.; Martienssen, R.A. The expanding world of small RNAs in plants. *Nat. Rev. Mol. Cell Biol.* **2015**, *16*, 727–741. [CrossRef]
51. Arikit, S.; Zhai, J.; Meyers, B.C. Biogenesis and function of rice small RNAs from non-coding RNA precursors. *Curr. Opin. Plant Biol.* **2013**, *16*, 170–179. [CrossRef] [PubMed]
52. Axtell, M.J.; Jan, C.; Rajagopalan, R.; Bartel, D.P. A two-hit trigger for siRNA biogenesis in plants. *Cell* **2006**, *127*, 565–577. [CrossRef] [PubMed]
53. Allen, E.; Howell, M.D. miRNAs in the biogenesis of trans-acting siRNAs in higher plants. *Semin. Cell Dev. Biol.* **2010**, *21*, 798–804. [CrossRef] [PubMed]
54. De Felippes, F.F.; Marchais, A.; Sarazin, A.; Oberlin, S.; Voinnet, O. A single miR390 targeting event is sufficient for triggering TAS3-tasiRNA biogenesis in Arabidopsis. *Nucleic Acids Res.* **2017**, *45*, 5539–5554. [CrossRef]
55. Dotto, M.C.; Petsch, K.A.; Aukerman, M.J.; Beatty, M.; Hammell, M.; Timmermans, M.C. Genome-wide analysis of *leafbladeless1*-regulated and phased small RNAs underscores the importance of the TAS3 ta-siRNA pathway to maize development. *PLoS Genet.* **2014**, *10*, e1004826. [CrossRef] [PubMed]
56. Park, W.; Li, J.; Song, R.; Messing, J.; Chen, X. CARPEL FACTORY, a Dicer homolog, and HEN1, a novel protein, act in microRNA metabolism in *Arabidopsis thaliana*. *Curr. Biol.* **2002**, *12*, 1484–1495. [CrossRef]
57. Parent, J.S.; Bouteiller, N.; Elmayan, T.; Vaucheret, H. Respective contributions of *Arabidopsis* DCL2 and DCL4 to RNA silencing. *Plant J. Cell Mol. Biol.* **2015**, *81*, 223–232. [CrossRef]
58. Xie, Z.; Johansen, L.K.; Gustafson, A.M.; Kasschau, K.D.; Lellis, A.D.; Zilberman, D.; Jacobsen, S.E.; Carrington, J.C. Genetic and functional diversification of small RNA pathways in plants. *PLoS Biol.* **2004**, *2*, e104. [CrossRef]
59. Xie, Z.; Allen, E.; Wilken, A.; Carrington, J.C. DICER-LIKE 4 functions in trans-acting small interfering RNA biogenesis and vegetative phase change in *Arabidopsis thaliana*. *Proc. Natl. Acad. Sci. USA* **2005**, *102*, 12984–12989. [CrossRef]
60. Margis, R.; Fusaro, A.F.; Smith, N.A.; Curtin, S.J.; Watson, J.M.; Finnegan, E.J.; Waterhouse, P.M. The evolution and diversification of Dicers in plants. *FEBS Lett.* **2006**, *580*, 2442–2450. [CrossRef]

61. Vaucheret, H.; Vazquez, F.; Crete, P.; Bartel, D.P. The action of *ARGONAUTE1* in the miRNA pathway and its regulation by the miRNA pathway are crucial for plant development. *Genes Dev.* **2004**, *18*, 1187–1197. [CrossRef]
62. Borges, F.; Pereira, P.A.; Slotkin, R.K.; Martienssen, R.A.; Becker, J.D. MicroRNA activity in the *Arabidopsis* male germline. *J. Exp. Bot.* **2011**, *62*, 1611–1620. [CrossRef] [PubMed]
63. Harvey, J.J.; Lewsey, M.G.; Patel, K.; Westwood, J.; Heimstadt, S.; Carr, J.P.; Baulcombe, D.C. An antiviral defense role of AGO2 in plants. *PLoS ONE* **2011**, *6*, e14639. [CrossRef]
64. Zilberman, D.; Cao, X.; Johansen, L.K.; Xie, Z.; Carrington, J.C.; Jacobsen, S.E. Role of *Arabidopsis ARGONAUTE4* in RNA-directed DNA methylation triggered by inverted repeats. *Curr. Biol.* **2004**, *14*, 1214–1220. [CrossRef] [PubMed]
65. Wu, J.; Yang, Z.; Wang, Y.; Zheng, L.; Ye, R.; Ji, Y.; Zhao, S.; Ji, S.; Liu, R.; Xu, L.; et al. Viral-inducible Argonaute18 confers broad-spectrum virus resistance in rice by sequestering a host microRNA. *eLife* **2015**, *4*. [CrossRef] [PubMed]
66. Zhai, L.; Sun, W.; Zhang, K.; Jia, H.; Liu, L.; Liu, Z.; Teng, F.; Zhang, Z. Identification and characterization of Argonaute gene family and meiosis-enriched Argonaute during sporogenesis in maize. *J. Integr. Plant Biol.* **2014**, *56*, 1042–1052. [CrossRef] [PubMed]
67. Marin, E.; Jouannet, V.; Herz, A.; Lokerse, A.S.; Weijers, D.; Vaucheret, H.; Nussaume, L.; Crespi, M.D.; Maizel, A. miR390, *Arabidopsis* TAS3 tasiRNAs, and their AUXIN RESPONSE FACTOR targets define an autoregulatory network quantitatively regulating lateral root growth. *Plant Cell* **2010**, *22*, 1104–1117. [CrossRef] [PubMed]
68. Montgomery, T.A.; Howell, M.D.; Cuperus, J.T.; Li, D.; Hansen, J.E.; Alexander, A.L.; Chapman, E.J.; Fahlgren, N.; Allen, E.; Carrington, J.C. Specificity of ARGONAUTE7-miR390 interaction and dual functionality in TAS3 trans-acting siRNA formation. *Cell* **2008**, *133*, 128–141. [CrossRef] [PubMed]
69. Xia, R.; Meyers, B.C.; Liu, Z.; Beers, E.P.; Ye, S.; Liu, Z. MicroRNA superfamilies descended from miR390 and their roles in secondary small interfering RNA Biogenesis in Eudicots. *Plant Cell* **2013**, *25*, 1555–1572. [CrossRef] [PubMed]
70. Sievers, F.; Higgins, D.G. Clustal Omega for making accurate alignments of many protein sequences. *Protein Sci.* **2018**, *27*, 135–145. [CrossRef]
71. Letunic, I.; Bork, P. Interactive Tree of Life v2: Online annotation and display of phylogenetic trees made easy. *Nucleic Acids Res.* **2011**, *39*, W475–W478. [CrossRef]
72. Zaratiegui, M.; Irvine, D.V.; Martienssen, R.A. Noncoding RNAs and gene silencing. *Cell* **2007**, *128*, 763–776. [CrossRef] [PubMed]
73. Cho, S.H.; Coruh, C.; Axtell, M.J. miR156 and miR390 regulate tasiRNA accumulation and developmental timing in *Physcomitrella patens*. *Plant Cell* **2012**, *24*, 4837–4849. [CrossRef]
74. Aukerman, M.J.; Sakai, H. Regulation of flowering time and floral organ identity by a microRNA and its *APETALA2*-like target genes. *Plant Cell* **2003**, *15*, 2730–2741. [CrossRef] [PubMed]
75. Jung, J.H.; Seo, Y.H.; Seo, P.J.; Reyes, J.L.; Yun, J.; Chua, N.H.; Park, C.M. The *GIGANTEA*-regulated microRNA172 mediates photoperiodic flowering independent of *CONSTANS* in *Arabidopsis*. *Plant Cell* **2007**, *19*, 2736–2748. [CrossRef] [PubMed]
76. Wu, G.; Poethig, R.S. Temporal regulation of shoot development in *Arabidopsis thaliana* by miR156 and its target *SPL3*. *Development* **2006**, *133*, 3539–3547. [CrossRef] [PubMed]
77. Chuck, G.S.; Tobias, C.; Sun, L.; Kraemer, F.; Li, C.; Dibble, D.; Arora, R.; Bragg, J.N.; Vogel, J.P.; Singh, S.; et al. Overexpression of the maize *Corngrass1* microRNA prevents flowering, improves digestibility, and increases starch content of switchgrass. *Proc. Natl. Acad. Sci. USA* **2011**, *108*, 17550–17555. [CrossRef] [PubMed]
78. Doebley, J.; Stec, A.; Hubbard, L. The evolution of apical dominance in maize. *Nature* **1997**, *386*, 485–488. [CrossRef] [PubMed]
79. Kidner, C.A.; Martienssen, R.A. Spatially restricted microRNA directs leaf polarity through ARGONAUTE1. *Nature* **2004**, *428*, 81–84. [CrossRef] [PubMed]
80. Palatnik, J.F.; Allen, E.; Wu, X.; Schommer, C.; Schwab, R.; Carrington, J.C.; Weigel, D. Control of leaf morphogenesis by microRNAs. *Nature* **2003**, *425*, 257–263. [CrossRef] [PubMed]
81. Schommer, C.; Palatnik, J.F.; Aggarwal, P.; Chetelat, A.; Cubas, P.; Farmer, E.E.; Nath, U.; Weigel, D. Control of jasmonate biosynthesis and senescence by miR319 targets. *PLoS Biol.* **2008**, *6*, e230. [CrossRef]

82. Jiao, Y.; Wang, Y.; Xue, D.; Wang, J.; Yan, M.; Liu, G.; Dong, G.; Zeng, D.; Lu, Z.; Zhu, X.; et al. Regulation of *OsSPL14* by OsmiR156 defines ideal plant architecture in rice. *Nat. Genet.* **2010**, *42*, 541–544. [CrossRef] [PubMed]
83. Lu, Z.; Yu, H.; Xiong, G.; Wang, J.; Jiao, Y.; Liu, G.; Jing, Y.; Meng, X.; Hu, X.; Qian, Q.; et al. Genome-wide binding analysis of the transcription activator ideal plant architecture1 reveals a complex network regulating rice plant architecture. *Plant Cell* **2013**, *25*, 3743–3759. [CrossRef] [PubMed]
84. Moreno, M.A.; Harper, L.C.; Krueger, R.W.; Dellaporta, S.L.; Freeling, M. *liguleless1* encodes a nuclear-localized protein required for induction of ligules and auricles during maize leaf organogenesis. *Genes Dev.* **1997**, *11*, 616–628. [CrossRef] [PubMed]
85. Liu, L.; Du, Y.; Shen, X.; Li, M.; Sun, W.; Huang, J.; Liu, Z.; Tao, Y.; Zheng, Y.; Yan, J.; et al. *KRN4* controls quantitative variation in maize kernel row number. *PLoS Genet.* **2015**, *11*, e1005670. [CrossRef] [PubMed]
86. Chuck, G.; Meeley, R.; Irish, E.; Sakai, H.; Hake, S. The maize *tasselseed4* microRNA controls sex determination and meristem cell fate by targeting *Tasselseed6/indeterminate spikelet1*. *Nat. Genet.* **2007**, *39*, 1517–1521. [CrossRef] [PubMed]
87. Banks, J.A. MicroRNA, sex determination and floral meristem determinacy in maize. *Genome Biol.* **2008**, *9*, 204. [CrossRef] [PubMed]
88. Peng, T.; Qiao, M.; Liu, H.; Teotia, S.; Zhang, Z.; Zhao, Y.; Wang, B.; Zhao, D.; Shi, L.; Zhang, C.; et al. A resource for inactivation of microRNAs using Short Tandem Target Mimic technology in model and crop plants. *Mol. Plant* **2018**, *11*, 1400–1417. [CrossRef] [PubMed]
89. Donaire, L.; Barajas, D.; Martinez-Garcia, B.; Martinez-Priego, L.; Pagan, I.; Llave, C. Structural and genetic requirements for the biogenesis of tobacco rattle virus-derived small interfering RNAs. *J. Virol.* **2008**, *82*, 5167–5177. [CrossRef] [PubMed]
90. Tsuji, H.; Aya, K.; Ueguchi-Tanaka, M.; Shimada, Y.; Nakazono, M.; Watanabe, R.; Nishizawa, N.K.; Gomi, K.; Shimada, A.; Kitano, H.; et al. GAMYB controls different sets of genes and is differentially regulated by microRNA in aleurone cells and anthers. *Plant J. Cell Mol. Biol.* **2006**, *47*, 427–444. [CrossRef] [PubMed]
91. Archak, S.; Nagaraju, J. Computational prediction of rice (*Oryza sativa*) miRNA targets. *Genom. Proteom. Bioinform.* **2007**, *5*, 196–206. [CrossRef]
92. Nogueira, F.T.; Madi, S.; Chitwood, D.H.; Juarez, M.T.; Timmermans, M.C. Two small regulatory RNAs establish opposing fates of a developmental axis. *Genes Dev.* **2007**, *21*, 750–755. [CrossRef] [PubMed]
93. McConnell, J.R.; Emery, J.; Eshed, Y.; Bao, N.; Bowman, J.; Barton, M.K. Role of *PHABULOSA* and *PHAVOLUTA* in determining radial patterning in shoots. *Nature* **2001**, *411*, 709–713. [CrossRef] [PubMed]
94. Emery, J.F.; Floyd, S.K.; Alvarez, J.; Eshed, Y.; Hawker, N.P.; Izhaki, A.; Baum, S.F.; Bowman, J.L. Radial patterning of *Arabidopsis* shoots by class III HD-ZIP and KANADI genes. *Curr. Biol.* **2003**, *13*, 1768–1774. [CrossRef] [PubMed]
95. Nogueira, F.T.; Chitwood, D.H.; Madi, S.; Ohtsu, K.; Schnable, P.S.; Scanlon, M.J.; Timmermans, M.C. Regulation of small RNA accumulation in the maize shoot apex. *PLoS Genet.* **2009**, *5*, e1000320. [CrossRef] [PubMed]
96. Shkibbe, D.S.; Schnable, P.S. Male sterility in maize. *Maydica* **2005**, *50*, 367–376.
97. Wang, H.; Nussbaum-Wagler, T.; Li, B.; Zhao, Q.; Vigouroux, Y.; Faller, M.; Bomblies, K.; Lukens, L.; Doebley, J.F. The origin of the naked grains of maize. *Nature* **2005**, *436*, 714–719. [CrossRef]
98. Wang, H.; Studer, A.J.; Zhao, Q.; Meeley, R.; Doebley, J.F. Evidence that the origin of naked kernels during maize domestication was caused by a single amino acid substitution in *tga1*. *Genetics* **2015**, *200*, 965–974. [CrossRef]
99. Wu, L.; Zhang, D.; Xue, M.; Qian, J.; He, Y.; Wang, S. Overexpression of the maize *GRF10*, an endogenous truncated growth-regulating factor protein, leads to reduction in leaf size and plant height. *J. Integr. Plant Biol.* **2014**, *56*, 1053–1063. [CrossRef]
100. Yan, J.; Zhao, C.; Zhou, J.; Yang, Y.; Wang, P.; Zhu, X.; Tang, G.; Bressan, R.A.; Zhu, J.K. The miR165/166 mediated regulatory module plays critical roles in ABA homeostasis and response in *Arabidopsis thaliana*. *PLoS Genet.* **2016**, *12*, e1006416. [CrossRef]
101. Zhang, J.; Zhang, H.; Srivastava, A.K.; Pan, Y.; Bai, J.; Fang, J.; Shi, H.; Zhu, J.K. Knockdown of rice microRNA166 confers drought resistance by causing leaf rolling and altering stem xylem development. *Plant Physiol.* **2018**, *176*, 2082–2094. [CrossRef]

102. Hochholdinger, F.; Yu, P.; Marcon, C. Genetic control of root system development in maize. *Trends Plant Sci.* **2018**, *23*, 79–88. [CrossRef] [PubMed]
103. Du, Q.; Wang, K.; Zou, C.; Xu, C.; Li, W.X. The *PILNCR1*-miR399 regulatory module is important for low phosphate tolerance in maize. *Plant Physiol.* **2018**, *177*, 1743–1753. [CrossRef]
104. Duvick, D.N. The contribution of breeding to yield advances in maize (*Zea mays* L.). *Adv. Agron.* **2005**, *86*, 83–145. [CrossRef]
105. Buckler, E.S.; Holland, J.B.; Bradbury, P.J.; Acharya, C.B.; Brown, P.J.; Browne, C.; Ersoz, E.; Flint-Garcia, S.; Garcia, A.; Glaubitz, J.C.; et al. The genetic architecture of maize flowering time. *Science* **2009**, *325*, 714–718. [CrossRef]
106. Lackey, E.; Ng, D.W.; Chen, Z.J. RNAi-mediated down-regulation of *DCL1* and *AGO1* induces developmental changes in resynthesized *Arabidopsis* allotetraploids. *New Phytol.* **2010**, *186*, 207–215. [CrossRef] [PubMed]
107. Moon, J.; Candela, H.; Hake, S. The Liguleless narrow mutation affects proximal-distal signaling and leaf growth. *Development* **2013**, *140*, 405–412. [CrossRef] [PubMed]
108. Luan, M.; Xu, M.; Lu, Y.; Zhang, Q.; Zhang, L.; Zhang, C.; Fan, Y.; Lang, Z.; Wang, L. Family-wide survey of miR169s and *NF-YAs* and their expression profiles response to abiotic stress in maize roots. *PLoS ONE* **2014**, *9*, e91369. [CrossRef]
109. Portwood, J.L., 2nd; Woodhouse, M.R.; Cannon, E.K.; Gardiner, J.M.; Harper, L.C.; Schaeffer, M.L.; Walsh, J.R.; Sen, T.Z.; Cho, K.T.; Schott, D.A.; et al. MaizeGDB 2018: The maize multi-genome genetics and genomics database. *Nucleic Acids Res.* **2019**, *47*, D1146–D1154. [CrossRef] [PubMed]
110. Yan, J.; Gu, Y.; Jia, X.; Kang, W.; Pan, S.; Tang, X.; Chen, X.; Tang, G. Effective small RNA destruction by the expression of a short tandem target mimic in *Arabidopsis*. *Plant Cell* **2012**, *24*, 415–427. [CrossRef]
111. Franco-Zorrilla, J.M.; Valli, A.; Todesco, M.; Mateos, I.; Puga, M.I.; Rubio-Somoza, I.; Leyva, A.; Weigel, D.; Garcia, J.A.; Paz-Ares, J. Target mimicry provides a new mechanism for regulation of microRNA activity. *Nat. Genet.* **2007**, *39*, 1033–1037. [CrossRef]
112. Jeong, D.H.; Park, S.; Zhai, J.; Gurazada, S.G.; De Paoli, E.; Meyers, B.C.; Green, P.J. Massive analysis of rice small RNAs: Mechanistic implications of regulated microRNAs and variants for differential target RNA cleavage. *Plant Cell* **2011**, *23*, 4185–4207. [CrossRef] [PubMed]

 © 2019 by the authors. Licensee MDPI, Basel, Switzerland. This article is an open access article distributed under the terms and conditions of the Creative Commons Attribution (CC BY) license (http://creativecommons.org/licenses/by/4.0/).

Review

Gene Regulation Mediated by microRNA-Triggered Secondary Small RNAs in Plants

Felipe Fenselau de Felippes

Science and Engineering Faculty, Queensland University of Technology, Brisbane, Australia; felipe.felippes@qut.edu.au

Received: 31 March 2019; Accepted: 24 April 2019; Published: 26 April 2019

Abstract: In plants, proper development and response to abiotic and biotic stimuli requires an orchestrated regulation of gene expression. Small RNAs (sRNAs) are key molecules involved in this process, leading to downregulation of their target genes. Two main classes of sRNAs exist, the small interfering RNAs (siRNAs) and microRNAs (miRNAs). The role of the latter class in plant development and physiology is well known, with many examples of how miRNAs directly impact the expression of genes in cells where they are produced, with dramatic consequences to the life of the plant. However, there is an aspect of miRNA biology that is still poorly understood. In some cases, miRNA targeting can lead to the production of secondary siRNAs from its target. These siRNAs, which display a characteristic phased production pattern, can act in *cis*, reinforcing the initial silencing signal set by the triggering miRNA, or in *trans*, affecting genes that are unrelated to the initial target. In this review, the mechanisms and implications of this process in the gene regulation mediated by miRNAs will be discussed. This work will also explore techniques for gene silencing in plants that are based on this unique pathway.

Keywords: tasiRNA; phasiRNA; miRNA; secondary siRNA

1. Introduction

MicroRNAs (miRNAs) are molecules that play pivotal roles in the control of gene expression, and together with small-interfering RNAs (siRNAs) they form the two major classes of regulatory small RNAs (sRNAs) in plants. Biogenesis of miRNAs relies on the activity of DICER-LIKE 1 (DCL1), an RNAse III enzyme that processes transcripts with imperfectly, self-complementary foldback structures to 21–22 nt long mature miRNA. The miRNA is then loaded into ARGONAUTE (AGO), conferring sequence specificity to the RNA-induced silencing complex (RISC), which promotes cleavage of the target transcript through the slicing activity of AGO. Alternatively, RISC-mediated gene downregulation can also be achieved via translation inhibition, a process still poorly understood in plants [1].

There are many examples of physiological and developmental pathways regulated by the direct action of miRNAs [2], yet there is an aspect of miRNA activity that only now has become more evident, and that is the ability of these molecules to indirectly regulate gene expression through the production of secondary siRNAs. In most cases, the outcome of miRNA-loaded RISC activity is cleavage and subsequent degradation of the target transcript, however, in a few cases, targeting can result in synthesis of a double-stranded RNA (dsRNA), having the target transcript as template for RNA-DEPENDENT RNA POLYMERASE 6 (RDR6) [3–7]. This step also requires the activity of SUPPRESSOR OF GENE SILENCING 3 (SGS3) and SILENCING DEFECTIVE 5 (SDE5) [3–9]. The newly synthetized dsRNA molecule is primarily the substrate for DCL4 to generate a new population of "secondary siRNAs", which have a phased pattern as their main characteristic, with siRNAs being produced in intervals 21 or 24 nt from the miRNA cleavage site [4,5,7,10–12]. These secondary siRNAs can have a dramatic impact on gene regulation mediated by miRNAs. They can act in *cis*, amplifying the silencing effect on their targets, or in *trans*, promoting downregulation of genes that otherwise would not be targeted by the trigger miRNA. Moreover, if the secondary siRNA precursor is a member of a gene family or shares sequences with other transcripts, silencing could spread to other genes, creating a regulatory cascade initially triggered by a single miRNA targeting event [13,14]. In addition to these features, generation of secondary siRNAs by miRNAs can add several other advantages to the regulation of gene expression, which will be discussed in more detail in this review.

The first miRNA-triggered secondary siRNA-producing loci were initially identified and characterized in *Arabidopsis thaliana*, where non-coding RNA molecules were found to give rise to siRNAs suppressing the expression of genes that were unrelated to their precursor molecules; therefore, they were referred as *trans*-acting siRNAs (tasiRNAs) [3,4]. To date, four families of tasiRNA-producing loci (*TAS1-4*) have been described in *A. thaliana*. *TAS1* and *TAS2* are both targeted by miR173, with *TAS3* and *TAS4* tasiRNA biogenesis being dependent on miR390 and miR828, respectively [3–6,15]. Additional *TAS* genes (*TAS5-10*) have been described or predicted in species other than *A. thaliana*, suggesting that many secondary siRNA-producing loci are yet to be discovered [16–19]. Indeed, with the advance of genomic-scale analyses, several phased, secondary siRNA-producing loci were recently identified in various plant species. As for *TAS* transcripts, one or more miRNAs were shown or predicted to target the precursor RNA molecule. However, different from classic tasiRNAs, generation of these newly identified secondary siRNAs can also be associated with protein-coding genes, and their activity in promoting cleavage of their target in *trans* is often not shown. Therefore, these secondary siRNAs are called phased siRNAs (phasiRNAs), and the loci where they come from is referred to as a *PHAS* gene [20,21]. In summary, phased, miRNA-triggered secondary siRNAs are generally referred to as phasiRNAs, while tasiRNAs are a specialized subclass of phasiRNAs for which function has been demonstrated to occur in *trans*. In addition, *TAS* loci are usually considered as noncoding with no function other than being precursor molecules to secondary siRNAs [13,20,21]. This review will focus on the factors leading to miRNA-triggered production of secondary siRNAs as well as the main features and possible advantages of this system for control of gene expression in plants. In addition, different methods to trigger gene downregulation using this silencing mechanism will be discussed.

2. Biogenesis of miRNA-Triggered Secondary siRNAs

The question of how some miRNAs can trigger the production of secondary siRNAs from their targets has been one of the major subjects of phasiRNA research. Two hypotheses have been commonly used to explain this peculiar phenomenon, known simply as "one-hit" and "two-hit" models. However, recent findings have reshaped our understanding of how miRNA-triggered siRNAs are generated (Figure 1). It is worth noting that, despite most of what is known has originated from studies using *TAS* loci as a model, the mechanisms for miRNA-triggered secondary siRNA biogenesis seems to be valid for the majority, if not all, phasiRNAs.

2.1. "Two-Hit" Model

The first attempt to explain biogenesis of phasiRNAs came from the observation that *TAS3* in *Physcomitrella patens* (*PpTAS3*), unlike most of the plant miRNA targets, displayed two miR390 complementary sequences, and the majority of sRNAs produced from this transcript were confined between these two sites [22]. The authors also showed that this pattern in the *PpTAS3* gene could be extended to *A. thaliana* and several other species. Interestingly, the miR390 target site located 5′ to the tasiRNAs was not cleavable in *A. thaliana*; however, mutations disrupting this or the cleavable 3′ miR390 site resulted in plants showing phenotypes associated with the impairment of *TAS3* function [22]. This and the discovery that other secondary siRNA-producing loci are also flanked by sRNA complementary sites led to a model where dual sRNA hits would act as a trigger for recruitment of RDR6.

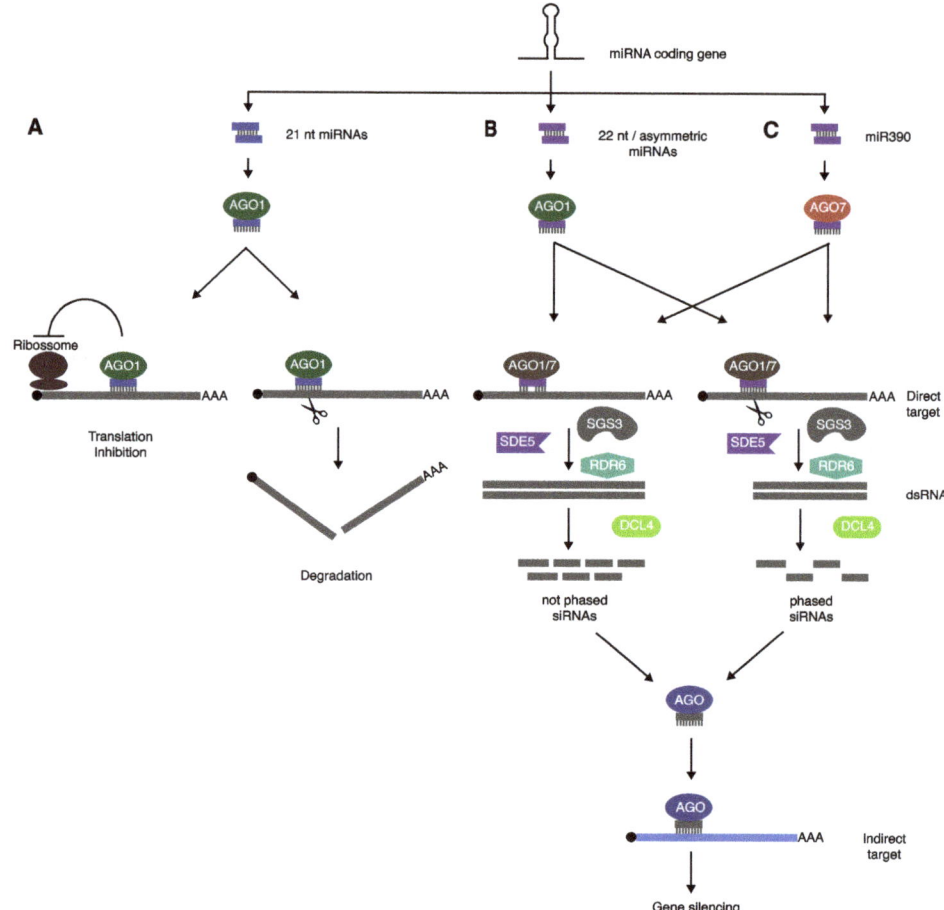

Figure 1. Biogenesis of miRNA-triggered secondary siRNAs. (**A**) Most plant miRNAs are processed into 21 nt long molecules, loaded into AGO1, and promote post-transcriptional gene silencing (PTGS) via translational repression or cleavage, followed by degradation of the target transcript. (**B**) Production of secondary siRNAs occurs when 22 nt long/asymmetric miRNA are bound to AGO1 to target a transcript. (**C**) Alternatively, 21 nt long miRNAs, such as miR390, can also initiate transitivity via interaction with AGO7. In both cases (B and C), cleavage is required for phasing, but not for generation of secondary siRNAs. Biogenesis of secondary siRNAs is dependent on the action of RDR6, SGS3, and SDE5, resulting in the synthesis of a dsRNA, which is mainly processed by DCL4. These siRNAs are loaded into AGOs and can drive gene silencing of their targets.

2.2. "One-Hit" Model

The "two-hit" model, however, was not sufficient to explain biogenesis of secondary siRNAs from other loci, such as *TAS1*, *TAS2* and *TAS4*, in which tasiRNA precursors were all cleaved at a single site upstream of the sRNA production region [3–6,15,22]. One of the initial insights into the mechanism behind tasiRNA generation from these transcripts came from experiments testing the requirements for secondary siRNA biogenesis from *TAS1* in *A. thaliana* [23,24]. These authors have shown that miR173-mediated targeting was not only necessary, but also sufficient to trigger tasiRNA production. This observation suggested that miRNAs, such as miR173, had unique features that differentiated them from the majority of other miRNAs that could not trigger the production of secondary siRNAs from their targets. Indeed, it was later demonstrated that miRNA length and the structure of the miRNA/miRNA* duplex (miRNA* refers to the sequence complementary to the predominant miRNA in the precursor molecule) were key determinants in triggering miRNA-dependent secondary siRNA production [25–27]. In plants, most miRNAs are processed as 21 nt long molecules. Interestingly, genome-wide analysis of sRNAs found that the majority of miRNAs and siRNAs associated with secondary siRNA production were 22 nt in length, indicating that sRNA size could be a crucial aspect for phasiRNA biogenesis. To test this hypothesis, Cuperus et al. [26] and Chen et al. [25] engineered miRNA precursors to produce mature miRNAs, either 21 or 22 nt of length, and tested their ability to initiate secondary siRNA production. For instance, miR173, which initiates tasiRNA production in *TAS1* and *TAS2*, is naturally found as a molecule 22 nt in length; however, its ability to trigger tasiRNA biogenesis was abolished when this molecule was 21 nt in length. Accordingly, turning the 21 nt miR319, which does not trigger secondary siRNAs production, into a 22 nt miRNA conferred this molecule the capacity to start siRNA generation from its target transcript. In the vast majority of cases, the generation of 22 nt miRNAs, instead of the more commonly found 21 nt variety, is caused by the presence of an asymmetric bulge in the pairing between miRNA and miRNA* in the precursor molecule, resulting in maturation, by DCL1, of an miRNA/miRNA* duplex with a 22/21 nt configuration [25,26]. Interestingly, 22 nt miRNAs can also be created by post-processing modification events, as shown in the case of the soybean miR1510. This sRNA, which is able to trigger phasiRNA production, is processed as 21 nt molecule, but accumulates as a 22 nt isoform via monouridylation [28]. Despite this, it is not only miRNA size that seems to account for the ability to initiate transitivity (another name for RDR6-dependent secondary siRNA production). MiRNAs that are 21 nt in length can also trigger transitivity when their miRNA* is found as a 22 nt molecule. It has been proposed that asymmetry in the miRNA/miRNA* duplex, which is also found in miRNAs processed as 22 nt molecules, is sufficient for the initiation of secondary siRNA production in target transcripts. This idea was confirmed by producing an asymmetric miR173/miR173* duplex, where both miRNA and miRNA* were produced as 21 nt entities, demonstrating that this configuration could efficiently trigger transitivity [27].

2.3. A Unified "One-Hit" Model

The models described above have been considered as two independent mechanisms leading to phasiRNA generation in plants. However, a recent study has shown that this might not be the case, and these two processes might be more similar than previously suspected. In *A. thaliana*, miR390 was recently shown to trigger tasiRNA production from the *TAS3* transcript even when only one targeting event occurred, similar to what happened with other *A. thaliana TAS* families [29]. Supporting the idea that "one-hit" is sufficient for tasiRNA production in *TAS3*, many dicots, conifers, and cycads carry a second *TAS3*-related gene, referred to as *TAS3-2*, which, in some species of citrus, chicories, and populous, possess only one miR390 target site [30,31]. In addition, *TAS3* in spruce has been characterized as a large family with 18 members, some of them carrying only one miR390 complementary sequence [32]. Taken together, these observations suggest that "one-hit" might be the basic system behind secondary siRNA production, and the "two-hit" configuration may have evolved as a regulatory mechanism to avoid possible off-targeting incidents by limiting the region from where secondary siRNAs are produced [29]. Another peculiarity of this unified model concerns cleavage of the precursor transcript as a requirement for secondary siRNA generation. Since the discovery of tasiRNAs, the slicing activity of AGO within RISC has been considered essential for the recruitment of RDR6 [13,14,33]. However, secondary siRNAs deriving from *TAS1* and *TAS3* can also be detected, even when the respective transcripts are not cleaved by AGO, indicating that a non-cleavable interaction of RISC with its target is sufficient to trigger efficient phasiRNA production [29,34]. Nonetheless, slicing of the target transcript is still crucial for the proper phase and, therefore, function of tasiRNAs, which is probably due to the lack of a well-defined end of the dsRNA molecule caused by the absence of cleavage.

Despite unification under a same "one-hit" process, tasiRNA biogenesis from *TAS3* and *TAS1/2/4* still differs regarding the initiation mechanism. While *TAS1*, *TAS2*, and *TAS4* give rise to secondary siRNAs after being hit by miRNAs that are 22 nt in length, miR390, which targets *TAS3*, is 21 nt long and does not show asymmetric structures in its precursor [3–6,15,25,26]. Another particularity of the *TAS3*/miR390 system is that miR390 is "assigned" with its own AGO protein. In *A. thaliana* there are 10 AGOs (AGO1-10), with most miRNAs loaded in AGO1, including the bulk of those that can trigger transitivity [33,35,36]. However, miR390 is not only preferentially found associated with AGO7, but it is also the specific ligand of this protein, which seems to select miRNA through recognition of its initial 5' adenosine residue and the central region of the miR390/miR390* duplex [33,37]. More interestingly, unlike when it is loaded into AGO7, the ability of miR390 to initiate tasiRNA production was abolished when this miRNA was found or forced to interact with AGO1 or AGO2 [33], highlighting the high degree of specialization found among members of this family.

In summary, in this unified model the production of secondary siRNA from miRNA targets is dependent on a targeting event, where the AGO involved is found in a competent status. This condition is achieved when AGO1, for instance, interacts with an miRNA that is 22 nt long and/or asymmetric as a duplex. AGO7, on the other hand, would be a specialized form of this protein that could continually promote the production of secondary siRNAs from its targets, allowing miRNAs neither 22 nt long nor asymmetric to initiate transitivity. How AGO1, loaded with 22 nt /asymmetric miRNAs or the miR390/AGO7 complex, can route its target to the RDR6 pathway is still unknown. One could speculate that once loaded with 22 nt /asymmetric miRNAs, AGO1 would suffer a change in its configuration allowing the onset of transitivity. Such a change in conformation is supported by crystal structure analysis of *Thermus thermophilus* AGO bound to DNA guide strands of different sizes [38]. In the case of AGO7, it is plausible that this protein has evolved to constitutively be found in this competent form. It is clear that further work will be required to test this hypothesis.

2.4. Other Elements Involved in Secondary siRNA Production

The subcellular location where miRNA-triggered secondary siRNAs are produced has also been the focus of investigation. Many components required for phasiRNA biogenesis, such as DCL4, RDR6, SGS3, and AGO7 accumulate in the cytoplasm [39–43]. More specifically, SGS3, RDR6, and AGO7 have been shown to co-localize in cytoplasmic foci, called siRNA bodies, which are distinct from processing-bodies (P-bodies) involved in mRNA turnover [41,42]. In addition, AGO7, miR390, and SGS3 were shown to be present in microsomal fractions and localized in the endoplasmic reticulum (ER), suggesting that phasiRNA production was connected to cytoplasmic membrane structures [40,42]. Indeed, it has been reported that miRNAs, including 22 nt ones, their target transcripts, and AGO1 are found associated to membrane-bound polysomes (MBPs) in the ER. Moreover, miRNA-guided cleavage could also be detected in MBP fractions [44]. Corroborating the view that proper subcellular localization was crucial for miRNA-triggered secondary siRNA production, phasiRNA generation was affected in *ago1-27* plants, most likely from a decrease in association between MBP and AGO1 [44]. Similarly, *TAS3*-tasiRNA biogenesis was impaired when AGO7 was forced to accumulate in the nucleus [42]. It is still unclear how these different components are brought together to the same subcellular compartment, and clearly more research will be necessary.

The role of SGS3 and SDE5 in the production of miRNA-triggered secondary siRNAs is another subject that still remains somewhat enigmatic. SGS3 has been identified from the beginning as an essential component of this system [3–7]. In vitro experiments have shown that SGS3 acts in conjunction with the cleaved transcript, protecting it against degradation and making it available for RDR6 [45]. However, it seems likely that SGS3 has other functions in addition to solely stabilizing the cleaved RNA. As discussed previously, tasiRNA production was shown to be independent of miRNA-mediated cleavage of the precursor transcript, yet SGS3 was still required for the synthesis of secondary siRNAs under non-slicing conditions [29,34]. This scenario is corroborated by the association of SGS3 with a slicing-defective RISC that binds uncut target RNAs [45]. A possible additional function of SGS3 in the production of secondary siRNAs could be in the proper placement of factors involved in transitivity in the same subcellular location, as suggested by the interaction of SGS3 with RDR6 and colocalization with

AGO7 in specialized cytoplasmic siRNA bodies [41,42]. As with *sgs3*, the accumulation of secondary siRNAs is also abolished in *sde5* mutants, suggesting a key, although, to date, unclear role for this protein in transitivity [8,9]. SDE5 encodes for a putative RNA export protein, and its role has been suggested to involve the traffic of mRNAs between the nucleus and cytoplasm and/or to route RNA to RDR6 [8,9]. Genetic experiments have placed SDE5 function downstream of SGS3, but upstream of RDR6 activity [46]. Nonetheless, the mode-of-action of these proteins still needs to be investigated in more detail.

In addition to the core components of the pathway, there are other elements that are not essential for the production of miRNA-dependent secondary siRNAs but still have an influence on the biogenesis of these molecules. Components of the THO/TREX complex, which is involved in the intercellular trafficking of mRNAs, have been shown to affect tasiRNA synthesis. In mutant plants where this complex had been disrupted, some tasiRNAs accumulated at lower levels when compared to the wild type [9,47]. It has been suggested that the THO/TREX complex is involved in the transport of *TAS* precursors from their production site to subcellular locations where secondary siRNA biogenesis takes place. Although tasiRNA precursors are considered to be non-coding transcripts, some *TAS* genes have short open reading frames (ORFs) located just upstream of the tasiRNA-producing region, which could potentially give rise to small peptides. Indeed, it has been described that some of these ORFs interact with ribosomes and are actually translated [44,46,48,49]. Interestingly, this process seems to be important for the proper accumulation of tasiRNAs from the transcripts involved. In *TAS2* and *TAS3*, mutations affecting these ORFs result in reduction of tasiRNA accumulation, most likely because of decreased stability of the *TAS* precursor caused by lower levels of association with ribosomes [46,49]. Alternatively, ribosome occupancy has been suggested as a factor that defines the regions of a transcript giving origin to secondary siRNAs [44]. The importance of translation in the production of secondary siRNAs is corroborated by the observation that production of synthetic tasiRNAs (syn-tasiRNAs) is improved with the introduction of a stop codon immediately before the miR173 target site [50].

3. Features and Advantages of miRNA-Triggered Secondary siRNA Gene Regulation

In *A. thaliana*, miRNA targeting events leading to transitivity are uncommon, with only a few cases described, and are better exemplified by the *TAS1–4* families. Nonetheless, gene regulation promoted by miRNA-triggered secondary siRNAs have an important impact on plant development and physiology. For instance, miR390 targeting of *TAS3* results in the production of tasiRNAs that can regulate the expression of different auxin response factor genes (*ARFs*), affecting important functions such as leaf morphology, the transition from juvenile to adult phase, and flower and root formation to mention a few [3,5,6,11,51–55]. In recent years, with the popularization and expansion of genomic-based studies, several loci that spawn phased, miRNA-triggered secondary siRNAs were identified in numerous other species. In many cases, these sRNA populations seem to have a role in a variety of pathways related to development, response to stresses, and disease resistance (for more details on these pathways and the miRNAs/phasiRNAs involved, please see this recent review [21]). But what would be the benefits of such an indirect role of miRNAs in the control of gene expression? In the second part of this review, some features and putative added values of indirect gene regulation by miRNAs via secondary siRNAs will be discussed.

The obvious consequence resulting from the production of tasi- and phasiRNAs is the amplification and potential enhancement of the silencing signal (Figure 2A). From a single miRNA targeting event, a population of secondary siRNAs is produced, all with the potential to silence in *cis*, multiplying the number of molecules that could cause downregulation of the precursor loci and, therefore, increase silencing pressure on the target. Indeed, evidence of secondary siRNA targeting in *cis* are quite common. For instance, in watermelon, *Medicago*, and citrus, several phasiRNAs were reported to target their precursor transcripts [20,56,57]. Despite their function mainly being associated with the silencing of unrelated genes in *trans*, tasiRNAs are also known to promote cleavage of the transcript of origin in *cis*. TasiRNA-5D2(-), one of the tasiRNAs emerging from *TAS3*, has been shown to cut its precursor transcript in different species [5,20,22,29,58,59]. In this case however, because it involves a non-coding transcript, it is possible that the re-attack of tasiRNA-5D2(-) acts more as a feedback regulatory mechanism, fine-tuning tasiRNA levels.

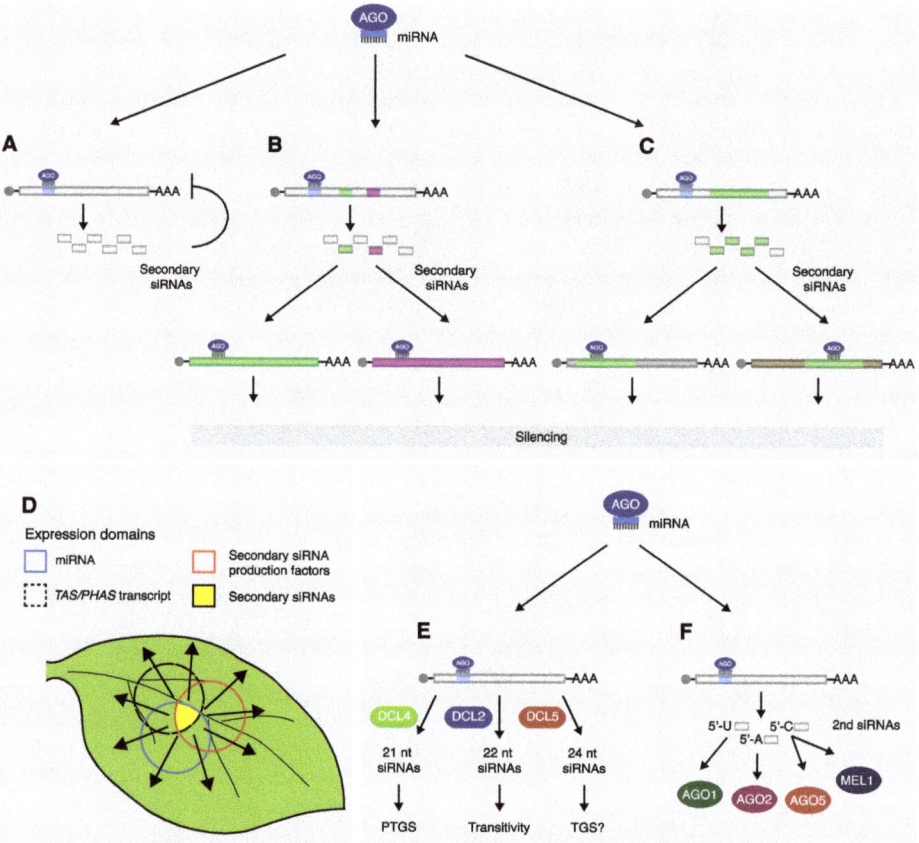

Figure 2. Features and advantages of gene regulation via miRNA-triggered secondary siRNAs. (**A**) By producing secondary siRNAs, miRNAs can increase the silencing pressure on their targets. (**B**) Secondary siRNAs targeting distinct transcripts can be produced from the same precursor, increasing the silencing range of the trigger miRNA. (**C**) The number of genes indirectly regulated by an miRNA can be increased if secondary siRNAs are produced from regions containing a conserved sequence shared by different loci. (**D**) Production of secondary siRNAs is restricted to regions where all elements participating in their biogenesis are present; however, they could later spread to neighboring cells to function non-cell autonomously (as indicated by the arrows). (**E**) The dsRNA synthetized by RDR6 is mainly processed by DCL4, generating 21 nt long siRNAs that are involved in PTGS. Alternatively, the dsRNA can also be the substrate for other DCLs, such as DCL2 and DCL5, resulting in the biogenesis of secondary siRNAs with different characteristics and functions. (**F**) Compared to most miRNAs, which have a 5' terminal uridine and are loaded into AGO1, secondary siRNAs show an increased diversity on their 5' extremity, allowing for sorting into different AGO proteins, with possible consequences to their activities.

Another interesting aspect of gene regulation mediated by secondary siRNAs is the possibility that one miRNA could affect the expression of several genes that otherwise would not have been targeted (Figure 2B). This is well illustrated in cotton, where cleavage of *MYB2* by miR828 results in the production of tasiRNAs that have been predicted to target several unrelated genes, such as sucrose synthase, histone acetyltransferase, and glutamate receptor, none of which are targeted by the triggering miRNA [60]. The *P. patens* miR390/*TAS3* system is an additional case where one miRNA can promote downregulation of several genes that are unrelated, sequence- and function-wise. In addition to the well-characterized, interspecies-conserved, tasiRNA targeting *ARF*-like mRNAs, *P. patens TAS3* transcripts also give origin to secondary siRNAs that can promote downregulation of three AP2 domain-containing transcripts [61]. Alternatively, phasiRNAs can also increase the number of genes regulated by a single miRNA if these secondary siRNAs are produced from conserved regions (Figure 2C). In this scenario, the newly generated siRNAs could function not only in *cis*, but also in *trans*, with the potential to affect any transcript that shares this same conserved region. The best example of such a regulatory network has been described in *Medicago* and involves the generation of phasiRNAs from nucleotide-binding leucine-rich repeat (*NB-LRR*) disease-resistance genes. In this species, three miRNA families (miR1507, miR2109, and miR2118) were described to target different NB-LRR-conserved motifs in 74 transcripts, leading to the biogenesis of phasiRNAs with the potential to regulate 60% of the estimated 540 *NB-LRR* genes [20]. Another mechanism resulting in the expansion of the miRNA activity range is through the production of secondary siRNAs that are 22 nt in size (Figure 2E). In peaches, two miR7122-triggered tasiRNAs, which are predominantly 22 nt in length, have been described to initiate phasiRNA generation from their targets [62], similar to what has been reported for the miR173/*TAS2* pathway in *Arabidopsis* [25].

As discussed previously, mechanisms leading to the formation of phasiRNAs require elements of the miRNA pathway as well as new components, such as RDR6, SGS3, SDE5, and AGO7. This increase in complexity brings new possibilities of regulation with interesting consequences to the indirect function of miRNAs in controlling gene expression. In *Arabidopsis*, tasiRNA production from *TAS3* is dependent on the activity of miR390 and AGO7 [5,33]. Interestingly, the *TAS* transcript, the AGO protein, and miRNA have distinct expression patterns and as a consequence; synthesis of secondary siRNAs from *TAS3* is restricted to cells where all these elements are present (Figure 2D) [33,54,63,64]. This spatio-temporal coordination has been shown to be important for the proper development of leaves. Abaxial/adaxial fate specification is a result of asymmetric expression of *ARFs* in the leaf, caused by the polarized accumulation pattern of *TAS3*-tasiRNAs [63–65]. This localized accumulation of tasiRNAs is only possible because of a delimited presence of AGO7 and *TAS3* to the adaxial side, restricting the biogenesis of secondary siRNAs to this region, despite the broader miR390 expression domain [63,64].

With few exceptions, most miRNA precursors are processed by DCL1 into mature molecules 21/22 nt in length, loaded into AGO1, and promote post-transcription gene silencing [1]. However, a whole new level of plasticity can be added to the control of gene expression when miRNA-triggered secondary siRNAs are employed (Figure 2E). The dsRNA molecule synthetized by RDR6 from *TAS* and *PHAS* transcripts are primarily processed by DCL4 into 21-nt-long siRNAs, which like miRNAs, act post-transcriptionally [7,10,11]. Nevertheless, grasses possess an additional DCL enzyme, DCL5 (formerly known as DCL3b), which is responsible for the production of phasiRNAs 24 nt in length from transcripts targeted by miR2275 [66,67]. Interestingly, this is the same size of siRNAs that interact with AGO4, the main effector of the transcriptional gene silence (TGS) pathway, which results in DNA methylation and subsequent silencing [1]. The implications of this discovery are still elusive. These 24 nt phasiRNAs were first described in rice and maize reproductive tissues. They accumulated in meiotic-stage anthers and, therefore, were believed to be involved in reproduction [12,68]. More importantly, given the size of these siRNAs, it is tempting to speculate that 24 nt long, DCL5-dependent phasiRNAs can be associated with AGO4 to promote DNA methylation, adding a new layer to gene regulation mediated by miRNAs. Supporting this view, Xia and colleagues [69] found that the miR2275/24 nt phasiRNAs pathway is not only present in monocots but also in eudicots plants. However, differently from the former group, miR2275-dependent, 24 nt long phasiRNA production in eudicots does not rely on the activity of a specific protein, such as DCL5, but instead it most likely requires the action of DCL3. This is the same enzyme responsible for producing the 24 nt siRNA associated with AGO4 and involved in TGS [1]. Reflecting the high level of specialization and conservation of the pathway, the vast majority of mature miRNAs have a 5′ terminal uridine (U), which has been shown to be a key determinant for the sorting of sRNAs into AGO1 [33,35,36]. In contrast, AGO2 and AGO4 prefer sRNAs that contain an adenosine (A) at the 5′ end, while AGO5 is more often associated to molecules that have a cytosine (C). Many of the secondary siRNAs with conserved functions, such as tasiRNAs that target *ARF* genes, are similar to miRNAs, having an uridine at the 5′ extremity of the mature molecule and, thus, are loaded into AGO1 [5]. Nonetheless, many phasiRNAs do not follow this trend, with many of them found associated to other AGOs such as AGO2 [33,35,36]. In rice, MEL1 is a specialized protein ortholog of AGO5, and it has been described to preferentially bind phasiRNAs that begin with a cytosine [70]. The function of these secondary siRNAs is still poorly understood, but nevertheless, MEL1 has been shown to mediate sporophytic germ-cell development and meiosis, suggesting that these sRNAs might play a direct role in these processes [71]. In summary, the variability of features found among the different AGOs can also be explored when miRNAs that trigger transitivity are involved in gene regulation (Figure 2F).

In addition to the features discussed above, there may be other unknown or poorly understood characteristics of phasiRNAs that could add extra value to miRNA-regulated pathways. For instance, compared to other classes of sRNAs, tasiRNA have been described to display extended cell-to-cell mobility, suggesting that, by initiating transitivity, miRNAs could increase their range of activity (Figure 2D) [72]. Indeed, an artificial miRNA (amiRNA) designed to be produced as a molecule 22 nt in length was reported to start secondary siRNA biogenesis from its target, resulting in silencing in tissues that otherwise would not be affected by amiRNAs of regular size or siRNAs produced from a hairpin construct [73].

4. Utilizing miRNA-Triggered Secondary siRNAs to Promote Directed Gene Silencing

Silencing promoted by sRNAs is not only an important mechanism to control gene expression, but it has also been used as a powerful tool to downregulate transcripts in both academic and applied purposes [74]. Despite not being as popular as founding techniques, such as artificial miRNAs (amiRNAs) and hairpin RNA interference (hpRNAi), systems based on the ability of miRNAs to start transitivity also exist and are undoubtedly a valuable addition to the collection of methods aiming to control gene activity (Figure 3).

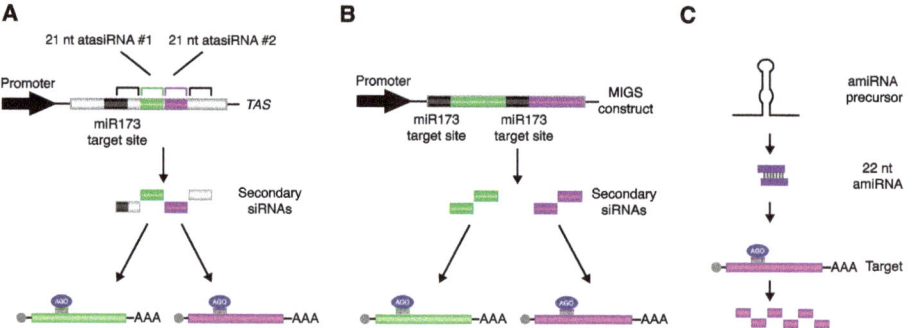

Figure 3. Methods for gene silencing based on miRNA-triggered secondary siRNAs. (**A**) Silencing using atasiRNAs consists of replacing one or more of the tasiRNAs in the *TAS* gene for a sequence designed to target the gene of interest. (**B**) miRNA-induced gene silencing (MIGS) constructs can be generated by placing the sequence recognized by an miRNA that can start transitivity in front of a fragment of the target gene (e.g., miR173). Downregulation of more than one gene using this technique can be easily accomplished by repeating the same pattern with different gene fragments. (**C**) By using specific precursors (such as *MIR173*), 22 nt long amiRNAs can be produced. These molecules can then initiate secondary siRNA synthesis from their targets, adding new features to the original method.

Artificial tasiRNA (atasiRNA), also known as synthetic tasiRNA (syn-tasiRNA), was the first method developed based on miRNA-triggered biogenesis of secondary siRNAs. It has been used to successfully reduce gene expression of endogenous sequences and to interfere with viroid infections [23,24,33,75–77]. This approach consists of replacing one or more tasiRNAs in a *TAS* transcript for sequences devised to target the gene of interest, in a process similar to the design of amiRNAs (Figure 3A) [75,78]. One of the most useful features of this technique is the possibility of having one precursor producing several atasiRNAs, each targeting different sequences, which could be located in the same or distinct transcripts [76]. AtasiRNAs share many of the advantages and limitations of amiRNAs. One distinct advantage is that high levels of specificity can be achieved, decreasing the chance of off-targeting. However, to efficiently design atasiRNA molecules that specifically downregulate one or just a few genes, with minimal chances of silencing unwanted transcripts, it is important that information about the entire genome is made available. Moreover, compared to approaches that make use of whole gene fragments for generation of the silencing construct, such as hpRNAi, this method is more susceptible to the effects of target accessibility that could reduce the effectiveness of the sRNA [79].

An additional system exploring the transitivity initiated by miRNAs is referred to as miRNA-induced gene silencing (MIGS), which has as its main characteristic the easiness of design [80,81]. With a single PCR step, the target site of an miRNA triggering phasiRNA production can be fused upstream of the fragment of a gene of interest. Upon transcript cleavage by RISC, the newly synthetized secondary siRNAs can subsequently promote silencing of related target sequences. MIGS is also a powerful tool to downregulate multiple genes using a single vector (Figure 3B). This is achieved simply by linking fragments of different targets, each with their own miRNA target site [80]. Since this method does not require genome-wide data, it is an interesting alternative to be used in species where this information is still lacking. Despite the aforementioned advantages, the risk of off-targeting needs to be considered when employing this system. Differently to atasiRNAs, MIGS constructs give rise to a population of siRNAs, all with the capacity to silence related sequences. Therefore, depending on the degree of conservation present in the fragment used, genes (other than the intended target) that share similar sequences could also become silenced. To date, MIGS has been shown to be an effective tool to silence genes in several species, including *Arabidopsis, Nicotiana benthamiana, Medicago*, soybean, rice, and petunia [80,82–86].

A common theme between atasiRNAs and MIGS is the requirement of a trigger miRNA. Therefore, it is important to take into consideration that the spatio-temporal expression pattern of the miRNA could influence the way in which atasiRNAs and MIGS-derived siRNAs are produced. In addition, some species might not code for the miRNA initiating the production of secondary siRNAs from the silencing constructs. To overcome this issue, a collection of plasmids has been created to allow the co-expression of the triggering miRNA and the MIGS/atasiRNA construct from a single vector [80,87]. Alternatively, designing an amiRNA to be produced as a 22 nt long molecule could also be a way to silence genes via secondary siRNAs, without the limitation of a two-component system (Figure 3C). McHale and colleagues [73] have demonstrated that a 22 nt long amiRNA targeting *CHALCONE SYNTHASE* (*CHS*) was able to cause widespread silencing due to the production of secondary siRNAs by RDR6.

5. Conclusion and Final Remarks

This review has discussed some of the crucial aspects related to the production of secondary siRNA triggered by miRNA and how this process can add valuable features to the control of gene expression mediated by sRNAs. Moreover, different methods to promote gene silencing in plants that are based on this unique ability of certain miRNAs were discussed, showing that they can be important alternatives to well-established systems, such as amiRNA and hpRNAi.

In recent years, our understanding of the mechanisms leading to miRNA-triggered secondary siRNA generation, and the importance of these molecules in plant physiology and development, has increased rapidly, yet this pathway is still one of the least understood among different processes involving sRNAs. It is still unknown, for example, how 22 nt long miRNA loaded into AGO1 or the miR390/AGO7 complex can lead to the recruitment of RDR6 to the target transcript. Also, the role and molecular mechanisms behind the activity of many phasiRNAs recently described in different plant species remain elusive. The elucidation of these and other aspects related to miRNA-triggered secondary siRNAs will greatly improve our understanding of how sRNAs impact the proper development of plants and the response to abiotic and biotic stresses. In addition, this new knowledge could be useful for the development of new technologies for biotechnological applications.

Funding: This work was funded by Australian Research Council (ARC), grant number FL160100155.

Acknowledgments: The author would like to thank Samanta Bolzan de Campos and Dr. Christopher Andrew Brosnan for critical reading and suggestions on the preparation of this review.

Conflicts of Interest: The author declares no conflict of interest.

References

1. Bologna, N.G.; Voinnet, O. The diversity, biogenesis, and activities of endogenous silencing small rnas in arabidopsis. *Annu. Rev. Plant Biol.* **2014**, *65*, 473–503. [CrossRef] [PubMed]
2. Li, S.; Castillo-González, C.; Yu, B.; Zhang, X. The functions of plant small rnas in development and in stress responses. *Plant J.* **2017**, *90*, 654–670. [PubMed]
3. Peragine, A.; Yoshikawa, M.; Wu, G.; Albrecht, H.L.; Poethig, R.S. Sgs3 and sgs2/sde1/rdr6 are required for juvenile development and the production of trans-acting sirnas in arabidopsis. *Genes Dev.* **2004**, *18*, 2368–2379. [CrossRef]
4. Vazquez, F.; Vaucheret, H.; Rajagopalan, R.; Lepers, C.; Gasciolli, V.; Mallory, A.C.; Hilbert, J.-L.; Bartel, D.P.; Crété, P. Endogenous trans-acting sirnas regulate the accumulation of arabidopsis mrnas. *Mol. Cell* **2004**, *16*, 69–79. [CrossRef]
5. Allen, E.; Xie, Z.; Gustafson, A.M.; Carrington, J.C. Microrna-directed phasing during trans-acting sirna biogenesis in plants. *Cell* **2005**, *121*, 207–221. [CrossRef] [PubMed]
6. Williams, L.; Carles, C.C.; Osmont, K.S.; Fletcher, J.C. A database analysis method identifies an endogenous trans-acting short-interfering rna that targets the arabidopsis arf2, arf3, and arf4 genes. *Proc. Natl. Acad. Sci. USA* **2005**, *102*, 9703–9708. [CrossRef]
7. Yoshikawa, M.; Peragine, A.; Park, M.Y.; Poethig, R.S. A pathway for the biogenesis of trans-acting sirnas in arabidopsis. *Genes Dev.* **2005**, *19*, 2164–2175. [CrossRef]
8. Hernandez-Pinzon, I.; Yelina, N.E.; Schwach, F.; Studholme, D.J.; Baulcombe, D.; Dalmay, T. Sde5, the putative homologue of a human mrna export factor, is required for transgene silencing and accumulation of trans-acting endogenous sirna. *Plant J.* **2007**, *50*, 140–148. [CrossRef] [PubMed]
9. Jauvion, V.; Elmayan, T.; Vaucheret, H. The conserved rna trafficking proteins hpr1 and tex1 are involved in the production of endogenous and exogenous small interfering rna in arabidopsis. *Plant Cell* **2010**, *22*, 2697–2709. [CrossRef] [PubMed]
10. Gasciolli, V.; Mallory, A.C.; Bartel, D.P.; Vaucheret, H. Partially redundant functions of arabidopsis dicer-like enzymes and a role for dcl4 in producing trans-acting sirnas. *Curr. Biol.* **2005**, *15*, 1494–1500. [CrossRef] [PubMed]
11. Xie, Z.; Allen, E.; Wilken, A.; Carrington, J.C. Dicer-like 4 functions in trans-acting small interfering rna biogenesis and vegetative phase change in arabidopsis thaliana. *Proc. Natl. Acad. Sci. USA* **2005**, *102*, 12984–12989. [CrossRef] [PubMed]
12. Johnson, C.; Kasprzewska, A.; Tennessen, K.; Fernandes, J.; Nan, G.-L.; Walbot, V.; Sundaresan, V.; Vance, V.; Bowman, L.H. Clusters and superclusters of phased small rnas in the developing inflorescence of rice. *Genome Res.* **2009**, *19*, 1429–1440. [CrossRef] [PubMed]
13. Fei, Q.; Xia, R.; Meyers, B.C. Phased, secondary, small interfering rnas in posttranscriptional regulatory networks. *Plant Cell* **2013**, *25*, 2400–2415. [CrossRef]
14. Vazquez, F.; Hohn, T. Biogenesis and biological activity of secondary sirnas in plants. *Scientifica* **2013**, *2013*, 783253. [CrossRef] [PubMed]
15. Rajagopalan, R.; Vaucheret, H.; Trejo, J.; Bartel, D.P. A diverse and evolutionarily fluid set of micrornas in arabidopsis thaliana. *Genes Dev.* **2006**, *20*, 3407–3425. [CrossRef]

16. Arif, M.A.; Fattash, I.; Ma, Z.; Cho, S.H.; Beike, A.K.; Reski, R.; Axtell, M.J.; Frank, W. Dicer-like3 activity in physcomitrella patens dicer-like4 mutants causes severe developmental dysfunction and sterility. *Mol. Plant* **2012**, *5*, 1281–1294. [CrossRef]
17. Li, F.; Orban, R.; Baker, B. Somart: A web server for plant mirna, tasirna and target gene analysis. *Plant J.* **2012**, *70*, 891–901. [CrossRef]
18. Zhang, C.; Li, G.; Wang, J.; Fang, J. Identification of trans-acting sirnas and their regulatory cascades in grapevine. *Bioinformatics* **2012**, *28*, 2561–2568. [CrossRef]
19. Zuo, J.; Wang, Q.; Han, C.; Ju, Z.; Cao, D.; Zhu, B.; Luo, Y.; Gao, L. Srnaome and degradome sequencing analysis reveals specific regulation of srna in response to chilling injury in tomato fruit. *Physiol. Plant* **2017**, *160*, 142–154. [CrossRef]
20. Zhai, J.; Jeong, D.-H.; De Paoli, E.; Park, S.; Rosen, B.D.; Li, Y.; González, A.J.; Yan, Z.; Kitto, S.L.; Grusak, M.A.; et al. Micrornas as master regulators of the plant nb-lrr defense gene family via the production of phased, trans-acting sirnas. *Genes Dev.* **2011**, *25*, 2540–2553. [CrossRef]
21. Deng, P.; Muhammad, S.; Cao, M.; Wu, L. Biogenesis and regulatory hierarchy of phased small interfering rnas in plants. *Plant Biotechnol. J.* **2018**, *16*, 965–975. [CrossRef]
22. Axtell, M.J.; Jan, C.; Rajagopalan, R.; Bartel, D.P. A two-hit trigger for sirna biogenesis in plants. *Cell* **2006**, *127*, 565–577. [CrossRef]
23. Montgomery, T.A.; Yoo, S.J.; Fahlgren, N.; Gilbert, S.D.; Howell, M.D.; Sullivan, C.M.; Alexander, A.; Nguyen, G.; Allen, E.; Ahn, J.H.; et al. Ago1-mir173 complex initiates phased sirna formation in plants. *Proc. Natl. Acad. Sci. USA* **2008**, *105*, 20055–20062. [CrossRef]
24. Felippes, F.F.; Weigel, D. Triggering the formation of tasirnas in arabidopsis thaliana: The role of microrna mir173. *EMBO Rep.* **2009**, *10*, 264–270. [CrossRef]
25. Chen, H.-M.; Chen, L.-T.; Patel, K.; Li, Y.-H.; Baulcombe, D.C.; Wu, S.-H. 22-nucleotide rnas trigger secondary sirna biogenesis in plants. *Proc. Natl. Acad. Sci. USA* **2010**, *107*, 15269–15274. [CrossRef]
26. Cuperus, J.T.; Carbonell, A.; Fahlgren, N.; Garcia-Ruiz, H.; Burke, R.T.; Takeda, A.; Sullivan, C.M.; Gilbert, S.D.; Montgomery, T.A.; Carrington, J.C. Unique functionality of 22-nt mirnas in triggering rdr6-dependent sirna biogenesis from target transcripts in arabidopsis. *Nat. Struct. Mol. Biol.* **2010**, *17*, 997–1003. [CrossRef]
27. Manavella, P.A.; Koenig, D.; Weigel, D. Plant secondary sirna production determined by microrna-duplex structure. *Proc. Natl. Acad. Sci. USA* **2012**, *109*, 2461–2466. [CrossRef] [PubMed]
28. Fei, Q.; Yu, Y.; Liu, L.; Zhang, Y.; Baldrich, P.; Dai, Q.; Chen, X.; Meyers, B.C. Biogenesis of a 22-nt microrna in phaseoleae species by precursor-programmed uridylation. *Proc. Natl. Acad. Sci. USA* **2018**, *115*, 8037–8042. [CrossRef]
29. de Felippes, F.F.; Marchais, A.; Sarazin, A.; Oberlin, S.; Voinnet, O. A single mir390 targeting event is sufficient for triggering tas3-tasirna biogenesis in arabidopsis. *Nucleic Acids Res.* **2017**, *45*, 5539–5554. [CrossRef] [PubMed]
30. Krasnikova, M.S.; Milyutina, I.A.; Bobrova, V.K.; Ozerova, L.V.; Troitsky, A.V.; Solovyev, A.G.; Morozov, S.Y. Novel mir390-dependent transacting sirna precursors in plants revealed by a pcr-based experimental approach and database analysis. *J. Biomed. Biotechnol.* **2009**, *2009*, 952304–952309. [CrossRef]
31. Xia, R.; Zhu, H.; An, Y.-Q.; Beers, E.P.; Liu, Z. Apple mirnas and tasirnas with novel regulatory networks. *Genome Biol.* **2012**, *13*, R47. [CrossRef]
32. Xia, R.; Xu, J.; Arikit, S.; Meyers, B.C. Extensive families of mirnas and phas loci in norway spruce demonstrate the origins of complex phasirna networks in seed plants. *Mol. Biol. Evol.* **2015**, *32*, 2905–2918. [CrossRef]
33. Montgomery, T.A.; Howell, M.D.; Cuperus, J.T.; Li, D.; Hansen, J.E.; Alexander, A.L.; Chapman, E.J.; Fahlgren, N.; Allen, E.; Carrington, J.C. Specificity of argonaute7-mir390 interaction and dual functionality in tas3 trans-acting sirna formation. *Cell* **2008**, *133*, 128–141. [CrossRef] [PubMed]
34. Arribas-Hernández, L.; Marchais, A.; Poulsen, C.; Haase, B.; Hauptmann, J.; Benes, V.; Meister, G.; Brodersen, P. The slicer activity of argonaute1 is required specifically for the phasing, not production, of trans-acting short interfering rnas in arabidopsis. *Plant Cell* **2016**, *28*, 1563–1580. [CrossRef]
35. Mi, S.; Cai, T.; Hu, Y.; Chen, Y.; Hodges, E.; Ni, F.; Wu, L.; Li, S.; Zhou, H.; Long, C.; et al. Sorting of small rnas into arabidopsis argonaute complexes is directed by the 5′ terminal nucleotide. *Cell* **2008**, *133*, 116–127. [CrossRef]

36. Takeda, A.; Iwasaki, S.; Watanabe, T.; Utsumi, M.; Watanabe, Y. The mechanism selecting the guide strand from small rna duplexes is different among argonaute proteins. *Plant Cell Physiol.* **2008**, *49*, 493–500. [CrossRef]
37. Endo, Y.; Iwakawa, H.O.; Tomari, Y. Arabidopsis argonaute7 selects mir390 through multiple checkpoints during risc assembly. *EMBO Rep* **2013**, *14*, 652–658. [CrossRef]
38. Wang, Y.; Sheng, G.; Juranek, S.; Tuschl, T.; Patel, D.J. Structure of the guide-strand-containing argonaute silencing complex. *Nature* **2008**, *456*, 209–213. [CrossRef]
39. Glick, E.; Zrachya, A.; Levy, Y.; Mett, A.; Gidoni, D.; Belausov, E.; Citovsky, V.; Gafni, Y. Interaction with host sgs3 is required for suppression of rna silencing by tomato yellow leaf curl virus v2 protein. *Proc. Natl. Acad. Sci. USA* **2008**, *105*, 157–161. [CrossRef]
40. Elmayan, T.; Adenot, X.; Gissot, L.; Lauressergues, D.; Gy, I.; Vaucheret, H. A neomorphic sgs3 allele stabilizing mirna cleavage products reveals that sgs3 acts as a homodimer. *FEBS J.* **2009**, *276*, 835–844. [CrossRef]
41. Kumakura, N.; Takeda, A.; Fujioka, Y.; Motose, H.; Takano, R.; Watanabe, Y. Sgs3 and rdr6 interact and colocalize in cytoplasmic sgs3/rdr6-bodies. *FEBS Lett.* **2009**, *583*, 1261–1266. [CrossRef] [PubMed]
42. Jouannet, V.; Moreno, A.B.; Elmayan, T.; Vaucheret, H.; Crespi, M.D.; Maizel, A. Cytoplasmic arabidopsis ago7 accumulates in membrane-associated sirna bodies and is required for ta-sirna biogenesis. *EMBO J.* **2012**, *31*, 1704–1713. [CrossRef] [PubMed]
43. Pumplin, N.; Sarazin, A.; Jullien, P.E.; Bologna, N.G.; Oberlin, S.; Voinnet, O. DNA methylation influences the expression of dicer-like4 isoforms, which encode proteins of alternative localization and function. *Plant Cell* **2016**, *28*, 2786–2804. [CrossRef] [PubMed]
44. Li, S.; Le, B.; Ma, X.; Li, S.; You, C.; Yu, Y.; Zhang, B.; Liu, L.; Gao, L.; Shi, T.; et al. Biogenesis of phased sirnas on membrane-bound polysomes in arabidopsis. *elife* **2016**, *5*, e22750. [CrossRef] [PubMed]
45. Yoshikawa, M.; Iki, T.; Tsutsui, Y.; Miyashita, K.; Poethig, R.S.; Habu, Y.; Ishikawa, M. 3′ fragment of mir173-programmed risc-cleaved rna is protected from degradation in a complex with risc and sgs3. *Proc. Natl. Acad. Sci. USA* **2013**, *110*, 4117–4122. [CrossRef] [PubMed]
46. Yoshikawa, M.; Iki, T.; Numa, H.; Miyashita, K.; Meshi, T.; Ishikawa, M. A short open reading frame encompassing the microrna173 target site plays a role in trans-acting small interfering rna biogenesis. *Plant Physiol.* **2016**, *171*, 359–368. [CrossRef] [PubMed]
47. Yelina, N.E.; Smith, L.M.; Jones, A.M.; Patel, K.; Kelly, K.A.; Baulcombe, D.C. Putative arabidopsis tho/trex mrna export complex is involved in transgene and endogenous sirna biosynthesis. *Proc. Natl. Acad. Sci. USA* **2010**, *107*, 13948–13953. [CrossRef] [PubMed]
48. Hou, C.Y.; Lee, W.C.; Chou, H.C.; Chen, A.P.; Chou, S.J.; Chen, H.M. Global analysis of truncated rna ends reveals new insights into ribosome stalling in plants. *Plant Cell* **2016**, *28*, 2398–2416. [CrossRef]
49. Bazin, J.; Baerenfaller, K.; Gosai, S.J.; Gregory, B.D.; Crespi, M.; Bailey-Serres, J. Global analysis of ribosome-associated noncoding rnas unveils new modes of translational regulation. *Proc. Natl. Acad. Sci. USA* **2017**, *114*, E10018–E10027. [CrossRef] [PubMed]
50. Zhang, C.; Ng, D.W.; Lu, J.; Chen, Z.J. Roles of target site location and sequence complementarity in trans-acting sirna formation in arabidopsis. *Plant J.* **2012**, *69*, 217–226. [CrossRef]
51. Adenot, X.; Elmayan, T.; Lauressergues, D.; Boutet, S.; Bouché, N.; Gasciolli, V.; Vaucheret, H. Drb4-dependent tas3 trans-acting sirnas control leaf morphology through ago7. *Curr. Biol.* **2006**, *16*, 927–932. [CrossRef] [PubMed]
52. Fahlgren, N.; Montgomery, T.A.; Howell, M.D.; Allen, E.; Dvorak, S.K.; Alexander, A.L.; Carrington, J.C. Regulation of auxin response factor3 by tas3 ta-sirna affects developmental timing and patterning in arabidopsis. *Curr. Biol.* **2006**, *16*, 939–944. [CrossRef]
53. Yoon, E.K.; Yang, J.H.; Lim, J.; Kim, S.H.; Kim, S.-K.; Lee, W.S. Auxin regulation of the microrna390-dependent transacting small interfering rna pathway in arabidopsis lateral root development. *Nucleic Acids Res.* **2009**, *38*, 1382–1391. [CrossRef] [PubMed]
54. Marin, E.; Jouannet, V.; Herz, A.; Lokerse, A.S.; Weijers, D.; Vaucheret, H.; Nussaume, L.; Crespi, M.D.; Maizel, A. Mir390, arabidopsis tas3 tasirnas, and their auxin response factor targets define an autoregulatory network quantitatively regulating lateral root growth. *Plant Cell* **2010**, *22*, 1104–1117. [CrossRef] [PubMed]

55. Matsui, A.; Mizunashi, K.; Tanaka, M.; Kaminuma, E.; Nguyen, A.H.; Nakajima, M.; Kim, J.-M.; Nguyen, D.V.; Toyoda, T.; Seki, M. Tasirna-arf pathway moderates floral architecture in arabidopsis plants subjected to drought stress. *BioMed Res. Int.* **2014**, *2014*, 303451. [CrossRef] [PubMed]
56. Liu, Y.; Ke, L.; Wu, G.; Xu, Y.; Wu, X.; Xia, R.; Deng, X.; Xu, Q. Mir3954 is a trigger of phasirnas that affects flowering time in citrus. *Plant J.* **2017**, *92*, 263–275. [CrossRef] [PubMed]
57. Liu, L.; Ren, S.; Guo, J.; Wang, Q.; Zhang, X.; Liao, P.; Li, S.; Sunkar, R.; Zheng, Y. Genome-wide identification and comprehensive analysis of micrornas and phased small interfering rnas in watermelon. *BMC Genomics* **2018**, *19*, 111. [CrossRef]
58. Jagadeeswaran, G.; Zheng, Y.; Li, Y.-F.; Shukla, L.I.; Matts, J.; Hoyt, P.; Macmil, S.L.; Wiley, G.B.; Roe, B.A.; Zhang, W.; et al. Cloning and characterization of small rnas from medicago truncatula reveals four novel legume-specific microrna families. *New Phytol.* **2009**, *184*, 85–98. [CrossRef] [PubMed]
59. Rajeswaran, R.; Aregger, M.; Zvereva, A.S.; Borah, B.K.; Gubaeva, E.G.; Pooggin, M.M. Sequencing of rdr6-dependent double-stranded rnas reveals novel features of plant sirna biogenesis. *Nucleic Acids Res.* **2012**, *40*, 6241–6254. [CrossRef]
60. Guan, X.; Pang, M.; Nah, G.; Shi, X.; Ye, W.; Stelly, D.M.; Chen, Z.J. Mir828 and mir858 regulate homoeologous myb2 gene functions in arabidopsis trichome and cotton fibre development. *Nat. Commun.* **2014**, *5*, 3050. [CrossRef]
61. Axtell, M.J.; Snyder, J.A.; Bartel, D.P. Common functions for diverse small rnas of land plants. *Plant Cell* **2007**, *19*, 1750–1769. [CrossRef]
62. Xia, R.; Meyers, B.C.; Liu, Z.; Beers, E.P.; Ye, S.; Liu, Z.; Liu, Z. Microrna superfamilies descended from mir390 and their roles in secondary small interfering rna biogenesis in eudicots. *Plant Cell* **2013**, *25*, 1555–1572. [CrossRef] [PubMed]
63. Chitwood, D.H.; Nogueira, F.T.S.; Howell, M.D.; Montgomery, T.A.; Carrington, J.C.; Timmermans, M.C.P. Pattern formation via small rna mobility. *Genes Dev.* **2009**, *23*, 549–554. [CrossRef]
64. Schwab, R.; Maizel, A.; Ruiz-Ferrer, V.; Garcia, D.; Bayer, M.; Crespi, M.; Voinnet, O.; Martienssen, R.A. Endogenous tasirnas mediate non-cell autonomous effects on gene regulation in arabidopsis thaliana. *PLoS ONE* **2009**, *4*, e5980. [CrossRef]
65. Pekker, I.; Alvarez, J.P.; Eshed, Y. Auxin response factors mediate arabidopsis organ asymmetry via modulation of kanadi activity. *Plant Cell* **2005**, *17*, 2899–2910. [CrossRef] [PubMed]
66. Margis, R.; Fusaro, A.F.; Smith, N.A.; Curtin, S.J.; Watson, J.M.; Finnegan, E.J.; Waterhouse, P.M. The evolution and diversification of dicers in plants. *FEBS Lett.* **2006**, *580*, 2442–2450. [CrossRef] [PubMed]
67. Song, X.; Li, P.; Zhai, J.; Zhou, M.; Ma, L.; Liu, B.; Jeong, D.-H.; Nakano, M.; Cao, S.; Liu, C.; et al. Roles of dcl4 and dcl3b in rice phased small rna biogenesis. *Plant J.* **2011**, *69*, 462–474. [CrossRef] [PubMed]
68. Zhai, J.; Zhang, H.; Arikit, S.; Huang, K.; Nan, G.L.; Walbot, V.; Meyers, B.C. Spatiotemporally dynamic, cell-type-dependent premeiotic and meiotic phasirnas in maize anthers. *Proc. Natl. Acad. Sci. USA* **2015**, *112*, 3146–3151. [CrossRef] [PubMed]
69. Xia, R.; Chen, C.; Pokhrel, S.; Ma, W.; Huang, K.; Patel, P.; Wang, F.; Xu, J.; Liu, Z.; Li, J.; et al. 24-nt reproductive phasirnas are broadly present in angiosperms. *Nat Commun* **2019**, *10*, 627. [CrossRef]
70. Komiya, R.; Ohyanagi, H.; Niihama, M.; Watanabe, T.; Nakano, M.; Kurata, N.; Nonomura, K.-I. Rice germline-specific argonaute mel1 protein binds to phasirnas generated from more than 700 lincrnas. *Plant J.* **2014**, *78*, 385–397. [CrossRef]
71. Nonomura, K.I.; Morohoshi, A.; Nakano, M.; Eiguchi, M.; Miyao, A.; Hirochika, H.; Kurata, N. A germ cell specific gene of the argonaute family is essential for the progression of premeiotic mitosis and meiosis during sporogenesis in rice. *Plant Cell* **2007**, *19*, 2583–2594. [CrossRef] [PubMed]
72. de Felippes, F.F.; Ott, F.; Weigel, D. Comparative analysis of non-autonomous effects of tasirnas and mirnas in arabidopsis thaliana. *Nucleic Acids Res.* **2011**, *39*, 2880–2889. [CrossRef] [PubMed]
73. McHale, M.; Eamens, A.L.; Finnegan, E.J.; Waterhouse, P.M. A 22-nt artificial microrna mediates widespread rna silencing in arabidopsis. *Plant J.* **2013**, *76*, 519–529. [CrossRef] [PubMed]
74. Pandey, P.; Senthil-Kumar, M.; Mysore, K.S. Advances in plant gene silencing methods. In *Plant Gene Silencing*; Mysore, K.S., Senthil-Kumar, M., Eds.; Springer: New York, NY, USA, 2015; Volume 1287, pp. 3–23.
75. de la Luz Gutiérrez-Nava, M.; Aukerman, M.J.; Sakai, H.; Tingey, S.V.; Williams, R.W. Artificial trans-acting sirnas confer consistent and effective gene silencing. *Plant Physiol.* **2008**, *147*, 543–551. [CrossRef] [PubMed]

76. Carbonell, A.; Takeda, A.; Fahlgren, N.; Johnson, S.C.; Cuperus, J.T.; Carrington, J.C. New generation of artificial microrna and synthetic trans-acting small interfering rna vectors for efficient gene silencing in arabidopsis. *Plant Physiol.* **2014**, *165*, 15–29. [CrossRef] [PubMed]
77. Carbonell, A.; Daròs, J.-A. Artificial micrornas and synthetic trans-acting small interfering rnas interfere with viroid infection. *Mol. Plant Pathol.* **2017**, *18*, 746–753. [CrossRef] [PubMed]
78. Schwab, R.; Ossowski, S.; Riester, M.; Warthmann, N.; Weigel, D. Highly specific gene silencing by artificial micrornas in arabidopsis. *Plant Cell* **2006**, *18*, 1121–1133. [CrossRef]
79. Ossowski, S.; Schwab, R.; Weigel, D. Gene silencing in plants using artificial micrornas and other small rnas. *Plant J.* **2008**, *53*, 674–690.
80. de Felippes, F.F.; Wang, J.; Weigel, D. Migs: Mirna-induced gene silencing. *Plant J.* **2012**, 541–547. [CrossRef]
81. de Felippes, F.F. Downregulation of plant genes with mirna-induced gene silencing. In *Sirna Design*; Taxman, D., Ed.; Humana Press: Totowa, NJ, USA, 2013; Volume 942, pp. 379–387.
82. Benstein, R.M.; Ludewig, K.; Wulfert, S.; Wittek, S.; Gigolashvili, T.; Frerigmann, H.; Gierth, M.; Flügge, U.-I.; Krueger, S. Arabidopsis phosphoglycerate dehydrogenase1 of the phosphoserine pathway is essential for development and required for ammonium assimilation and tryptophan biosynthesis. *Plant Cell* **2013**, *25*, 5011–5029. [CrossRef] [PubMed]
83. Imin, N.; Mohd-Radzman, N.A.; Ogilvie, H.A.; Djordjevic, M.A. The peptide-encoding cep1 gene modulates lateral root and nodule numbers in medicago truncatula. *J. Exp. Bot.* **2013**, *64*, 5395–5409. [CrossRef] [PubMed]
84. Han, Y.; Zhang, B.; Qin, X.; Li, M.; Guo, Y. Investigation of a mirna-induced gene silencing technique in petunia reveals alterations in mir173 precursor processing and the accumulation of secondary sirnas from endogenous genes. *PLoS ONE* **2015**, *10*, e0144909–e0144916. [CrossRef] [PubMed]
85. Jacobs, T.B.; Lawler, N.J.; LaFayette, P.R.; Vodkin, L.O.; Parrott, W.A. Simple gene silencing using the trans-acting sirna pathway. *Plant Biotechnol. J.* **2015**, *14*, 117–127. [CrossRef]
86. Zheng, X.; Yang, L.; Li, Q.; Ji, L.; Tang, A.; Zang, L.; Deng, K.; Zhou, J.; Zhang, Y. Migs as a simple and efficient method for gene silencing in rice. *Front. Plant Sci.* **2018**, *9*, 662. [CrossRef]
87. Baykal, U.; Liu, H.; Chen, X.; Nguyen, H.T.; Zhang, Z.J. Novel constructs for efficient cloning of srna-encoding DNA and uniform silencing of plant genes employing artificial trans- acting small interfering rna. *Plant Cell Rep.* **2016**, *35*, 2137–2150. [CrossRef] [PubMed]

© 2019 by the author. Licensee MDPI, Basel, Switzerland. This article is an open access article distributed under the terms and conditions of the Creative Commons Attribution (CC BY) license (http://creativecommons.org/licenses/by/4.0/).

MDPI
St. Alban-Anlage 66
4052 Basel
Switzerland
Tel. +41 61 683 77 34
Fax +41 61 302 89 18
www.mdpi.com

Plants Editorial Office
E-mail: plants@mdpi.com
www.mdpi.com/journal/plants

www.ingramcontent.com/pod-product-compliance
Lightning Source LLC
LaVergne TN
LVHW070641100526
838202LV00013B/857